Pan Pacific Microelectronics Symposium and Tabletop Exhibition 2010

Kauai, Hawaii, USA
26-28 January 2010

ISBN: 978-1-61567-947-8

Printed from e-media with permission by:

Curran Associates, Inc.
57 Morehouse Lane
Red Hook, NY 12571

Some format issues inherent in the e-media version may also appear in this print version.

Copyright© (2010) by Surface Mount Technology Association (SMTA)
All rights reserved.

Printed by Curran Associates, Inc. (2010)

For permission requests, please contact Surface Mount Technology Association (SMTA)
at the address below.

Surface Mount Technology Association (SMTA)
5200 Wilson Road
Suite 215
Edina, MN 55424

Phone: (952) 920-4682
Fax: (952) 926-1819

www.smta.org

Additional copies of this publication are available from:

Curran Associates, Inc.
57 Morehouse Lane
Red Hook, NY 12571 USA
Phone: 845-758-0400
Fax: 845-758-2634
Email: curran@proceedings.com
Web: www.proceedings.com

TABLE OF CONTENTS

SESSION TP1: FAILURES IN MANUFACTURING

Electrostatic Discharge (ESD)-Sources of Electrostatic Charge in a Production Line (SMT) 1
  Hartmut Berndt
Mechanical Failures in Pb-Free Processing: Evaluating the Effect of Pad Crater Defects on Process Strain Limits for BGA Devices 8
  John McMahon, Brian Gray
Quality and Reliability Analysis of Lead Free PCBs in Simulated Production Conditions and Long Term Use 14
  Sammy Shina, Greg Morose, Richard Anderson, Helena Pasquito, Bob Farrell, Don Longworth, Wendi Boger, Ken Degan, Donald Abbott, David Pinsky, Karen Ebner, Amit Sarkhel, Michael Miller, Louis Feinstein, Deb Fragoza, Eric Ren, Charlie Bickford

SESSION TP2: ELECTROCHEMICAL FAILURE MECHANISMS

Corrosion Driven Whisker Growth in SAC305 21
  Keith Sweatman, Junya Masuda, Takashi Nozu, Masuo Koshi, Tetsuro Nishimura

SESSION TP3: INSPECTION AND FAILURE ANALYSIS

A Catalogue of Failure Mechanisms in Flip Chip Devices Detected Using Acoustic Micro Imaging 25
  Janet E. Semmens
Modern 2D/3D X-Ray Inspection - Emphasis on BGA, QFN, 3D Packages, and Counterfeit Components 31
  Evstatin Krastev, David Bernard

SESSION WA1: SIP SOLUTIONS

Head Stack Assembly Percussion Swage Processing Morphologies for Hard Disk Drive Products 38
  Jeffry S. Bennin, Michael R. Tiller
New Packaging and Interconnect Technologies for Ultra Thin Chips 46
  Christine Kallmayer, Rolf Aschenbrenner, Julian Haberland, Herbert Reichl
Chip-last Interconnection to Thin Organic Packages with Advanced Dilectrics and Fine Pitch I/Os 52
  Rao R. Tummala, Venkatesh Sundaram, Nitesh Kumbhat, Nithya Sankaran
Advantages of a New Wafer Level Integration Concept Based on Direct Bonded Silicon on LTCC 58
  J. Müller, M. Fischer, H. Bartsch de Torres, B. Pawlowski, S. Barth
From the Single Chip to the Wafer Integration 64
  Gilles Poupon, Jean Charles Souriau, Hervé Boutry, Jean Brun, Nicolas Sillon

SESSION WA2: MANUFACTURING CHALLENGES

Molded Interconnect Devices-Progressive Approach for Mechatronic Products and Efficient Manufacturing Processes 71
  Christian Goth, Jörg Franke, Klaus Feldmann
Micron Level Placement Accuracy for High Aspect Ratio Die in Printing Products 78
  Zeger Bok, Daniel D. Evans Jr.
Important Considerations in Packaging: A Small Hall-Effect Sensor 84
  Ron Molnar, Jeff Wise
Capillary Underfill Physical Limitations for Future Packages 90
  Horatio Quinones, Tom Ratledge
High Brightness LED Assembly Using DPC Substrate and SuperMCPCB 94
  Nick Wang, Allen Hsu, Andy Lim, Jerry Tan, Charles Lin, Heinz Ru, Thompson Jiang, David Liao

SESSION WP1: PROCESS TECHNOLOGIES

High Thermal Mass, Very High Lead Count SMT Connector Rework Process: Process and Problem Resolution 99
Jim Bielick, Brian Chapman, Mitchell Ferrill, Michael Fisher, Phil Isaacs, Eddie Kobeda, Theron Lewis

High Speed Touch Screen Assembly Using Anisotropic Conductive Adhesives (ACAS) Vertical Ultrasonic Bonding Method 105
Kyung-Wook Paik, Seung-Ho Kim, Kiwon Lee

Lead-Free Flux Technology and Influence on Cleaning 109
Ning-Cheng Lee

SESSION WP2: OPTIMIZATION TECHNIQUES

DPBO-A New Control Chart for Electronics Assembly 116
Daryl L. Santos

Using Test Optimization to Improve Rework Effectiveness 121
Juan Coronado, Jorge Valle, Omar Garcia, Luis Manuel Zamora, Fernando Rodas, Mario Aguilar, Frederico Santos, Carlos Alberto Robles, Zhen (Jane) Feng, Dason Cheung, Murad Kurwa

How IT Can Enable Energy Optimization: A Case Study on Textile Factories 127
Eugenio Capra, Chiara Francalanci, Daniele Zagordi, Alex Zazzera

SESSION WP3: SUCCESS STRATEGIES

A Survey on IT Managers Green Awareness 133
Eugenio Capra, Alessandro Caleffii

On Improving Operations Scheduling in Electronics Manufacturing 139
Daryl L. Santos

Emerging Competitors in Emerging Markets: What Patent Applications Reveal 145
H. J. Neuhaus, R. A. Fillion, C. E. Bauer

SESSION THA1: TSV FOR 3D PACKAGING

Emerging Substrate Technologies for Packaging 151
Henry H. Utsunomiya

Quantitative Analyses of Interfacial Properties of Cu-Cu Bonds for TSV Integration 158
Bioh Kim, Thorsten Matthias, Markus Wimplinger, Paul Lindner, Eun-Jung Jang, Jae-Won Kim, Young-Bae Park

Cost Effective Cu-TSV Interconnects by EMC3D and Technical Challenges with Cu-TSV 161
Paul Siblerud

Optimization of Copper Plating Parameters for Fast Filling of Through Silicon Vias (TSV) 162
Su Wang, S. W. Ricky Lee

Silicon Interposer for Heterogeneous Integration 171
M. Juergen Wolf, Kai Zoschke, Robert Weiland, Mathias Klein, Klaus-Dieter Lang, Herbert Reichl

KEYNOTE ADDRESS I

Nanotechnology Is Now Starting to Find Applications in Electronics 176
Alan Rae

KEYNOTE ADDRESS II

Solar Cells on Ultra-Thin Crystalline Silicon 181
Robert Mertens

Author Index

Sheraton Kauai Resort
Island of Kauai
January 26-28, 2010

Pan Pacific 2010 Exhibitors

Advanced Packaging/SMT Magazine/Pennwell

Circuits Assembly Magazine

Cobar/Balver Zinn

NordsonDAGE

* Global SMT & Packaging

* Kyzen Corporation

NordsonAsymtek

Semitool, Inc.

* Sonoscan Inc.

* STI Electronics, Inc.

* SMTA Corporate Members

2010 Sponsors

Libra Industries – Luau Sponsor

Kyzen Corporation – Golf Outing
(Two Hole Sponsorships)

KYZEN
CORPORATION

A special thanks to the SMTA Board of Directors for their support of the 2010 Pan Pacific Symposium!

Officers
Dan Baldwin, Ph.D., President
Engent, Inc.

Bill Barthel, VP Tech Programs
Plexus Corporation

Tom Forsythe, VP Communications
Kyzen Corporation

Roy Starks, VP Membership
Libra Industries

Kola Akinade, Ph.D., Secretary
Cisco Systems

Marie Cole, Treasurer
IBM Corporation

Planning Committee
Jeff Kennedy, Chair
Celestica, Inc.

Denis Barbini, Ph.D.,

Joseph Belmonte
BOSE Corporation

Hal Hendrickson
Dage Precision Industries

Ning-Cheng Lee, Ph.D.
Indium Corporation of America

Laura Turbini, Ph.D
Research In Motion

Thank you to the Pan Pacific Technical Committee for organizing the 2010 symposium

GENERAL CHAIR
Charles Lin, Ph.D.
Bridge Semiconductor, Taiwan

NORTH AMERICAN COORDINATOR
Charles E. Bauer, Ph.D.
TechLead Corporation, USA

TECHNICAL PROGRAM CO-CHAIRS
Soren Norlyng
MICRONSULT, Denmark

M. Juergen Wolf
Fraunhofer Institute, Germany

TECHNICAL PROGRAM COMMITTEE
Australia
Keith Sweatman, Nihon Superior
China
Tom Chung, Ph.D., ASTRI
Sang Liu, Ph.D., Huawei Technologies Co., Ltd.
Abby Tsoi, Kasion Automation

Japan
Yoshikazu Nakamura, Ph.D.,
Future Package Technology
Masahide Tsukamoto, Ph.D.,
Masashi Engineering, Inc.
Henry Utsunomiya,
Interconnection Technologies, Inc.
Korea
Kyung Wook Paik, Ph.D., KAIST
Kim Il Ung, Ph.D., Samsung Electronics Co., Ltd
Malaysia
Teng Hoon Ng, Celestica, Inc.
Singapore
Charles Lin, Ph.D., Bridge Semiconductor
Peter Pooh, ConnectCounty Holdings Bhd.
Taiwan
Shen-Li Fu, Ph.D., I-Shou University
Yu-Jung Huang, Ph.D., I-Shou University
Charles Lin, Ph.D., Bridge Semiconductor
USA
Bill Barthel, Plexus Corporation
Horatio Quinones, Ph.D., Asymtek
Paul Wang, Ph.D., Microsoft (USA)

ELECTROSTATIC DISCHARGE (ESD) – SOURCES OF ELECTROSTATIC CHARGE IN A PRODUCTION LINE (SMT)

Hartmut Berndt
B.E.STAT European ESD competence centre
Kesslesdorf, Saxony, Germany
hberndt@bestat-esd.com

ABSTRACT

The number of failures caused by electrostatic discharges (ESD) has been increasing for some time now. So, it is necessary for everyone, who handles electrostatic sensitive devices (ESDS), to know the reasons of such failures. This presentation will give an overview about possible causes for ESD in a SMT production line.

Particularly automated production lines have some processing steps, where electrostatic charges are increasingly generated. So far one has been focused on the human being. This is controllable. Measurements in production lines show electrostatic charges at the following processing steps: application of soldering paste (printer), assembling (automated and manual pick and place), and labelling as well as optical and electrical tests (ICT). The electronic components are always assembled directly and without any covering on the PCBs. Thus, the wire bonding process leads to a damage of the electronic components. This process step is a very critical part in the production line. The electronic devices will be directly contacted with a metal needle.

The process steps, where the PCBs are covered with enclosures must be inspected either. Such enclosures are mostly made of isolating materials, like plastics. Thus, those can be electrostatic charged highly, while assembling.

An optimized ESD Control System for machines with the emphasis on cost-effectiveness will be presented.

Key words:
Automation SMT production line
Automation handling equipment
Electrostatic discharge
ESD Control System
Product quality

INTRODUCTION

All electronic components and assemblies are to be at risk of electrostatic discharges. Producers, suppliers, distributors and users have to realize the ESD control system during the whole manufacturing process, during the measurements as well as during the application. All active electronic components, beginning with simple diodes, transistors or complex inner circuits, require an extern ESD control system. In the next step, SMD resistors and condensers, and prospectively NEMS and MEMS will be included in this danger category. Tests show, that these passive components can be damaged through electrostatic discharges.

The structures of electronic components become smaller. Already 5 volts of an electrostatic charge are enough to change the structures in small electronic components. The structures will achieve such small dimensions, so electrostatic charges can cause permanent damages. In the year 2022 the sizes of the electronic components will be less than 10 nm. Electrostatic charges of 0,1 nC and electrostatic fields of 10 V/cm will be enough then to damage ESDS permanently.

PROCESS STEPS IN A SMT PRODUCTION LINE

Every time PCBs are handled, electrostatic charges are generated. A SMT production line has different process steps, where such charges may be generated. As a matter of principle a PCB can always be charged by any movements. The isolating plastic, which is used as basic material, is mostly the main reason. The material is electrostatically charged by friction, p. e. by conveyor belts, although these are mostly made of conductive material.

Soldering Printing

One of these processes is the so called soldering printing of PCBs with soldering paste. This procedure and the following slitting process PCB - printing colander leads to high charges. This would not be critical, unless ESDS exist on PCBs. Usually PCBs are assembled on both sides. That means that electronic components already exist during the second print or the backside-print. Very high electrostatic charges may arise while separating the printing colander from the PCB. This slitting process is typical example for the generation of electrostatic charges. It does not matter if the colander is made of metal or plastic.

AOI

Afterwards an optical/vision inspection, so called AOI, follows. This process does not generate any electrostatic charges by itself, but the transportation does.

Optical test procedures are probably the only processes, which do not cause any electrostatic charges.

Pick-and-Place Machine

The PCBs arrive at the machine, which is electrostatically charged on the surface. Now a charge exchange happens inside the machine. Electronic components are electrostatically charged and are assembled with the PCB. The PCB is charged either. While placing the ESDS on the PCB the charge exchange takes place. This discharge current damages the ESDS.

Electronic components/ESDS are charged through the process „removing them from a tray or blister". Electrostatic charges are generated during this slitting process. The ESDS are picked by the placement head and placed on the several PCBs. In the past one had experienced with the material of such placement head. Nevertheless, electrostatic charges cannot be avoided or even discharged by these. The reason therefore is the ESDS' enclosure, which is generally made of plastic (isolating).

ICT

PCBs may be electrostatically charged during the transport between two process steps. The following ICT (integrated circuit test machine) leads to a sudden discharge of the existing electrostatic charges on the PCB or on the single electronic component. The reason therefore is the direct contact of the metal needle (measurement probe) with the component's pins. A series resistor would not be any solution, because the discharge happens directly at the contact point between needle and pin.

Assembly Processes

Different assembly processes cause the contact of isolating enclosure parts with static control sensitive components. Thus, an influence of the ESDS happens by the electrostatic field of isolating plastic parts. A charge transfer on the ESDS effected, which probably can cause discharges during the production process or at the customer.

Wire Bonding Process

A very critical process is the wire bonding process, during the handling of ESDS (naked chips) as well as during the wire bonding of whole assemblies. Mostly, PCBs are electrostatically charged by the enclosures or through the transport process. During the wire bonding process a direct contact between a metal needle and an ESDS occurs again. Thus, a sudden discharge is provoked and the ESDS is damaged.

Further Processes

Labeling processes, transport machines or systems, cutting systems or other steps can produce electrostatic potential differences. These differences can damage electronic parts:

- Isolating parts: plastic glass, plastic covers
- Pneumatic lines and cables: rubber transportation system, plastic rolls
- Anodized surfaces: aluminum
- Pick-up mechanisms: nozzles
- Vacuum cups
- Grippers

FAILURE MODELS (CDM, CBM UND FICBM)

Different failure models are used for the analyses of humans and machines. The HBM (Human Body Model) is always used for the electrostatic charge of a person. Otherwise the CDM (Charged Device Model) is applied for the charge analysis of machines or production lines. Nevertheless, both will not be sufficient in the future. New failure models like the CBM (Charged Board Model) or the FICBM (Field Induce Charged Board Model) become necessary. The CDM only considers a single electronic component; however the CBM is applied to analyze the whole PWB.

Reflecting the following considerations, single failure models are caused:

- A person touches an electronic component and the stored electrostatic charges are transported from the person to the electronic component. These charges are grounded by the connection between the electronic component and the earth potential.
- An electronic component or an electronic device acts as capacitor plate and stores electrostatic charges. While contacting the earth potential, damages are caused by a discharge pulse.
- A charged object is in an electric field. A potential is generate over the gateoxid or the pn-junction of an electronic component. Electrostatic charges are

generated and discharges cause damages (break down).

Already known failure models:
- HBM (Human Body Model)
- MM (Machine Model)
- CDM (Charged Device Model)
- FIM (Field Induced Model)

The first failure model only considers the charged person. The second one is a specialization of the HBM (Human Body Model). The third failure model assumes that the electronic component charges itself electrostatically and discharges itself suddenly by contacting metal. However, a person does not influence directly the charge and discharge process any more at this third one.

Reflecting the fact that quick discharges must be considered more and more in the future, all statements about electrostatic charges so far won't be sufficient any more. Very fast discharges in very short terms already exist. Those are really energy loaded and damage ESDS of course.

Nowadays, it is not enough to consider every single electronic component. One has to analyze the complete electronic assembly either. However, a suitable failure model is still missing. Two models are in preparation: the CBM (Charged Board Model) and the FICBM (Field Induce Charged Board Model). Both assume that the board (PWB) is electrostatic charged. The board has a higher capacity, so that it may store much more electrostatic charges. Those can also be grounded by a single electronic component. The high energy causes an early damage of the whole electronic component.

Charged Device Model (CDM)
The electronic component acts as a condenser. It gathers charges, such caused while sliding through a magazine or while contacting another charged object. Additionally, electrostatic charges are generated by removing the electronic component from a conductive tray or a belt. Electrostatic charges are generally caused by taking an electronic component out of conductive material, because it is not equipped with a conductive enclosure. Thus, an electronic component is always electrostatic charged after every mechanical process, independent of its actual handling like the movement in a pick-and-place-machine or another production line. However, just only a discharge damages the electronic component. The discharge can be realized directly or indirectly via further processes. It just

enough to bring the electronic component in the near of a dischargeable point or object. So, an electrical or electrostatic field may already provoke such a discharge. Damages of pn-junction, dielectric and other components are caused by a discharge impulse and its discharge current, depending on the grounding via the enclosure or the chip.

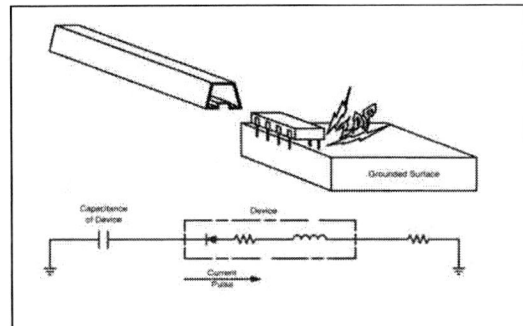

Picture 1 Typical CDM discharge

An electronic component can store energies up to 100 µJ. However, at a very low contact resistance ($< m\Omega$) and a conduction inductivity of 10 nH, such energy, depending on the charge amount, can realize a direct or an indirect connection to the earth potential. An output per impulse of several 100 W to 1000 W is reached by an increase of the discharge current impulse of several ns. Such outputs are enough to change the component parameters considerably or to destroy the electronic component finally (see Wunsch und Bell [6]).

Charged Board Model (CBM)
The previous models HBM and CDM are not enough to describe ESD failures. There will be almost breakdowns caused by humans, but the most ESDS are moved in automatically handling equipments. There is no direct influence of the human. The electronic components and assemblies charge themselves electrostatic. The capacity conditions of a PWB are absolute different in comparison to humans. That's why one has been searched for a new model for some years because of starting with new conditions. One solution is the charged board model (CBM). Here the capacity proportions are very difficult. The value of the capacity is higher as the body of the person. The result is a bigger electrostatic charge on the board. We have a bigger energy at the discharge. This leads to the influence of ESDS and it damages the ESDS.

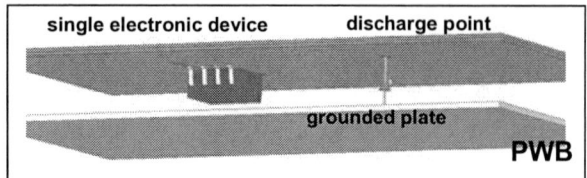

Picture 2 Typical CBM discharge

Field Induce Charged Board Model (FICBM)
A different model is the FICBM. The influence of the electrical field on PCBs has been neglected or has been thought that it is not important, until now. An electrical field is able to produce electrostatic charges on a PCB, like demonstrated. These electrostatic charges will be stored by the bigger capacity of the PCB or by the electronic components on it. The discharge process is not predictable. It can always happen, when the PCB is grounded or contacted. That's why the description of FICBM is not easy, but the effect is very important. Electrical fields can appear everywhere, where machines, motorized plants stand or where circuit processes are produced.

STEPS FOR THE MACHINE GROUNDING
The first and only requirements are the demand for a grounding of all metal parts as well as the demand for the avoidance of plastic usage, which could generate electrostatic charges and fields. He experiences show, that it is not enough for the protection of ESDS in automated machines and systems. ESDS won't be damaged by the operator, but through the machines. The transport operation of an ESDS in a machine can happen as following:

1. Removal of the ESDS out of packaging. This is the first partition act. The ESDS has an isolating case, so it will be electrostatic charged during the removal out of the reel or the tray.

2. The electrostatic charged ESDS will be transported to the PWB. Thereby a further electrostatic charge can happen. The movement at High speed Pick-and-Place System should be enough of the generation of electrostatic charges.

3. Through the placing on the PWB, different potential between the ESDS and the PWB exist. So the potential difference leads to a discharge, which will damage the ESDS.

These examples show, that electrostatic charges always develop, when ESDS are parted or moved. Electrostatic charges will always generate because of the reason, which the components as well as the PWBs are made of an isolating material. Other acts and production steps show, that this is not the only possibility for the generation of electrostatic charges in a production process. Further critical steps are for example: the printing of PWBs, the labeling of PWBs and assemblies as well as test constructions.

Manual handling of individual components is not common anymore. PWB assemblies are handled mainly by equipment and the final phases of mechanical assembly are done by both humans and robots. In consequence of this the Human Body Model (HBM) is not valid ESD simulation model as much as previous. The main electrostatic risk during automated manufacturing is with Charged Device Model (CDM) type of electrostatic discharges. The additional model, but not standardized yet is Charged Board Model (CBM).

In the CBM type of ESD the assembled Printed Wiring Board (PWB) or some of the mechanics parts can be charged during handling and the discharge to ground or between the objects can happen. CBM type of discharge is typically more severe than other models for components due to high capacitance and high stored charge of PWB assemblies or mechanics. There are some main ESD control principles which are important in ESD Protected Area (EPA) as well as in automated process equipment:

1. All conductive and dissipative items are grounded.

2. Materials or parts to be contacted with ESDS are made of electrostatic dissipative material.

3. Non-essential insulating materials are excluded.

4. Where insulating materials or parts are needed, the possible charges shall be minimized by special measures, like ionization, shielding or coating.

Enclosures of machines are normally made of conductive material. The conductive enclosure should have a straight and reliable connection to ground and the distance of the insulating parts should be long enough in order not to create high electrostatic fields close to ESDS. Special attention should be paid for grounding of parts which are separated from the enclosure or are movable, like adjustable conveyor.

Type of measurement	ESD item	Limits[1]
Resistance to ground (only metal parts)	equipment body (enclosure)	$R_G < 2\ \Omega$
	moving parts	$R_G < 10^6\ \Omega$
	all surfaces, in contact with the ESDS	$10^6\ \Omega < R_G < 10^9\ \Omega$
Point-to-point resistance or surface resistance	All surfaces, in contact with the ESDS	$10^6\ \Omega < R_G < 10^9\ \Omega$
Electrostatic potential or electrostatic voltage	Conductive parts of equipment	V < 5 volts
	Surfaces in contact with ESDS items or closer than 15 cm of them	V < 50 volts
	ESDS, PWB and mechanical parts [3]	V < 100 volts
Electrostatic field	ESDS, PWB and mechanical parts [2,3]	electrostatic field strength < 10000 volts/m measured at the position of ESDS or
	ESDS and PWB assemblies [3]	electrostatic charge $Q < 5 * 10^{-9}$ Coulomb

Table 1 Requirements for AHE (automated handling equipments) [4]

[1] Electrostatic potential, field and charge values are absolute values, so they can have either positive or negative readings.

[2] If the potential of process essential insulators exceeds 2000 volts the item must be kept a minimum of 30 cm from ESDS.

[3] Stored charges should be less than required to cause CDM or CBM type of damage for the device.

There are a lot of materials which can be in contact with ESDS items. Components to be placed are stored in reels with plastic tapes covered and nozzle picks the component from reel. Components are placed on the PWB and PWB is contacted with conveyor belts and possible support pins, gripper, clamps etc. All these materials should be made of electrostatic dissipative material at least in contact area and a resistance to ground value shall be between 10^6 and $10^9\ \Omega$.

Components and PWB material have plastic, insulating material and they can become charged by tribocharging, e.g. by rubbing against conveyor belt, touching on other product parts or in routing process. The charged ESDS item can subject to CDM or CBM risk. All rotating and sliding elements form an ESD risk. The tribocharging during automated manufacturing shall be minimized and metal contact to ESDS shall be prevented. Normally it is not enough, an ionizer shell be installed in the area of rotating material.

Ionizers are applied sometimes to remove electrostatic charges from machines. Electronic components and PWBs cannot be grounded. Thus, ionization is the only method minimizing electrostatic charges at the moment.

Ionization is just one opportunity decreasing electrostatic charges. Intelligent ionizers are able to detect electrostatic charges in machines and to generate equivalent charges for their decrease either. The limits are shown in table 1.

The mentioned ESD control steps are in common with the today's knowledge.

MEASUREMENT METHODS AND RESULTS
The following measurements are recommended to perform in order to evaluate the capability of automated equipment:

1. Resistance to ground
2. Point to point resistance
3. Electrostatic potential
4. Electrostatic field
5. Accumulated charge
6. EMI (Electromagnetic Interference)

Resistance to ground measurement is one of the most important measurements in automated equipment. Each individual part is measured, like equipment body, conveyor reel, gripper, nozzle, jig, support table/pins etc. In additional to this point-to-point or surface resistance have to measure from all surfaces which take a contact with ESDS items. The appropriate probes can be used.

Electrostatic potential is measured from ESDS and PWB assemblies by qualified meter. Electrostatic field is measured from insulating materials according to manufacturer's instructions of meter. The large conductive and grounded area can affect to the measurements of potential and field.

For evaluation of real ESD risk in automated processes, the handled ESD sensitive devices or products have to be analyzed when production is ongoing. The methods are to measure the potential and charge of the devices. The charge can be measured by individual charge meter or by measuring the discharge curve from the charged device. From the discharge curve the discharged current, energy and charge can be calculated. There are no exact acceptance levels; they must be analyzed according to ESD sensitivity of the device in case. Some requirements are in the table 1.

So called contact voltmeters (CVM) offer another opportunity to detect electrostatic charges. These CVMs are electrostatic voltmeters with a high input impedance ($> 1 * 10^{14}$ Ω, better 10^{15} Ω) and a low input capacity. Thus, electrostatic charges can be measured directly on the ESDS of the PWB without any damage. Those CVMs are new, so just a few tests have been realized. Electrostatic charges of about 200 V were measured on ESDS in the SMT process. Further practical tests will be realized in the next time to detect possible electrostatic charges.

First Measurement results

The main focus laid on measurements of electrostatic charges or on voltages on electronic components or PCBs. Picture 3 shows some measurement results, which were recorded directly in front of a pick-and-place machine. Further measurements were realized behind the soldering print machine, AOI and ICT. All measurements show, that electrostatic charges or electrostatic voltage are mostly above the limit of 100 volts.

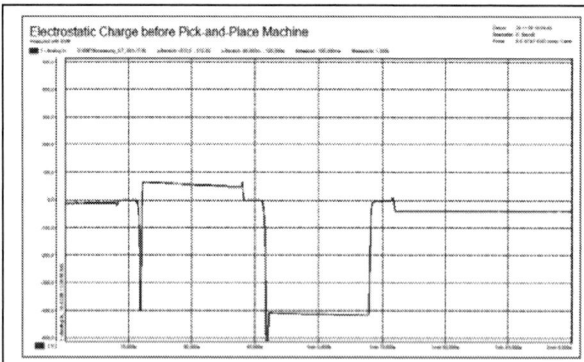

Picture 3 Electrostatic voltage of a PCB, in front of a pick-and-place machine (sample)

Further on, the electrostatic charge of different reels/blisters was measured. Therefore, the charge at various processing steps was determined (see table 2).

Carrier material Reels	Cover tape	Carrier	ESDS
Paper	- 50 V - 200 V	- 500 V - 700 V	+ 30 V + 50 V
Thermoformed plastic	+ 800 V	+ 6000 V	+ 20 V
Thermoformed plastic	- 100 V	+ 3000 V	+ 100 V

Table 2 Measurement results on complete reels with ESDS

Picture 4 shows a possible measurement arrangement at a tape-and-reel machine.

Picture 4 Blister on a feeder with ESDS, electrostatic voltage measured with a small field sensor

CONCLUSION

In conclusion, it can be said, that there are many sources for electrostatic charges in a SMT production line. The different steps lead to high or low electrostatic charge. First measurements were realized to determine the true charge. A great problem is that all machines must be stopped for the measurements. In the future it is very important, that we can measure in a normal machine, while operating.

Presently, the only possibility is ionization. The grounding of all parts does not suffice; it is just a basic requirement. The moving parts CB and ESDS cannot be grounded. Nevertheless, these are the parts, which are electrostatically charged and which cause the damages.

FAILURES OF PCB AND ESDS LEAD TO A REDUCTION OF THE PRODUCT QUALITY!

REFERENCES

[1] IEC 61340-5-1 Electrostatics - 08.2007: Part 5: Specification for the protection of electronic devices from electrostatic phenomena, Section 1: General requirements

[2] IEC 61340-5-2 Electrostatics – 08.2007: Part 5: Specification for the protection of electronic devices from electrostatic phenomena, Section 2: User guide

[3] ANSI/ESD S20.20-2007 ESD Association standards for the Development of an Electrostatic Discharge Control Program for – Protection of Electrical and Electronic Parts, Assemblies and Equipments

[4] ESD SP10.1-2000 ESD Association standard practice for Protection of Electrostatic Discharge Susceptible Items – Automated Handling Equipment (AHE)

[5] J. Paasi, P. Tamminen, H. Salmela, J-P. Leskinen, T. Viheriäkoski, "ESD Control in Automated Placement Process", Proc, EOS/ESD Symposium EOS-27 (2005)

[6] Wunsch, D.; Bell, R. R.: Determination of threshold failure levels of semiconductor diodes and transistors due to pulse voltages. IEEE Transactions on Nuclear Science, NS-15, 1

MECHANICAL FAILURES IN PB-FREE PROCESSING: EVALUATING THE EFFECT OF PAD CRATER DEFECTS ON PROCESS STRAIN LIMITS FOR BGA DEVICES

John McMahon and Brian Gray
Celestica Inc.
Toronto, ON, Canada
jmcmahon@celestica.com

ABSTRACT

The increased temperatures associated with Pb-free processes have produced significant challenges for PWB laminates. Newly developed laminates have different curing processes, are commonly filled with ceramic particles or micro-clays and can have higher T_g values. These changes which are aimed at improving the materials resistance to thermal excursions and maintaining electrical integrity through primary attach and rework operations have also had the effect of producing harder resin systems with lower fracture toughness.

Industry guidelines for mechanical stress limits were developed for materials processed using eutectic SnPb solders. A series of design and material implementations have gained wide acceptance by the industry to address mechanical failures at the corners of area array packages. Current accepted levels of process strain were established when the dominant and limiting failure mode was interfacial fracture (IFF) in complex intermetallic compound (IMC) layers at the solder / package interface. Changes in packaging processes, conversion to Lead (Pb) free solders and the subsequent compensation by laminate suppliers have produced significant shifts in failure mode occurrence and the "Pad-crater" failure mode has become far more common than IFF. By conducting a testing program that focuses on materials and geometries consistent with high complexity "Enterprise Computing and Telecomm" electronic assemblies, and evaluating the results against the current industry guidelines it is possible to determine whether the dominance of the "Pad-crater" defect mode will require revision of process strain guidelines. Test methods, test results, failure analysis and likely mitigation techniques are discussed.

Key words: Pad crater, spherical bend test, mechanical failure mode, process strain.

INTRODUCTION

The transition to lead free assembly of medium and low complexity products is essentially complete. Products designed for sale directly into the retail marketplace such as handsets, smart phones, gaming systems, PCs and Net books, are all routinely built with lead free solder alloys required to comply with widely enacted environmental legislation. The vast majority of these products are built

with alloys in the Tin Copper (Sn-Cu) and Tin Silver Copper (Sn-Cu-Ag) alloy systems. There are a wide variety of choices available but the common impact is that the minimum solder joint temperature for proper re-melting of solder spheres and powders is in the 230 to 240C range. The industry struggled initially. While there have been some very extensive field failure issues related to material selection, in general the materials required for robust assemblies in these less critical reliability, high replacement rate product categories are now readily available. Proper analysis and material qualification and product design will result in "wear out" failure mechanisms and warranty exposures that are in line with those experienced with eutectic tin lead (SnPb) assembly processes.

Less is known about the mechanical robustness of lead free systems. Considerable effort has expended on drop shock improvements for handheld devices and there are a plethora of new additions to low silver alloys attempting to enhance the energy absorption of these alloys, but comparatively little work has been documented on mechanical failures of larger assemblies. Laminate suppliers have over time implemented changes to reduce the Z-axis expansion and raise decomposition temperatures to make their materials robust in extended thermal excursions. Unfortunately these implementations have had a detrimental effect on some mechanical properties of laminates, particularly toughness.

Roggerman et al.[1] and others have published "cold ball pull" and "hot pin pull" testing results which identify that filled phenolic cured FR4 epoxy laminate systems fail at lower loads and absorb less energy to failure than unfilled resins from the same group. These micro particle fillers have been introduced to reduce Z-axis expansion and are widely implemented. We believe that this category of resin system represents the limiting case for mechanical integrity.

Mechanical failures in BGA solder joint systems have been categorized into ten modes to simplify cross company discussion and acceptance of standardized testing. There is wide industry agreement that Mode 3, failure at the package NiP/SnNi IMC layer was the limiting case when current procedures were designed. There has been some movement in the industry to convert to Cu-Sn based interfaces at the package side and this has reduced the number of maverick lot incidents considerably. This change combined with conversion to lead free processing has produced a new dominant failure mode. Anecdotally mechanical failures in lead free systems are almost always reported as Mode 10, pad lift/pad crater unless some significant process defect is present in the solder.

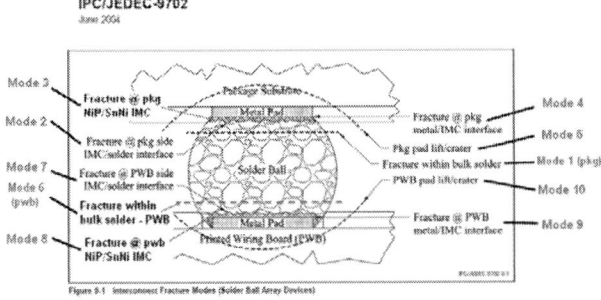

Figure 1. Failure modes in BGA solder joint systems [2]

However, the conversion to Lead free assembly is not complete, The challenges facing the electronics industry over the next few years are more formidable than those that have been recently overcome. With the expiry of the "lead in solder" exemption now targeted at 2014, Enterprise computing and Telecommunications equipment must convert to lead free assembly processes. These server room and backbone type products are designed to the maximum area that can be processed through standard SMT, wave solder and test equipment. Outline dimensions of 16 x 20 inches (40 x 50 cm) are typical. Current products typically have 20 to 30 Cu layers and thicknesses of 0.100 to 0.130 inches (2.5 to 3.5 mm) but certainly higher layer counts and thicknesses of 0.25 inches (6 mm) are on the roadmaps for these products. The high thermal mass associated with this type of assembly can drive a 5X to 10X increase in the heat energy requirements when compared to PC and consumer products. These increased energy requirements translate into longer thermal profiles and extended exposure times at high temperatures for all processing steps.

(ASICs) that usually exist as high I/O BGA packages. Power and I/O characteristics usually require that these are flip chip devices based on built up substrates with metal heat spreaders. The combination of thick laminate structure and large stiff package design almost certainly defines the maximum stress condition and therefore the lower boundary condition for mechanical integrity under flexure in high complexity assembly.

Thermal analysis of a typical high complexity product processed through standard 10 zone and 12 zone SMT ovens (Figure 1.) clearly indicates that proper reflow of large BGA devices on this type of product will expose the laminate to temperatures near the 260°C limit.

Evaluations intended to optimize process parameters and materials targeted for high complexity assembly must be performed on representative laminate stacks, package types and must be based on extended thermal profiles.

The intent of the current work is to validate that the current material sets defined for lead free solder processing can with stand equivalent mechanical stress excursions relative to those defined for Eutectic tin lead systems. This work will be conducted on test vehicles constructed and processed to be consistent with high complexity assembly.

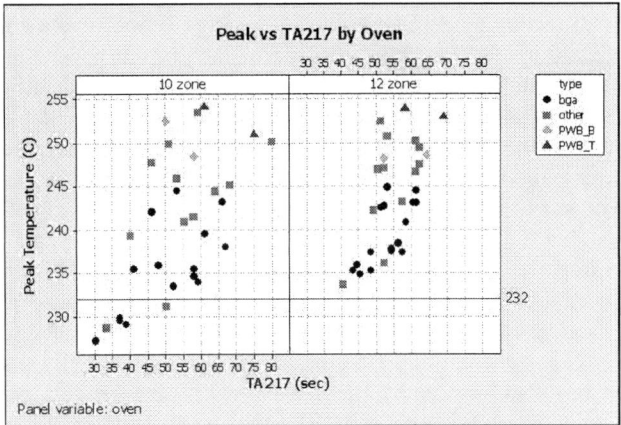

Figure 2. Thermal Analysis of High Complexity Reflow

EXPERIMENTAL DESIGN
The experiment was designed to assess the mechanical strain limits for lead free high complexity assembly and characterize the effect of lead free alloys and extended thermal requirements on safe working limits for board flexure in terms of peak strain and rising strain rate.

Surface strain analysis in PCBAs is the method by which we normalize a whole group of sub-parameters. Surface strain in uniform slabs is a function of deflection or curvature and board thickness (distance from the neutral axis of the slab). In electronic assembly it is complicated by non uniform reinforcement of the system by soldered components. Our interest is actually in the stress/strain concentrations that are inherent in the soldered connections. Specifically the stresses that are imposed on the solder, the interfaces of the solder joint and the resin systems that are directly in contact with the solder pads. In this work we have introduced two factors which are "designed" to generate variation in the results. These are board thickness and strain rate. The other three factors: sphere alloy, laminate and pad plating are under study. The expectation is that each combination of strain rate and board thickness will generate a separate distribution of failures. The experimental design is actually to produce six separate evaluations of the reduced factor list outlined in Table 1.

Factor	Levels
Sphere alloy	SnPb / SAC105 SAC305 / SAC405
Laminate	Standard Filled Phenolic / Toughened Filled Phenolic
Pad Plating	OSP / ENIG

Table 1. Material Factors

Strain Rate

The strain rate dependence of fracture in epoxy resin systems is well documented in basic materials research, and has been widely incorporated into existing industry specifications and guidelines, such as IPC/JEDEC-9702 - Monotonic Bend Characterization of Board-Level Interconnects[2] and IPC/JEDEC-9704 - Printed Wiring Board Strain Gage Test Publication [3]. A proper treatment of this topic is beyond the scope of the current paper, but testing for this experiment was targeted at three specific Principal Strain rates intended to cover the acceptable ranges of all major assembly processes. Those three targeted Principal strain rates are 1000, 3500, 7000 micro strain per second (μe/s).

Materials

Eight sub-lot variations of a mechanical test vehicle were procured. Solderable surfaces were plated in both OSP and ENIG to generate interfaces based on both Cu_6Sn_5 IMC and Ni_4Sn_3 IMC systems. The PWBs were obtained in two laminates provided by Isola. The first was a standard filled phenolic cured FR4 and the second was a non-commercial variant of the first which had been modified to reduce room temperature Young's modulus by approximately 40% in an attempt to toughen the resin system.

Two versions of physical design were generated, each with identical footprints and outlines but with two distinct laminate stacks to represent incremental levels of assembly complexity. The first version was made up of 20 copper layers and had a nominal thickness of 0.100 inches (2.54 mm), the second contained 26 copper layers and had a nominal thickness of 0.130 inches (3.3 mm). The 185 x 185 mm test vehicles have a single BGA footprint for a 40 x 40 mm – 1.0 mm pitch device.

Figure 3. Package / board construction - 20 layer stack

The 40 mm 1517 I/O built up flip chip packages were daisy chain devices provided by LSI. The package substrates were all plated with SAC305 over copper before spheres of the various alloys were attached. The BGA spheres were provided in Sn37Pb, SAC105, SAC305 and SAC405.

SAMPLE PREPARATION

Thermal profiles were prepared for each of the sphere alloys in the testing program. The SnPb devices were attached with SnPb paste while all of the Lead free devices were processed with SAC387 paste. This induces minor modifications to all of the SAC alloys in the final solder joints but it is typical of production assembly.

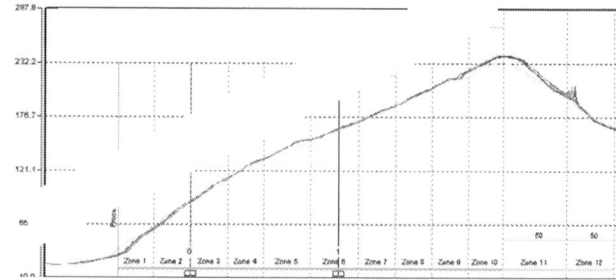

Figure 4. Typical SMT profile - 10 zone - SAC alloy

Assemblies were preconditioned by one pass through the SMT oven prior to BGA attachment to account for the fact that these large devices usually exist on the top side of double sided assemblies. Where forced rework was required assemblies were processed through two further hot gas cycles to simulate removal and replacement of the BGA device. Solder joints from these processes were properly formed with acceptable voiding and typical microstructure.

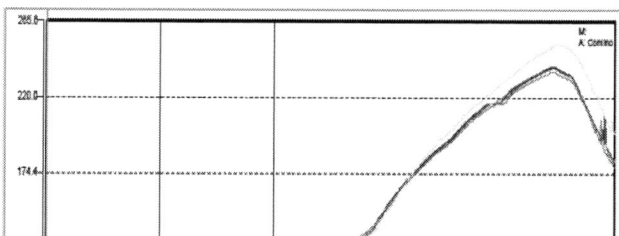

Figure 5. Typical Rework profile - SAC alloy

METTALURGY

XRF examination of the as received BGA devices with spheres attached indicates that silver levels are near the low end of the specification for all of the SAC alloy lots.

Element	Alloy	N	Mean (ppm)	SE
Ag	SAC105	50	8,411	64
	SAC305	50	25,458	80
	SAC405	50	36,374	95
	SnPb	50	491	41

Table 2. Silver content in attached solder spheres

Optical microscopy was performed on cross sections from all of the process lots. There were no remarkable results from primary attach. Solder conformation was normal, very little voiding was evident and the all of the standard phase compositions were measured. Packages are pre-plated with SAC305 directly over Cu so OSP boards produce Cu_6Sn_5 interfaces at both top and bottom interfaces. ENIG boards produce Cu6Sn5 at the top interface and a much more complicated Ni_3Sn / Ni_4Sn_3 / $(Cu-Ni)_4Sn_3$ interface at the board side pad.

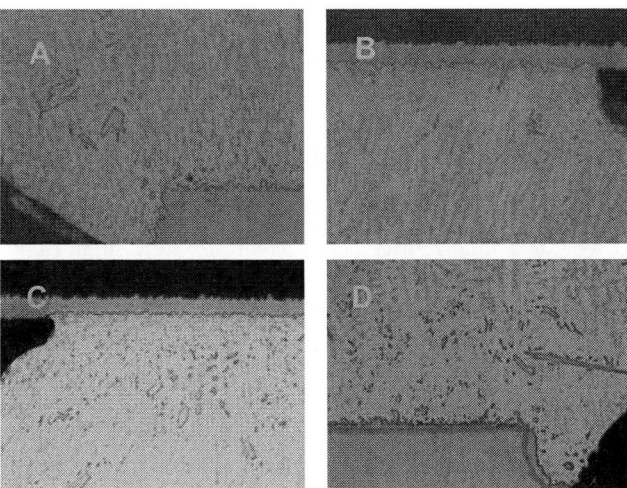

Figure 6. Forced Rework: A & B - OSP; C&D - ENIG

As expected, the extended thermal exposures generated by forced rework produce a fine dendritic network and distributions of primary Cu_6Sn_5 near the interfaces and thicker boundary IMC layers. There is significant consumption of the Cu pad on the package side. We also noted a significant portion of Cu containing IMC over the Ni layer at the board side of the ENIG assemblies.

FLEXURAL TEST BENDING MODE
Various flexural test bending modes have been employed in the electronics industry. IPC-9702 is based on four-point bend with the package aligned parallel to the bending direction. This mode reduces scatter in the results primarily because individual solder joints are reinforced by near neighbors at the same stress level. IPC-9702 was intended to reduce repetitive package qualification testing by standardizing methods. The results are easily compared between packages, but they do not represent the extreme condition and therefore are difficult to translate into safe working limits for tool qualifications and process characterizations. Orienting the package at 45 degrees to the bending direction increases the stress in the corner solder joints and provides a more conservative estimate of flexural limit. Hsieh & McAllister [4] have published an excellent comparative study of the various flexural options and identify spherical bend testing as the mode that generates the highest tensile stress in the corner solder joint for a given displacement from the as built condition. Celestica has selected the spherical bend test setup for this work for three primary reasons.

- It represents the most conservative estimate of Flexural limit.
- It most closely mimics the conditions imposed by "bed of nails" test equipment which is a standard process step for all most product and primary source for board deflection.
- It generates data at all four corners of the package because they are loaded equally.

TEST SETUP AND DATA COLLECTION
The spherical bend test fixture is based on a support plate with eight spherically ground pins evenly spaced on a 120 mm circle. The sample under test is centered on the circle and the load is applied from the back side in the center of the package footprint by another spherically ground pin.

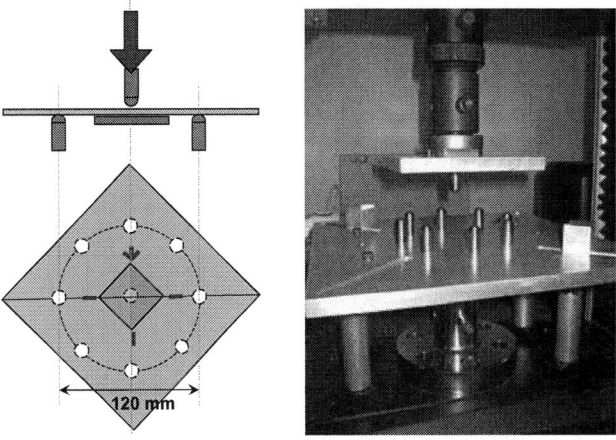

Figure 7. Spherical Bend Test Stand

The effect is to force the flat assembly into an area segment of a sphere whose radius is directly related to the displacement of the loading pin. The attached package acts as a stiffener in the center of the slab and stress is imposed in a manner directly related to the diagonal distance from the center of the package. The effect is to load the corner solder joints to failure. In fracture mechanics engineering terms the loading is mixed mode I & II.

Mode I: A tensile component of the load as the stiffener (package) resists being deformed by flexure imposed on the board. Mode 1 is opening mode when describing a horizontal crack in the solder joint.

Mode II: An in-plane shear stress component as the package resists being stretched as curvature is imposed on the system.

The loading system is described by arrows in Figure 8(A).

The principal strain on the board surface is by design coincident with the diagonal of the package and strain gauges are located on the board at the corners of the package. Where the assembled test unit is relatively compliant, gauges attached to the package corner in the same orientation also provide information. However the heavy metal heat spreader and its attachment to the package substrate make that information undecipherable for this sample set.

Data collection is setup to simultaneously record resistance in the daisy chain, strain in the six gauges attached to the sample, displacement of the test head and the load induced by that displacement. The six strain gauges record diagonal strain and rising strain rate at each corner and the additional two at a single corner allow calculation of the principal strain and principal strain rate.

RESULTS

Previous work with more compliant systems has allowed us to record either the change in resistance that signifies failure at one of the solder interfaces or a localized minimum in the strain profile that indicates a laminate failure in either the board or the package substrate. Preliminary trials on this sample set demonstrated that these very stiff "high complexity typical" systems do not store enough energy to create these very small events at low strain levels. Strain events do occur at higher levels of displacement but they are related to catastrophic failures and do not represent the first damage to the system. Alternatively events may occur at low strain levels but the systems produce enough signal noise from vibration effects to make the events undetectable.

These preliminary results also determined that Mode 10, "Pad crater" was not only the dominant failure mode under this test method, it was the only failure mode detected in assemblies from standard "Primary Attach" processes.

A more intensive destructive evaluation of samples subjected to increasing levels of displacement (flexure) determined that for this test setup three distinct displacement zones can be defined.

- Zone 1: A safe zone where no damage occurs
- Zone 2: A mixed zone where package corners both fail and survive.
- Zone 3: A zone where all package corners fail.

The intent of further testing was to define the boundary conditions for Zone 2 and generate distributions that would allow extrapolation to safe working limits. We defined a "step stress" procedure to test groups of samples to progressively higher peak strains at fixed displacement rates. For each data point peak strain, rising strain rate and outcome were recorded. The distributions of these estimates of survivable strain were used to generate working limits.

This work also required that we redefine our criteria for failure. In compliant systems where "strain events" are recorded, failed samples inspected after penetrating dye has been applied exhibit complete separation of the BGA structure from the board. The crack path is very consistent; it involves a cone shaped failure in the "butter coat" of the laminate and follows the top surface of the glass bundles in the first reinforcing layer. We have yet to see any work in the literature that investigates the crack path. We normally assume that cracks initiate at the surface but it has been postulated that there could be considerable internal separation and coalescence of micro cracks below the surface before this catastrophic failure occurs.

As result of our analysis we have defined two distinct portions of the Pad crater crack, which represent two different levels of failure categorized as cohesive and adhesive separation.

- Cohesive Separation: The crack initiates at the surface of the laminate in close proximity to the solder pad and travels at approximately 45 degrees to the first level of reinforcement.

- Adhesive Separation: The crack subsequently follows the top surface of the first reinforcement layer until it finds a short path back to the surface inboard of the solder pad.

We understand now that in these stiffer systems the cohesive portion of the crack occurs without generating any discernable disturbance in the strain profile, but we believe that it represents a significant risk to product shipping into service environments.

If we conceptualize a crack tip as an infinitely small radius then it represents an infinitely high stress concentration. We should accept that once the crack has initiated it will require much less energy for that crack to grow. Even benign service environments have significant amounts of thermal or vibration energy present which could over time grow these cracks. There are also the issues with moisture migration and condensation when the internal glass bundles are in contact with atmosphere.

To summarize, any damage to the laminate resin system discernable by dye penetration and optical microscopy is considered to be a failure.

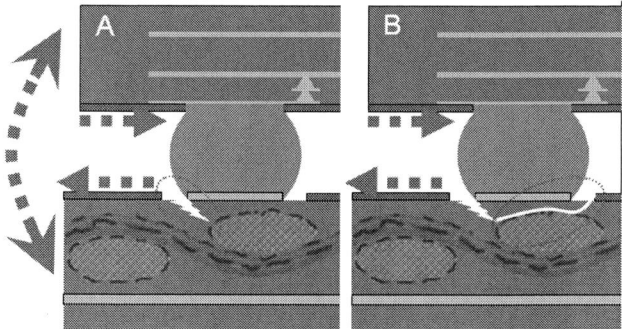

Figure 8. Pad Crater A- cohesive separation, B – Adhesive separation

Failure distributions were best modeled using Weibull distributions for two reasons.

- The failure rate is expected to be constantly increasing as the stressing factor is increased. In fact, there is appoint at which no further survivors will be detected. This is the boundary between Zone 2 and Zone 3.
- The data includes both failures and survivors. In statistical terms the data is right censored.

As expected the scatter in the results increases as the testing strain rate is increased. In general the distributions of all data at the various strain rates are consistent with the behavior currently accepted by the industry. The survivable peak strain decreases as rising strain rate increases.

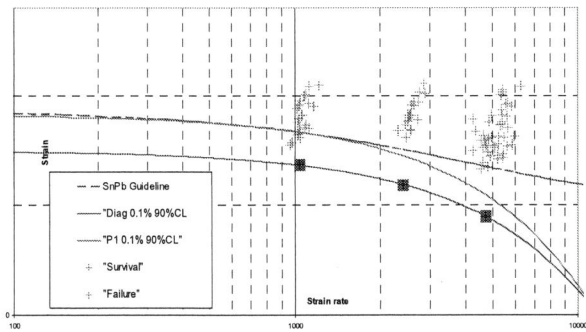

Figure 9. Typical Strain / strain rate behavior in FCBGA devices for a single board thickness

Distributions are combined for individual board thicknesses. Data and calculations of safe limits were plotted in the common strain / strain rate format on log linear graphs. Basic regression techniques were used to generate safe working limits over the range strain rates associated with manufacturing processes. Figure 9 is a typical output from this process and displays individual data points, and the two limit curves based on diagonal and principal strains respectively. This process was repeated for all board thicknesses under study. This data was compared against the current estimate of the board thickness relationship and seems to correlate well. There is some evidence that the strain rate dependency for this fracture mode is different than the relationship currently accepted by the industry for SnPb assembly.

The significant additional thermal exposure of laminates and solder pads caused by forced rework of large BGA devices produces changes in the board side solder interfaces and where ENIG pad plating is present introduces a second failure mode. Mode 8 – Fracture at the NiP / NiSn IMC layer at the board side was identified as a second failure mechanism but Mode 10 remains dominant. The occurrence of two failure modes did not significantly affect the distribution of the data or the calculated limits.

CONCLUSIONS

1. Mode 10 – "Pad Crater" is the dominant fracture mode for solder ball array devices when assemblies are fabricated using common Lead free process capable materials.
2. The Pad crater failure mechanism follows the strain rate dependent nature well established for brittle materials.
3. Safe working limits for monotonic failure in solder ball array devices can be established by board thickness for these materials.
4. Secondary fracture modes can be introduced by extended thermal exposures typical for hot gas rework of BGA devices.

FUTURE WORK

There are three additional test phases currently planned.

- We are continuing the test program to include more sub-lots that have been processed though extended thermal exposures to validate that automated rework processes do not produce a separate population across the range of strain rates associated with assembly processes.
- We are pursuing a cyclic test program to determine if multiple stress/strain excursions at what we now classify as safe levels can initiate Pad Crater type cracks.
- We are evaluating design modifications to mitigate this fracture mode.

ACKNOWLEDGMENTS

The authors gratefully acknowledge the contributions from Edward Kelly of Isola and Zafer Kutlu of LSI for their contributions in time and the material contributions from their corporations that made this testing program possible.

REFERENCES

[1] Roggeman B, Borgesen P, Li J, Godbole G, Tumne P, Srihari K, Levo T, Pitarresi J. Assessment of PCB Pad Cratering Resistance by Joint Level Testing, Electronics Components and Technology Conference, 2008.

[2] IPC Standard 9702 Monotonic Bend Characterization of Board Level Interconnects

[3] IPC/JEDEC-9704 - Printed Wiring Board Strain Gage Test Publication

[4] Hsieh G, McAllister A, Flip Chip Ball Grid Array Component Testing Under Board Flexure, EC&TC 2005

[5] He MY, Hutchinson JW. Asymmetric four-point crack specimen. J Appl Mech 2000;67:207–9.

[6] Mukadam M, Long G, Butler P, Vasudevan V. Impact of Cracking Beneath Solder Pads in Printed Circuit Boards on Reliability of Ball Grid Array Packages,

[7] Hornmug R, Smith M, Welch D, Sherman L, Goldman P. IPC Meeting Minutes, Laminate and Prepreg Materials Subcommittee. 22-Sep-2008 (http://members.ipc.org/committee/Minutes/3-11_m_0809.pdf)

QUALITY AND RELIABILITY ANALYSIS OF LEAD FREE PCBs IN SIMULATED PRODUCTION CONDITIONS AND LONG TERM USE

Sammy Shina, Ph.D., University of Massachusetts, Lowell, MA, USA
Greg Morose, Sc.D., Massachusetts Toxics Use Reduction Institute (TURI), Lowell, MA, USA
Richard Anderson Ph.D., and Helena Pasquito, Cobham, Lowell, MA, USA
Bob Farrell, Benchmark Electronics, Hudson, NH, USA
Don Longworth and Wendi Boger, DDi Inc., Newburyport, MA, USA
Ken Degan, Teradyne Inc., North Reading, MA, USA
Donald Abbott, Ph.D., Texas Instruments, Attleboro, MA, USA
David Pinsky, Karen Ebner, and Amit Sarkhel,
Raytheon Company, Tewksbury, MA, USA
Michael Miller and Louis Feinstein, Textron Systems, Wilmington, MA, USA
Deb Fragoza and Eric Ren, EMC, Hopkinton, MA, USA
Charlie Bickford, Wall Industries, Exeter, NH, USA

The New England Lead-free Electronics Consortium is a collaborative effort of New England companies spanning the electronics supply chain, created by the University of Massachusetts Lowell in 1999 and sponsored by the Toxics Use Reduction Institute and the U.S. EPA. The consortium has completed and published the results of three phases of manufacturing and testing of lead-free Printed Wiring Boards (PWBs) with the goal of achieving zero-defect lead-free soldering processes with comparable or superior reliability to that of leaded solder processes.

In this fourth phase of testing, which began in 2007, several simulated conditions of assembly and rework processes were evaluated in a matrix of multiple levels of components, PWB lead free surface finishes and solders, and compared to a baseline of leaded equivalent materials and processes. Both Through Hole (THT) and Surface Mount (SMT) Technologies were evaluated.

All quality and reliability testing was performed with industry standard methodologies, using specially trained production inspectors for the quality evaluation, and extreme thermal cycling and vibrations for reliability testing.

Our results indicate that with proper selection of currently available (2009) materials and finishes and careful control of the assembly processes, successful lead free assembly and rework can be achieved. Comparison of different strategies for rework, and recommendations for least copper dissolution for THT technology processes are discussed in chapter 6 of a book [1] published by the authors. Reliability testing showed inflection points for leaded versus lead free reliability that is currently being investigated through means detailed later in this paper

Key words: Lead Free, Visual Inspection, Rework, Reliability testing, Thermal Cycling, Design of Experiments.

INTRODUCTION

In January 2003, The European Union (EU) published Directives 2002/96/EC on Waste Electrical and Electronic Equipment (WEEE) and 2002/95/EC on the Restriction of the use of certain Hazardous Substances in electrical and electronic equipment (RoHS). These emerging directives have been the primary drivers for a global movement toward lead free electronics.

NEW ENGLAND LEAD-FREE CONSORTIUM

The consortium has been very successful in maintaining a synergistic and close working relationship between its ever changing and expanding member companies since 1999. The unique success of its endeavors is due to the fact that the member companies volunteer expertise and knowledge of their personnel and supply materials, production and test equipment towards the consortium projects. In return, they get to select particular component geometries, materials and technologies that they are interested in, jointly decide on the testing methodologies and share the results. In addition, member companies are able to work collaboratively with competing or potential customer companies through the consortium framework. The University of Massachusetts Lowell (UML) and the Toxics Reduction Institute (TURI) have provided project management, communications and leadership throughout the different phases of testing. Funding was initially provided by TURI for Phases I and II, and from the EPA for phases III and IV.

MATERIAL SELECTION FOR LEAD FREE EVALUATION

The selection of the materials was constrained by the resources available and the amount of testing parameters to be evaluated. The technique of Design of Experiments (DoE) was used to try to separate the effect of each parameter on the overall performance of quality and reliability of the PWBs. The test parameters were as follows:

1. Components. There were 886 SMT components (BGAs, microBGAs, resistors, TSOPs, PQFPs, PQFN, and MLFs), and 21 THT components (connectors, LEDs, DC/DC convertors, and capacitors) provided for assembly on each side of each test vehicle.

2. Solder types. There were 24 PWBs that were assembled with lead free materials and solders, 8 that were assembled with leaded solder and 3 Halogen free PWBs that were assembled with lead free solders.

3. PWB finishes. There were 4 types of surface finishes:
- Electroless Nickel Immersion Gold (ENIG). This surface finish involves using both electroless and immersion technologies to deposit the metallic surface finish.
- Hot Air Solder Leveling (HASL). Lead free alloy Sn100C was used. It is comprised of mostly tin, but also includes 0.6% copper, 0.05% nickel, and 0.0055% germanium.
- Organic Solderability Peservatives (OSP).
- Nano materials surface using nanosilver particles dispersed in a polymer (polyaniline), with a thickness between 45 to 65 nm. This was selected because it has the potential of addressing major lead free implementation challenges such as copper dissolution during rework and process improvement for assembly of lead free THT components.

4. Solder Compositions. There were 4 different solder pastes for assembly of the SMT components:
- Tin/silver/copper alloy (SAC305) with no clean chemistry flux (from two different suppliers).
- Tin/silver/copper alloy (SAC305OA) with organic acid chemistry flux.
- Tin lead alloy with no clean chemistry flux for baseline data source.

Three (3) different solder alloys were used in for the assembly of the THT components:
- Tin/silver/copper alloy (SAC305).
- Tin/copper alloy (Sn100C) at two different operation settings, 295 and 310 temperatures
- Tin/lead alloy for baseline purposes.

5. Laminates. Two different laminate materials were used:
- FR-4 laminate material was designed for use in lead free assembly environments (32 test vehicles) and has a glass transition (T_g) temperature of 180°C.
- Halogen-free flame retardants laminate with a glass transition (T_g) temperature of 180°C was used for three (3) test vehicles.

6. Test Vehicles. Thirty-five (35) test vehicles were assembled at two locations and shown in Figure 1. The test vehicles were 8" (inches) wide by 10" long, contained 20 layers and are 0.110" thick.

7. Experiment Matrices: These matrices are based on DoE principles and were used to selectively determine individual contribution of each parameter. They are shown in the following Tables 1, 2 and 3 for the 35 test vehicles

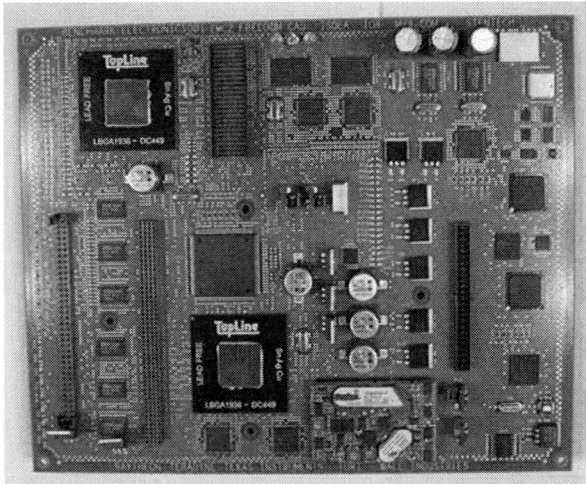

Figure 1. New England Consortium phase IV Vehicle

Table 1. Lead free Test Vehicles, Phase IV DoE

Test Vehic.	SMT Solder	TH Solder	Surface Finish	Lam- inate
1	SAC305 - 1	SAC305	ENIG	FR4
2	SAC305 - 1	SAC305	ENIG	FR4
3	SAC305 - 1	SAC305	HASL	FR4
4	SAC305 - 1	SAC305	HASL	FR4
5	SAC305 - 1	SAC305	OSP	FR4
6	SAC305 -1	SAC305	OSP	FR4
7	SAC305 - 1	SAC305	Nano	FR4
8	SAC305 - 1	SAC305	Nano	FR4
9	SAC305OA	Sn100C 1	ENIG	FR4
10	SAC305OA	Sn100C 1	ENIG	FR4
11	SAC305OA	Sn100C 1	HASL	FR4
12	SAC305OA	Sn100C 1	HASL	FR4
13	SAC305OA	Sn100C 1	OSP	FR4
14	SAC305OA	Sn100C 1	OSP	FR4
15	SAC305OA	Sn100C 1	Nano	FR4
16	SAC305OA	Sn100C 1	Nano	FR4
17	SAC305- 2	Sn100C 2	ENIG	FR4
18	SAC305- 2	Sn100C 2	ENIG	FR4
19	SAC305- 2	Sn100C 2	HASL	FR4
20	SAC305- 2	Sn100C 2	HASL	FR4
21	SAC305- 2	Sn100C 2	OSP	FR4
22	SAC305- 2	Sn100C 2	OSP	FR4
23	SAC305- 2	Sn100C 2	Nano	FR4
24	SAC305- 2	Sn100C 2	Nano	FR4

Table 2. Tin Lead Test Vehicles, Phase IV DoE

Test Vehicle	SMT Solder	TH Solder	Surface Finish	Lam- inate
25	Tin Lead	Tin Lead	ENIG	FR4
26	Tin Lead	Tin Lead	ENIG	FR4
27	Tin Lead	Tin Lead	HASL	FR4
28	Tin Lead	Tin Lead	HASL	FR4
29	Tin Lead	Tin Lead	OSP	FR4
30	Tin Lead	Tin Lead	OSP	FR4
31	Tin Lead	Tin Lead	Nano	FR4
32	Tin Lead	Tin Lead	Nano	FR4

All three (3) Halogen free test vehicles were made with OSP laminate finish, as shown in Table 3. All were soldered with SAC305 (2 NC, one soldered with Organic acid). The Halogen Free laminates were made with FR4.

Table 3. Halogen free Test Vehicles, Phase IV

Test Vehicle	SMT Solder	TH Solder	Surface Finish	Lam-inate
33	SAC305-1	SAC305	OSP	HF
34	SAC305-1	SAC305	OSP	HF
35	SAC305OA	SAC305	OSP	HF

SMT ASSEMBLY PROCESS

SMT component assembly was performed using an electroformed nickel stencil with thickness at the bottom of 0.005", and the top thickness of 0.004". A printer was used with print speed of 0.51 ips, front and rear blade pressure of 19.4 lbs, a separation speed of 0.055 ips, and a separation distance of 0.098". Reflow was performed using an oven with ten (10) heating zones and three (3) cooling zones, with a ramp to peak thermal profile. Three (3) thermal profiles were used as shown in Table 4.

Table 4. Thermal Profiles for Test Vehicles, Phase IV

Thermal Profile	Peak Temp °C	TAL seconds
Tin/Lead Test PCBs	210-218	60-90
Tin/Lead LF BGA (top)	222-230	60-90
Lead Free Test PCBs	240-248	60-90

THT ASSEMBLY PROCESS

The Soldering equipment used had robotic multiwave and selectwave soldering capabilities. The multiwave process uses a robot system to pick up, hold, and dip the test vehicle onto multiple nozzles that are mounted on a product specific nozzle plate; hold time being referred to as "dwell time". In order to reduce thermal stress, a target preheat temperature was used in the range of 110-115°C for all 35 test vehicles. A summary of the parameters used for soldering THT technology test vehicles is given Table 5.

Table 5. THT Soldering Parameters

Parameter	SAC 305	Sn100 (1)	Sn100 (2)	Tin Lead
Vehicle Number	1 – 8	9 - 16	17 - 24	25-32
Multiwave Pot Temp.	295°C	295°C	310°C	270°C
Multiwave Dwell Time	13 sec.	13 sec.	16 sec.	7 sec.
Selectwave Pot Temp.	300°C	300°C	310°C	270°C
Selectwave Nozzle Size	8 mm	8 mm	8 mm	4 mm

QUALITY INSPECTION RESULTS

Inspection was performed using associates at Benchmark Electronics. They were trained in IPC-A-610 Revision D standard for Class 3 High Performance Electronic Products by a training leader from Cobham. In addition, an X-ray inspection machine was also used for automatic inspection. All defects were identified on a component lead basis, so that a single component can have multiple defects. A total of 4,689 defects were identified for all 35 test vehicles. The tabulation of defects by solders, component technology and defects per test vehicle is given in Table 6, showing no significant differences in total defects/test vehicle type.

Table 6. Summary of Visual Defects

Component Type	LF Test Vehicles (1-24)	Tin/lead Vehicles (25-32)	HF Lamin. (33-35)
SMT	112	24	9
THT	3,062	1,091	391
Tot Defects/TV	132	139	133

SMT INSPECTION ANALYSIS

The inspection data for the SMT and THT component defects were analyzed using the MINITAB ® software. Graphs for Mean effects and interactions are shown in Figures 2 and 3. While all parameters of solder types and surface finish were not significant, the SAC305 OA solder paste had a higher mean defect rate (8.0 defects/TV) than the overall average of 4.1 defects/TV. Conversely, the nano surface finish had the lowest mean defect rate (2.75 defects/TV), compares with the other finishes that ranged between 4.0 and 5.5 defects/test vehicle.

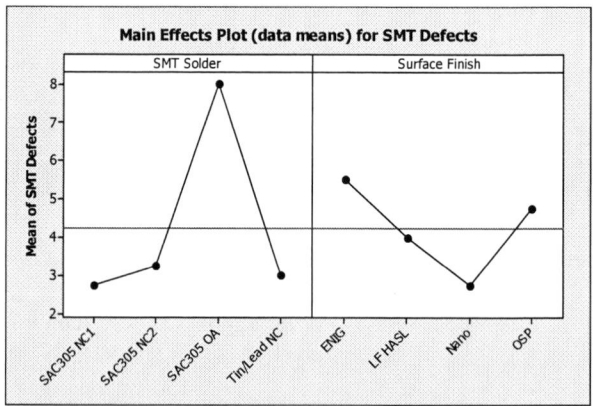

Figure 2. Main effects plot for the SMT defects

The parameter interaction plot, shown in Figure 3, indicates uneven performance between the different surface finish parameters when compared against multiple solder compositions. The combination with the highest defect level was the SAC 305 OA solder paste and the OSP finish with 10 defects/PWB. The lowest mean defect level was the tin/lead solder paste and the nano surface finish with zero defects/PWB.

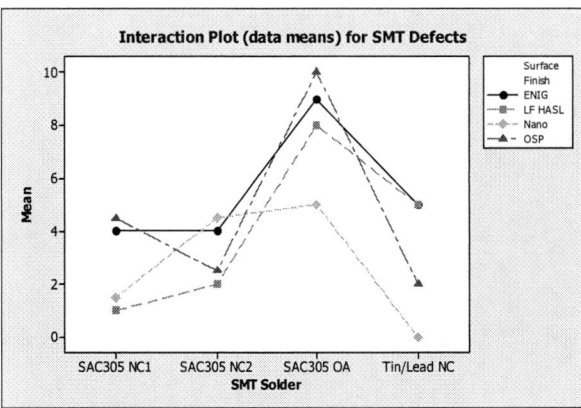

Figure 3. Interaction Plot for SMT Defects

The nano surface finish, had the lowest overall defect rate for all types of solders tested, while it had the highest defect rate for the SAC305 2 solder paste. This indicates a possible interaction in this combination. This plot also reveals that the HASL surface finish had the lowest defect rate for both the SAC305 1 and SAC305 2 solder pastes.

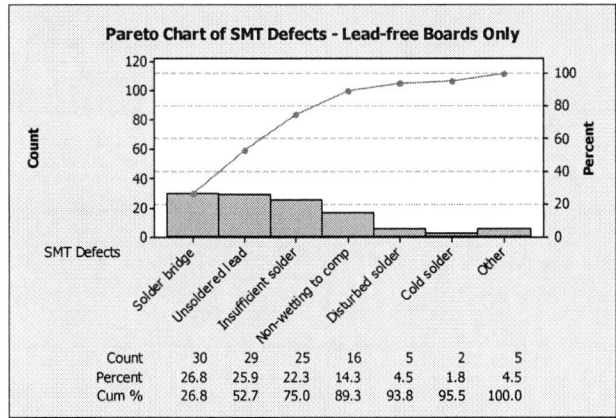

Figure 4. Pareto Chart of SMT defects

Figure 4 is a Pareto Chart for the SMT defects for lead free TVs only. The chart reveals that solder bridges, unsoldered leads, insufficient solder, and non-wetting to component were the most prevalent defect categories.

The three (3) Halogen free test vehicles had a much lower defect rate (3.0 defects/TV), when compared to the FR4 equivalent (7.25 defects/TV), all having OSP surface finish and soldered with two types of lead free solders.

THT INSPECTION ANALYSIS
THT analysis for Mean effects and interaction are shown in Figures 5 and 6. Due to drift that occurred during the multiwave operations, only sixteen test vehicles were analyzed (one replicate only). The interaction Plots in Figure 6 indicate the need for careful selection of materials to obtain good quality visual results. HASL exhibited the best overall performance, while ENIG had the worst variability. More testing is required as solder suppliers tune their materials.

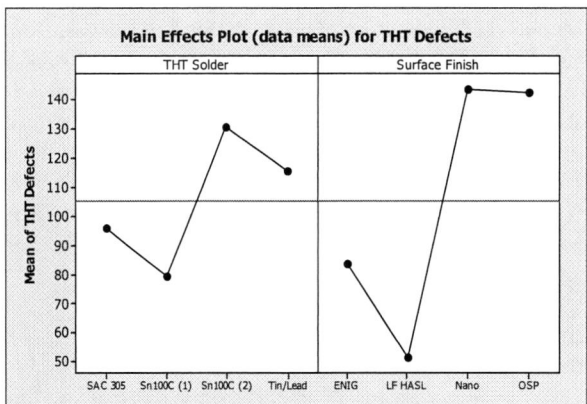

Figure 5. Main effects plot for the THT defects

Unlike the SMT interactions (Figure 3), the Nano and OSP finishes in THT testing showed similar interactions versus solders as shown in Figure 6.

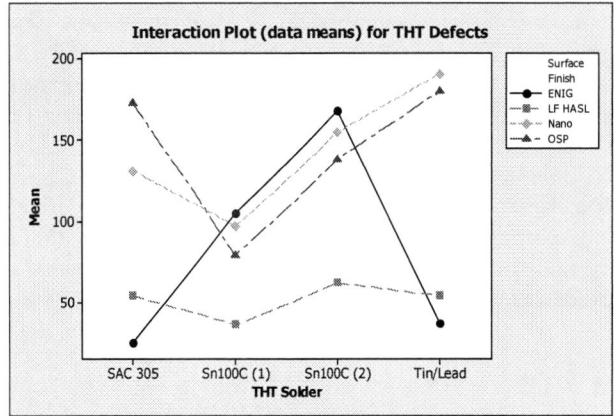

Figure 6. Interaction Plot for THT Defects

For the tin/lead test vehicles, 98.5% of the defects were either `insufficient solder" or "solder bridging." For the assembly of THT components with lead-free solder, the most prevalent defect type was insufficient solder (53.6%), the second most prevalent was solder bridging (37%), and the third was solder splatter (5.5%).

The three (3) Halogen free test vehicles soldered with tin/lead had a much lower defect rate (130 defects/TV), when compared to their equivalent FR4 test vehicles (177 defects/TV), all having OSP surface finish and soldered in two types of lead free solders (SAC 305 and NC OA).

VISUAL INSPECTION CONCLUSIONS
The defect rate was much higher for THT than for SMT components in lead free soldering, indicating the need for further THT process optimization. For SMT, no statistically significant difference was found for the 3 solders or within the 4 surface finishes used.

For THT components, there was no statistically significant difference for the solder types, but there was significance due to surface finishes. ENIG exhibited the most variability when used with various solders.

The highest THT defect level was at OSP and nano surface finishes. The nano finish was applied by stripping the OSP finish, then applying the nano finish. In addition, there was a time delay between the SMT and THT assembly. Therefore, it is recommended to try to minimize this time delay by conducting both operations in the same day. In summary, high quality assembly of components with lead free solder is achievable with careful selection of solder and finish materials.

RELIABILITY TESTING PLANS

Reliability testing was completed in two methods: Half of the test vehicles were tested to failure by thermal cycling performed at two locations in MA (Textron Systems and Cobham). The other half were tested to failure due to vibration testing at the Raytheon facility in Towson MD.

Each of the thermally cycled 16 DoE and the two halogen free test vehicles had fourteen daisy chains for continuous monitoring during cycling, using an Agilent 34980A data logger. The thermal cycling test adhered to the IPC-9701, with a dwell time at the temperature extremes of 15 minutes. The authors believed that IPC-9701 was developed to address tin/lead solders, and the SAC solders have a lower creep rate than tin/lead solders, limiting the amount of SAC solder damage during short dwell times. Failure was defined as a maximum of 20% nominal resistance increase within a maximum of five consecutive reading scans. The thermal profile used is shown in Figure 7.

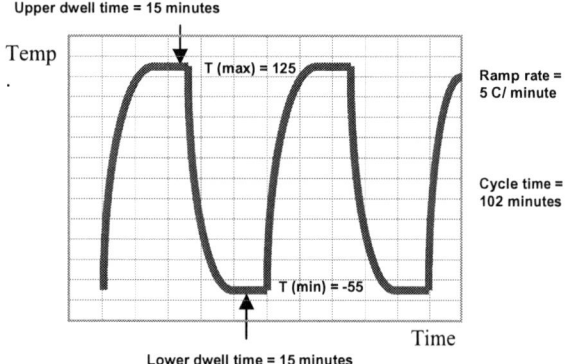

Figure 7. Thermal profile for Temperature cycling

Four of the 14 chains on the test vehicles were connected to discrete components (i.e. 0404, 0603, and 0805 resistors). The other ten daisy chains are connected to one component per daisy chain. Figure 8 shows the actual test setup in the environmental chamber.

THERMAL CYCLING RESULTS

The two test vehicles with halogen-free laminate material had early failures for all components at 220 thermal cycles. All other test vehicles daisy chains had initial component failures, then all failed after 1718 hours. Table 7 lists the component types and failure cycle numbers.

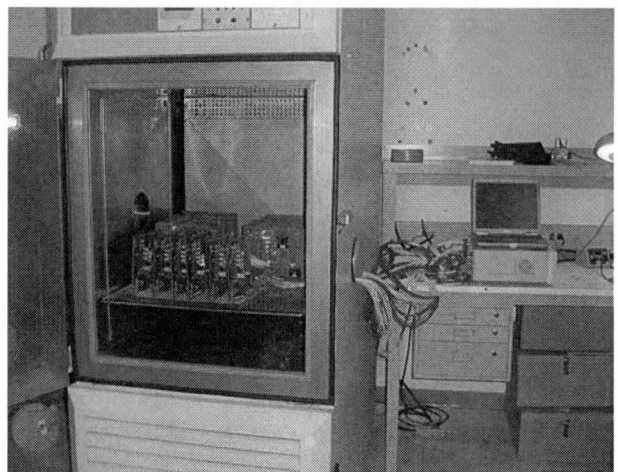

Figure 8. Thermal cycling in the Chamber

The complete failures of all components indicate a common problem that needed to be investigated. The same phenomena also occurred on the vibrations testing.

Table 7. Summary of Component Failures on TV12

Component Type	Failure Cycle
Resistor 0805	801
Resistor 0805	809
Resistor 0603	929
TSOP 48 Pins	972
Resistor 0402	1,129
MicroBGA 84	1,490
CABGA 100	1,718
MicroBGA 64	1,718
PBGA 256	1,718
PBGA 256	1,718
PQFP 208	1,718
TSOP 48 Pins	1,718
TSOP 48 Pins	1,718
TSOP 48 Pins	1,718

WEIBULL DISTRIBUTION PLOTS

A two parameter Weibull distribution has a shape and a scale. A shape value of 3 approximates a normal curve. A shape value lower than two low describes a right-skewed curve, and a value greater than four describes a left-skewed curve. A scale parameter is the 63.2 percentile ($N_{63.2}$) of the data, sometimes referred to as characteristic life. After the plot is generated for a component type, then various points of interest can be calculated such as the number of cycles to 1, 5 and 63.2 % cumulative failure N.

After 1,470 thermal cycles, Nine out of the twelve lead-free test vehicles (75%) have experienced failures for the U16 (ceramic chip array BGA, 100 balls,10 mm pitch with SAC soldered balls)) and three out of the four tin lead PWBs have experienced failures for the same component. Figures 9 and 10 show the Weibull distribution for the U16 component on a lead free and tin/lead Test Vehicles, and Table 8 shows % failures.

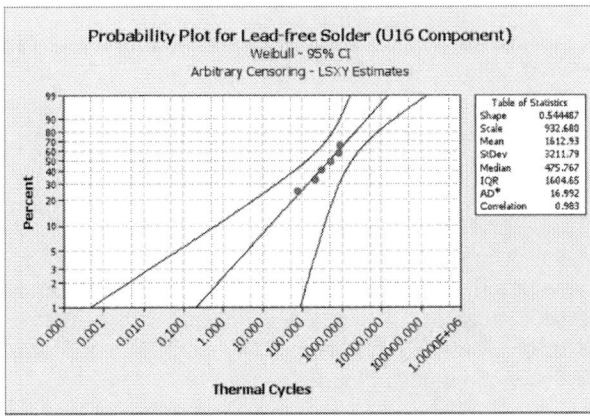

Figure 9. Weibull plot for U16 on a lead Free TV

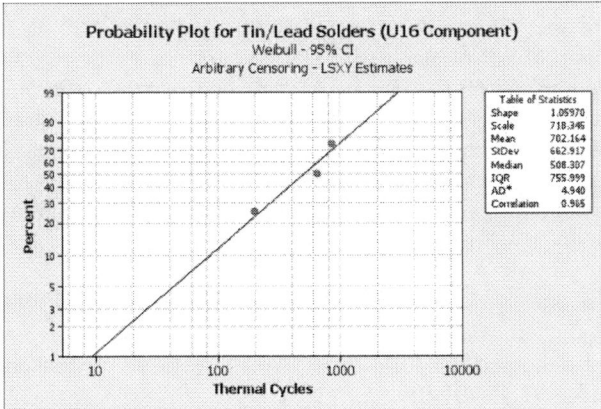

Figure 10. Weibull plot for U16 on a tin/lead TV

Table 8. Percentiles Failures of U16 using Weibull Plots

Percent Failures	Designation	Lead-free Test Cycles	Tin/lead Test Cycles
1	N_1	0.2	9.4
10	N_{10}	15	85.9
20	N_{20}	59.3	174.4
30	N_{30}	140.4	271.5
40	N_{40}	271.6	381.1
50	N_{50}	475.8	508.3
60	N_{60}	794.3	661.5
63.2	$N_{63.2}$	932.7	718.3
70	N_{70}	1,311.60	855.9
80	N_{80}	2,235.20	1,125.60
90	N_{90}	4,314.90	1,578.10

For U16, the tin/lead TVs appear more robust from N_1 through N_{50}, with a crossover point from N_{60} through N_{90} when the lead free TVs are more robust. This might indicate that there are two different failure modes: one is possibly an infant mortality related failure mode and the other is possibly a wear out mechanism failure mode.

The same phenomena occurred with 0805 resistors, the only other component to fail at the 1470 test cycle as shown in Table 9. However it seems that the inflection point, at which the reliability levels of the lead free and the tin/lead test vehicles switch are at different designation points for the two components.

Table 9. Percentiles Failures of 0805 using Weibull Plots

Percent Failures	Designation	Lead-free Cycles	Tin/lead Cycles
1	N_1	370.8	244.3
10	N_{10}	578.8	650.3
20	N_{20}	667.3	889
30	N_{30}	729.4	1,080.80
40	N_{40}	780.8	1,255.30
50	N_{50}	827.3	1,425.50
60	N_{60}	872.2	1,601.30
63.2	$N_{63.2}$	886.8	1,660.70
70	N_{70}	918.6	1,794.30
80	N_{80}	970.5	2,024.90
90	N_{90}	1,038.70	2,350.80

VIBRATION TEST RESULTS

Half of the test vehicles (16 from Tables 1 and 2) and two halogen free TVs were subjected to a rigorous vibration testing, shown in Figure 11.

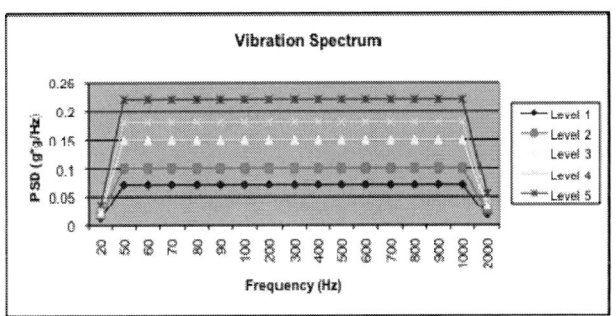

Figure 11. Vibration Spectrums for Test Vehicles

The results were similar to thermal cycling, with all the TVs failing at about the same time. As an example, in TV15, there was one failure (CABGA 100) at 161 minutes, a second failure (Micro BGA 84) at 184 minutes, and then all the remaining eleven components in the daisy chains (Table 7) failed between 232 – 233 minutes.

ANALYSIS OF CATASTROPHIC FAILURES

After making sure that the test setups were accurate, it was decided to perform failure analysis to confirm the causes of catastrophic failures that occurred around the same time for all test vehicles (around 1718 thermal cycles and 233 vibration cycles). The obvious reason could be that the interconnections in the test vehicles gave way under the thermal and mechanical test strains. The procedure for determining the failures is as follows:

I. Identify the network location of the defect as defined by the data logger through impedance measurement of each daisy chain. Using a Fluke 45 Dual display multi-meter with a precision power supply, possible defects in the TVs were located by measuring the resistance between the following 3 locations of each component in the chain:

1. From the power input to the 1st pin of the component
2. From the first to last pin of the component
3. From the last pin of the component back to the output.

The multi-meter was used to measure the resistance of the components and vias in all 34 test vehicles in the DoE.

II. Once these possible defects were located, a thermal Image Flir system (Thermovision A20) camera was used to validate and pinpoint the location of defects. The components were then powered, causing an increase in the resistance in the defect area, resulting in an increase of temperature, which can be captured by the camera. An example of this technique is given in Figure 12.

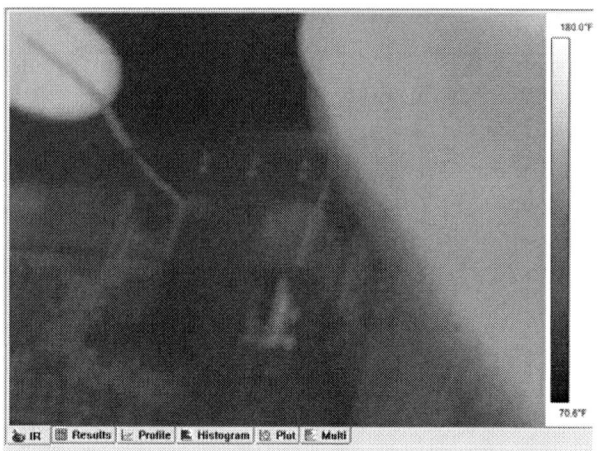

Figure 12. Thermal Camera Image of a possible IC defect

III. The suspect components can then be cross-sectioned to validate the source of the failures, whether it is the solder joint, the internal component wiring or the inter-connects vias on the test vehicles. Based on the data resulting from the cross sections, further analysis can be made as to the sources of the catastrophic failures.

RELIABILITY CONCLUSIONS

Thermal cycling was terminated at around 1700 cycles, after catastrophic failures of all test vehicles. Vibration testing was similarly terminated after 230 cycles. Several issues were concluded while others need to be resolved by further failures analysis of the failed components:

- Halogen free test vehicles had early failures for all components. Further development is needed in Halogen free before it is viable as a bromide replacement for fire retardant functions.
- Test vehicles with High Tg FR4 laminate material are robust with only 2 component types (ceramic chip BGA and 0805 resistor) surpassing 63% failure threshold after 1,470 cycles of severe thermal conditions.
- Ceramic chip BGA component (U16) showed that tin lead is more reliable than lead-free for early

failures, but less reliable for wear out failures. This is a crossover mechanism that indicates multiple failure modes.
- Resistor 0805 resistor showed reverse reliability properties than ceramic chip BGA.

FURTHER STUDY

The consortium plans to complete the reliability and subsequent failure analysis in 2010. The plan is to finish the cross sectioning of all component locations identified by thermal imaging. All results will be published in future conferences and industry journals.

ACKNOWLEDGEMENTS

The authors would like to acknowledge the contributions from the following individuals and corporations for their support of this research:

Freedom CAD for design of the test vehicle, AIM for providing solder for assembly of through hole components, Isola for providing laminate materials for the test vehicles, Ormecon for providing the nano surface finish, International Rectifier for providing components, Stentech for providing stencils, PWB Interconnect Solutions for IST testing.

In addition, the authors wish to acknowledge the EPA Region 1 for providing 2 grants for funding for this project during the period 2006 – 2009 (Award Nos. NP 971586010 and X9-97182701-0)

REFERENCES

[1] Shina, Sammy G., "Green Electronics Design and Manufacturing", McGraw-Hill, New York, April 2008.

[2] Shina Sammy G., "Six Sigma for Electronics Design and Manufacturing", McGraw Hill, New York, April 2002.

[3] Morose, G., Shina S., et al. "Visual and Reliability Testing Results of Circuit Boards Assembled with Lead Free Components, Soldering Materials and Processes in a Simulated Production Environment", proceedings of the APEX 2006 Conference.

[4] Minitab, Interpreting the Shape, Scale, and Threshold on a Weibull Probability Plot, www.minitab.com/support/ answers, August 2008.

[5] Manock, John, et al., Effect of Temperature Cycling Parameters on the Solder Joint Reliability of a Pb-free PBGA Package, SMTA Journal, Volume 21 Issue 3, 2008.

CORROSION DRIVEN WHISKER GROWTH IN SAC305

Keith Sweatman, Junya Masuda, Takashi Nozu, Masuo Koshi, and Tetsuro Nishimura
Nihon Superior Co., Ltd
Osaka, Japan

ABSTRACT
Corrosion has been identified as one source of the compressive stress that is a driver of whisker growth in high tin lead-free solders and in this paper the authors report a study directed at identifying the relationship between the extent of corrosion and the concomitant whisker growth. Printed circuit coupons with an OSP finish were soldered with SAC305 solder using wave, reflow, and hand soldering methods with "no-clean" fluxes typical of current commercial practice. These coupons were exposed to conditions of 40°C/95%RH, 60°C/90%RH and 85°C/85%RH for up to 5000 hours. As well as recording the location of whiskers, their density, and length as a function of time, the extent of corrosion was measured on cross-sections through the solder. The highest incidence and fastest growth rate occurred on test pieces exposed to 85°C/85% RH. The incidence and growth rate of whiskers was found to vary with the soldering method.

Key words: Whiskers, Corrosion, SAC305

INTRODUCTION
Although there are few confirmed reports of equipment failures due to whisker growth from solder joints rather than from electrodeposited tin finishes the possibility remains a concern for the electronics industry. It is now widely accepted that the primary driving force for whisker growth is relief of compressive stress and while it is well established that such stress can occur in electrodeposited coatings, particularly when the process is out of specification, a solder joint formed by unconstrained solidification from the molten state tends to be naturally stress-free. There are, however, ways in which compressive stress sufficient to induce whisker growth can be introduced to a solder joint. One such source of compressive stress recognized in tests such as JESD22A121, "Measuring Whisker Growth on Tin and Tin Alloy Surface Finishes" is corrosion induced by exposure to heat and humidity.

The study reported in this paper investigated the relationship between whisker growth in conditions of heat and humidity and surface corrosion accelerated by the residues of the fluxes typically used in the three common soldering processes, wave soldering, hand soldering and reflow soldering.

EXPERIMENTAL METHOD
Alloy
Sn-3.0Ag-0.5Cu (SAC305)

Test Vehicle
Interdigitated comb pattern, electrodeposited 35μm thick copper traces at 0.3mm spacing (Figure 1).

Soldering Methods
Solder was applied to the test vehicle by dip soldering, hand soldering and reflow soldering with a variety of commercially available wave soldering fluxes, flux cored solder wires and solder pastes using process parameters recommended for these materials (Table 1)

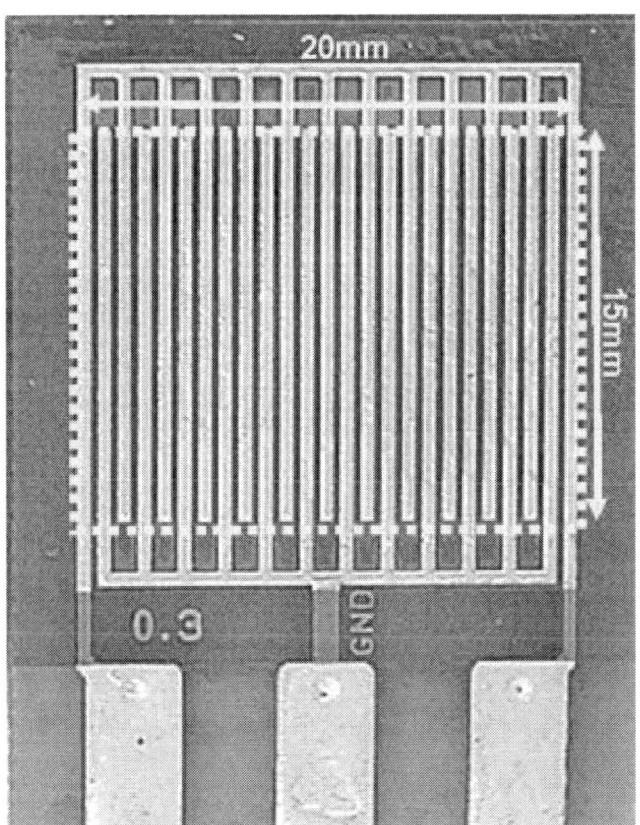

Figure 1. Test vehicle.

Exposure Environments
    40°C/95%RH
    60°C/90%RH
    85°C/85%RH

Test vehicles were inspected for whiskers at 500h and 1000h and then at 1000h intervals up to 5000h.

Whisker Measurement
The area defined by the yellow dotted line in Figure 1 was inspected for whiskers. Whisker densities were determined by superimposing grids on SEM images of the traces and

counting the squares in which whisker occurred. The longest whiskers in each field of view were noted and the length estimated in the SEM.

Corrosion Measurement

Figure 2 is typical of a cross-section of a soldered trace exposed to heat and humidity. Figure 3 is a magnified view of the edge of the trace where the solder is most exposed to flux residue and where most corrosion occurs. Solder corrosion was quantified by measuring on cross-sections such as this the total area of corrosion and expressing that as a percentage of the total cross-section area of the solder coating excluding the intermetallic compound at the solder/copper interface and in the matrix.

Table 1. Soldering Conditions

	Soldering Method		
	Hand	Dip	Reflow
Fluxes	A.B,C,D,E	F,G	H,K
Soldering Parameters	Tip 300°C Continuous	Solder 250°C	Ramp Profile 1.5°C/s 50s>227°C
A,B,C,D Halogenated Core Fluxes			
F,G: Halogenated Liquid Fluxes			
H, K Halogenated Paste Medium			

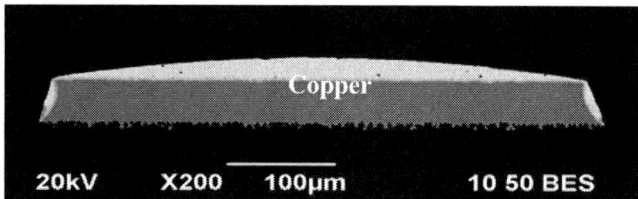

Figure 2. Cross-section of typical soldered trace

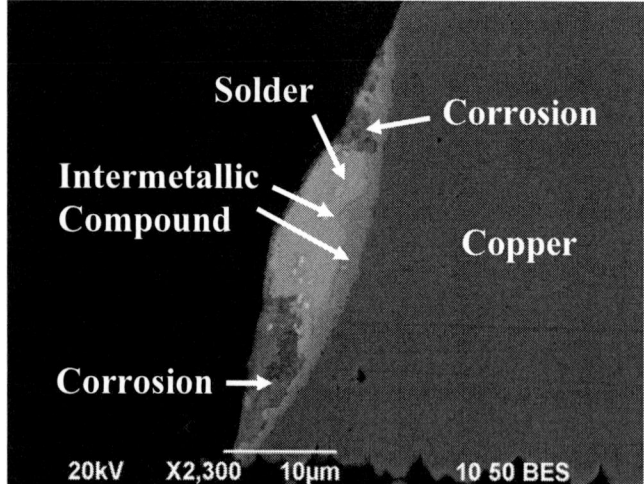

Figure 3. Cross-section of solder coating

RESULTS

As expected the extent of corrosion increased with increasing temperature and humidity (Figure 4). Under the same environmental conditions the extent of corrosion

varied with the soldering method with, on average, corrosion being greatest on the test vehicles that had been reflow soldered and least on those that had been hand soldered.

The typical distribution of whiskers is indicated schematically in Figure 5 with the greatest concentration of whiskers of the greatest length occurring on the edges.

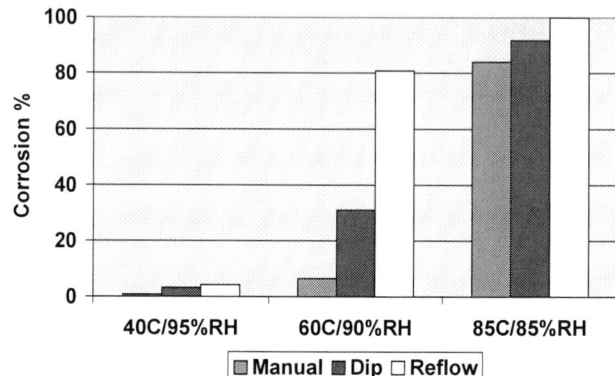

Figure 4. Corrosion at 3000h as a function of environment and soldering method.

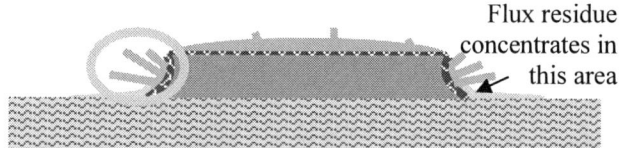

Figure 5. Schematic indication of pattern of whisker growth.

Significant whisker growth occurred only under the conditions of 60°C/90%RH and 85°C.85%RH and there is a relationship between the extent of corrosion and the maximum whisker length (Figure 6).

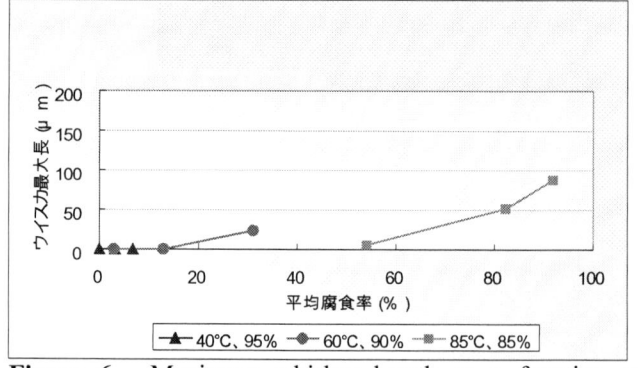

Figure 6. Maximum whisker length as a function of corrosion %

The environmental condition had a strong effect on the time to the occurrence of the first whisker (Table 2)

Table 2. Effect of environment on time to first whisker

Condition	Time	Acceleration Factor
40°C/95%RH	5000h	1
60°C/90%RH	1000-2000h	~3
85°C/85%RH	~500h	~10

The soldering method had a small but significant effect on the time to the occurrence of the first whisker (Table 3)

Table3. Effect of soldering method on time to first whisker

Method	Time	Acceleration Factor
Dip	500h	1
Hand	500-1000h	~x 1.5
Reflow	1000-2000h	~x 3

Maximum whisker length as a function of time in each of the three environments is reported on the basis of location in Figure 7. Maximum whisker length as a function of time at 85C.85%RH and location is reported on the basis of soldering method in Figure 8.

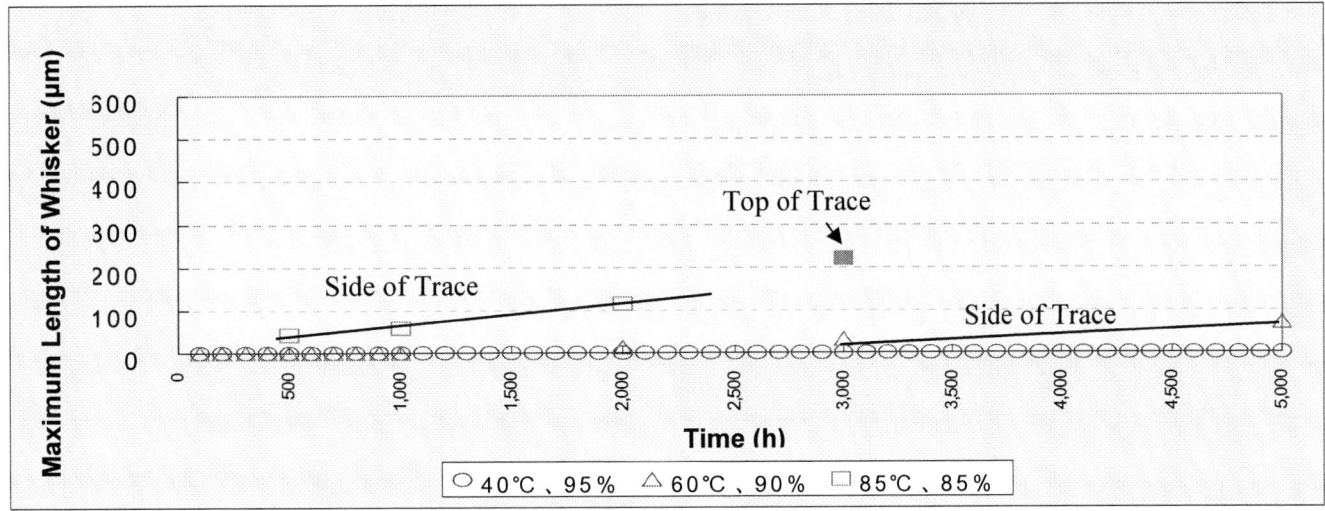

Figure 7. Maximum whisker length as a function of environment and time

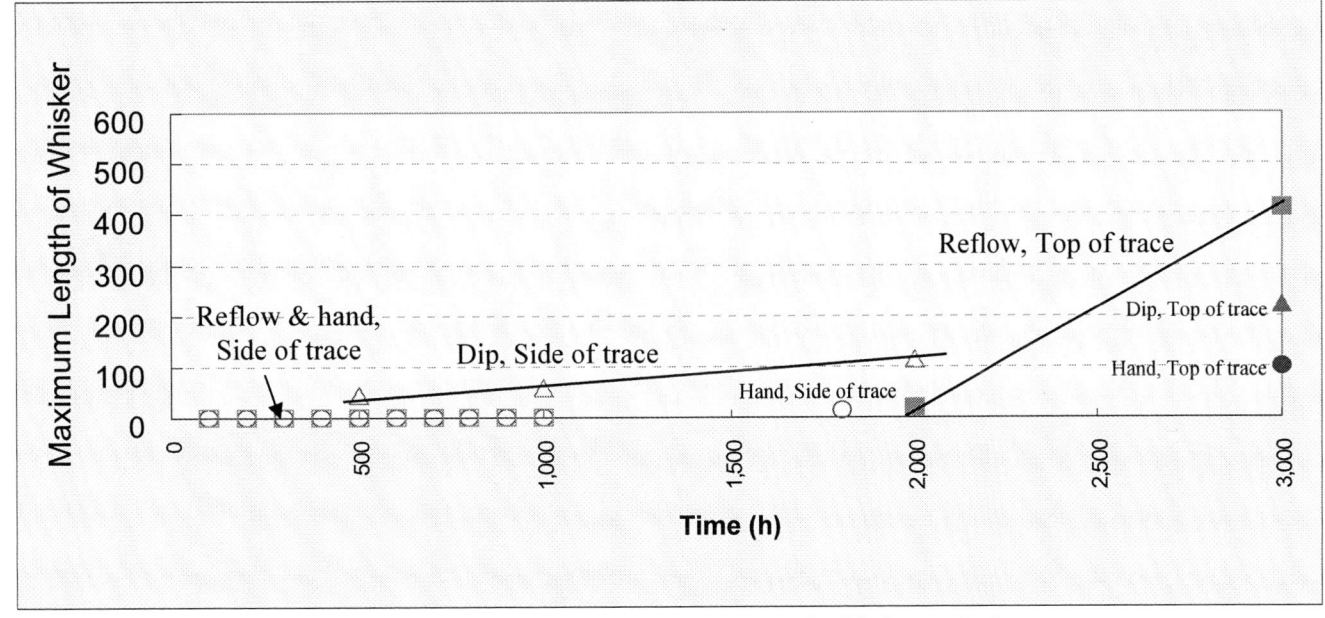

Figure 8. Maximum whisker length as a function of time at 85C/85%RH and soldering method

DISCUSSION
With no other obvious factors contributing to whisker grow the results confirm that corrosion induced by high temperature and humidity can induce of whisker growth. It can be hypothesised that the driver for the whisker growth is the compressive stresses generated by the increase in volume that occurs as the solder is converted to corrosion products.

It is generally considered that a whisker of 50µm represents a reliability risk since it has the potential to short typical circuitry. Under the conditions of this test whiskers of that length did develop in less than 1000h at 60°C/90%RH and 85°C/85%RH.

However, even at 95%RH a temperature of 40°C was not sufficient to generate whiskers of a length that could be considered a reliability risk.

The relationship between soldering method and whisker growth is presumably related to the extent to which the residue of the fluxing system used can contribute the ions that drive the corrosion process in the humid condition. There does not appear to be any obvious way in which the susceptibility of the solder itself could be affect by the method by which it is applied to the substrate as in all cases is solidifies unconstrained from the molten state.

The fact that most whisker growth occurs at the sides of the traces is presumably related to the fact that most of the flux residue ends up being concentrated in that area.

CONCLUSIONS
Under conditions of 60°C/90%RH and 85°C/85%RH corrosion that appears to be related to the character of the residues used in the soldering process can cause SAC305 solder to produce whiskers long enough to compromise circuit reliability. Where circuitry vulnerable to failure by shorts caused by whiskers is likely to be exposed to such conditions .consideration should be given to effective removal of flux residues or the selection of fluxes with residues that do not support the sort of corrosion that seems to drive whisker growth.

FUTURE WORK
Given the apparent relationship between flux residue and whisker growth under conditions of heat and humidity the possibility of formulating effective fluxes that have residues that do not promote whisker-inducing corrosion is being investigated.

A CATALOGUE OF FAILURE MECHANISMS IN FLIP CHIP DEVICES DETECTED USING ACOUSTIC MICRO IMAGING

Janet E. Semmens
Sonoscan, Inc.
Elk Grove Village, IL, USA
jsemmens@sonoscan.com

ABSTRACT

Acoustic Micro Imaging (AMI) has been used over the past years to successfully evaluate the quality of flip chip underfill and interconnect bonds. Acoustic micro imaging uses high frequency ultrasound (5 to 500 MHz) to image the internal features of samples. Ultrasound is sensitive to variations in the elastic properties of materials and is particularly sensitive to locating air gaps (delaminations, cracks and voids).

Flip chip technology is constantly progressing toward smaller devices and higher IO count which leads to thinner chips, and smaller bumps and bonds which in turn requires higher resolution in the acoustic images. This is driving AMI technology to develop higher frequency transducers to increase the available resolution in both the spatial (x,y) and axial (z) dimensions. Low-k and extra low-k dielectrics are being used in flip chips and these materials present additional challenges for acoustic analysis due to their material properties. However acoustic signal analysis and imaging techniques are available to compensate. As a result of advancements in AMI technology and the refinement of analysis methods more can be determined concerning the nature of the failure mode. For example the failure may be laminar cracking of the passivation or low-k dielectric layers, or failures of the solder bump interconnect. This paper will present a catalogue of flip chip failure modes encountered to date and describe the analysis methods used to detect and define the failure modes.

Key words: Acoustic Micro Imaging (AMI), Acoustic Microscopy, Flip Chip

INTRODUCTION

AMI (Acoustic Micro Imaging) is a non-destructive test method that utilizes high frequency ultrasound in the range of 5 MHz to 500 MHz. Ultrasound is sensitive to variations in the elastic properties of materials and is particularly sensitive to locating air gaps (delaminations, cracks and voids). There is a direct relationship between frequency and resolution in AMI. Higher frequencies have shorter wavelengths and, therefore, provide higher resolution. Lower frequencies, which have longer wavelengths, provide better penetration of the ultrasound energy through attenuating materials, thicker materials or multiple layer assemblies. Generally a compromise is found between sufficient resolution and maintaining satisfactory

penetration and working distance for a given application. More recently methods such as Frequency Domain Imaging have been used to improve the resolution/detectability of features in acoustic images.

ANALYSIS METHODS
Reflection Mode A-Scan, B-Scan, and C-Scan

In reflection mode Acoustic Micro Imaging the fundamental information is contained in what is called the A-Scan. The A-Scan displays the echo depth information in the sample at each x, y coordinate. Echoes displayed in the A-Scan correspond to different interfaces in the device being examined. The amplitude and phase (polarity) information of the echoes is used to characterize the condition at the interface and is dependant on the acoustic impedance value of the materials involved. The equation that describes the pulse reflection at an interface between materials is as follows:

$$R = I \frac{Z2 - Z1}{Z2 + Z1}$$

Where R is the amplitude of the reflected pulse, I is the amplitude of the incident pulse, Z1 is the intrinsic acoustic impedance of the material through which the pulse is traveling and Z2 is that of the next material which is encountered by the pulse.

There is a time/distance relationship between the echoes based on the acoustic velocities in the materials that can be used to predict the positions of the interface echoes for the various levels.

$$Velocity = 2 \times distance/time$$

In addition to separately gating the echoes to scan one level specific x–y area the different depth levels can be displayed in a cross sectional (x–z) view called a B-scan.

More recently many flip chips are back thinned which causes the layers to be very thin relative to the wavelength of the frequency needed for inspection. In some instances the echoes from the various levels may not be completely separated from one another on the A-scan and this causes interference effects that can be difficult to interpret. However, information from the multiple reflections (resonances) can be used to examine the specific levels and methods such as frequency domain analysis can be used to extract the information from complex echoes.

Frequency Domain (FFT) Imaging

Frequency Domain analysis has been used to enhance the resolution of features in acoustic images. In addition it has been observed that in some instances the detectability of certain internal features or defects is dependant on the frequency content of a specific echo.

Currently a method is used that stores the A-scan information for each x-y point in a scan. From the stored information images of depths within the device not included in the original gate for the image can be recreated and/or waveforms (echoes) can be viewed for analysis without rescanning the sample. In addition to this the echoes can be digitally processed, frequency filtered, etc., to extract further information about the condition of the sample, or extract information at or slightly beyond the limits of conventional AMI.

Frequency Domain imaging is one method that can extract further information by using the frequency content of the signal. In this technique each A-scan of the image relates to the localized frequency response of the corresponding pixel in the sample. For reference, the conventional image is a time domain image in which each pixel relates to the magnitude of a return echo [1].

The transducers used in AMI range in center frequency from 5 MHz to 300 MHz and above. In conventional acoustic imaging, the choice of transducer determines the spatial resolution, penetration, and other parameters. These transducers typically have highly damped waveforms in order to achieve better resolution, both spatial and axial, using time domain imaging. Figure 6 displays an A-scan with typical echoes (pulses) as seen in the time domain. However these highly damped waveforms contain broad-spectrum frequency information that can be displayed in the Fourier (frequency) domain. When using time domain imaging at a center frequency of 50 MHz, one can not manipulate the image data to produce a 30 MHz image or a 75 MHz image, because in the time domain the acoustic pulses themselves do allow frequency separation. However, the data file including the stored A-scans makes this kind of manipulation possible, within limits.

·

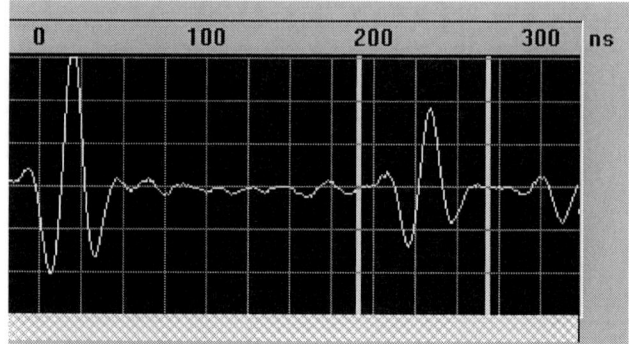

Figure 1: A-scan displaying typical waveforms (pulses) in the time domain.

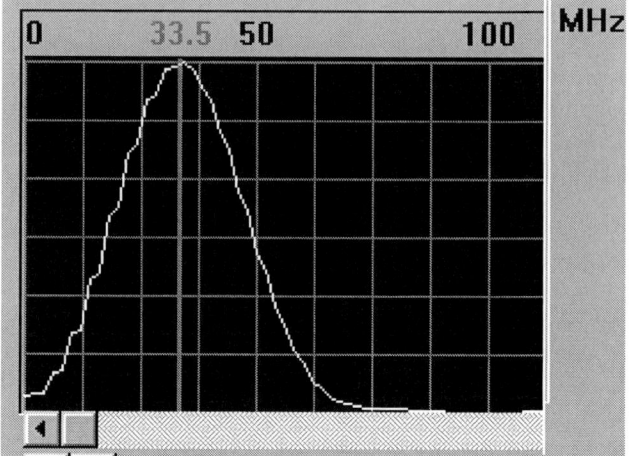

Figure 2: Broadband pulse content in the frequency domain for the echo within the gate on the A-scan in Figure 6.

Because the A-scans for each point in the image are collected with the image changes in frequency that may occur during reflection can be analyzed. For example, a pulse of 15 MHz ultrasound launched toward a material interface (such as molding compound to die) may be reflected with a different frequency content than originally pulsed, and this change – not otherwise detectable – may be indicative of the interface condition. The gated echo(es) from the stored A-scans can be filtered by means of a Fast Fourier Transform (FFT), also called a Frequency Domain algorithm, to isolate a given frequency. The Fourier transform decomposes the selected waveform(s) into sinusoids of different frequencies. The FFT identifies the different frequency sinusoids and their respective amplitudes. Figure 2 shows the frequency content distribution of the gated echo shown in Figure 6 in the frequency domain. Images can then be reconstructed from components of the frequency information. Specific features may yield more information at one frequency than another. Therefore, FFT filtering of the echo can bring out image detail that may not be visible with conventional time domain imaging.

Advancements have been made in recent years with respect to higher resolution in the acoustic images by increasing the frequency/design of the transducers. However, there is a point where the thickness and type of material in the packaging will limit the use of even higher frequency ultrasound even though package features are becoming increasingly smaller and internal layers increasingly thinner. Frequency domain (FFT) analysis of the waveforms has been used in the past to measure bond line thickness and has shown the capacity of measuring to thickness well beyond the axial resolution limit of a given frequency [2]. Frequency domain imaging can be used to improve detection/resolution in the lateral dimensions by removing the low frequency component from the image. Conversely by selecting a lower frequency component of the bandwidth features that were masked by the high frequency portion of the signal have been detected.

FLIP CHIP FAILURE MODES
Silicon Chip Cracks
Cracks can occur in the silicon chip due to some problem with the manufacturing process or handling of the parts. In some instances vertical cracks will extend to the back surface of the chip where they can be seen using optical methods though the crack opening dimension may make it difficult to see. In the acoustic image vertical cracks appear as dark lines in the image. The width of the line however is not restricted to the size of the opening but rather is dependent on the depth of the crack. The larger shadow caused by the crack depth renders the crack more detectable in the acoustic image. In addition there can be laminar cracks in the silicon near or in the active layers of the die that are not detected when viewing the back surface. These cracks are visible in the C-scan acoustic image (similar to a delamination) and further depth information on the crack can be seen using B-scan analysis. Figure 3 shows the B-scan through a laminar crack and vertical cracks in a flip chip device. The cross sectional view shows the laminar crack to extend upward from the chip/bump and underfill interface however this portion of the crack network does not reach the back surface of the device. The vertical segment of the crack does show a small opening at the back surface but this does not give a true indication of the extent of cracking below the surface.

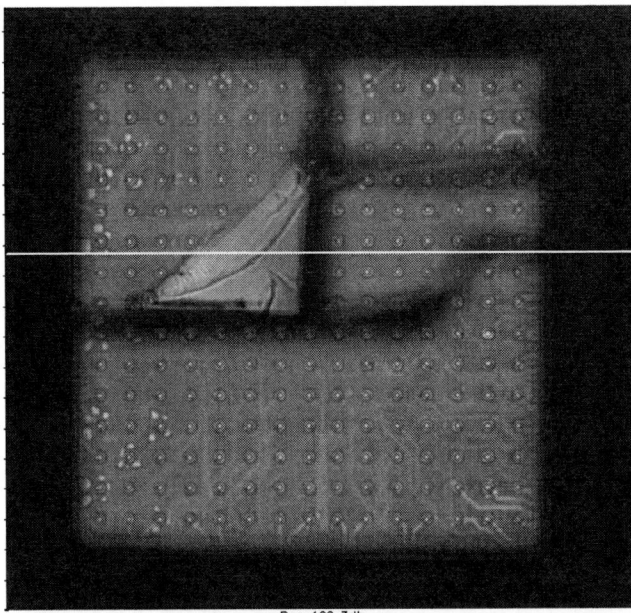

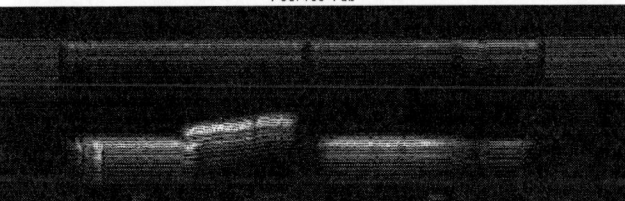

Figure 3: The upper part of the display shows the C-scan image of a flip chip containing large vertical cracks as well as a laminar portion of the crack. A bright line on the image indicates the position of the B-scan cross sectional view shown in the lower portion of the display.

Under Bump Metallization and Passivation Delamination
Defects have been detected in flip chips in the region closest to the silicon, just prior to the solder bump [3]. In many cases the features can be identified from their morphology in the acoustic image as they can appear larger than the size of the solder joint as a result of the thin laminar cracks extending into the surrounding passivation or dielectric layers. The position of the signals on the A-Scan indicates that the defects are present in the layers preceding the bond pad to bump level. Using a narrow gate at the leading edge of the chip to bump interface the echoes for the under bump layers can be isolated for the image and the defects clearly detected as seen in Figure 4.

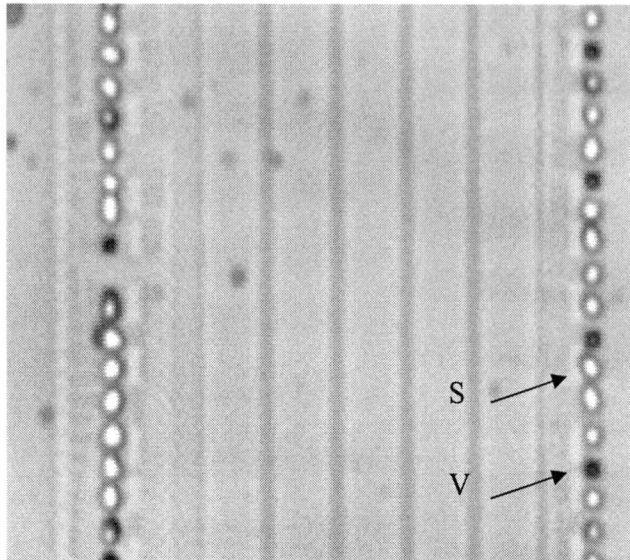

Figure 4: Acoustic image of chip/bond pad level using level specific imaging technique. Variations in the signal intensity are apparent from one bond site to another. the strong signal reflections characteristic of voids and delaminations are shown in white. The white bumps also are the ones which display the larger, irregular shapes. Dark spots (low signal reflections) indicate the bonded pads at this level. Two of the sites are indicated by the arrows for reference in the corresponding A-scan and DPA images.

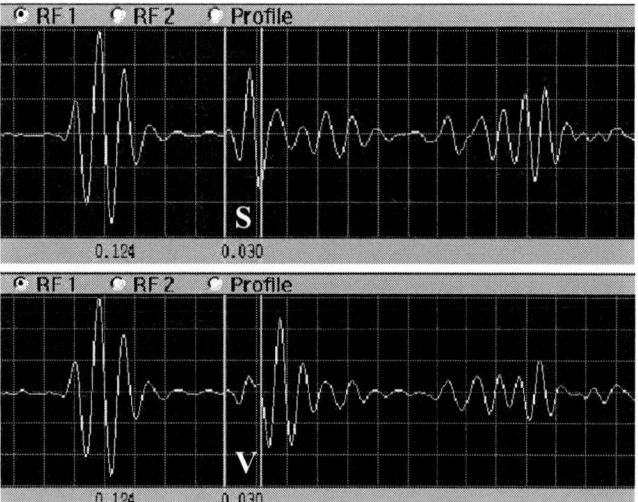

Figure 5: A-Scans showing the location of the gate (designated by the 2 vertical lines over the trace) to obtain the images of the bond pad and passivation separations from the silicon. The echoes corresponding to the defects show significantly higher amplitude than the echoes corresponding to the bonded pads.

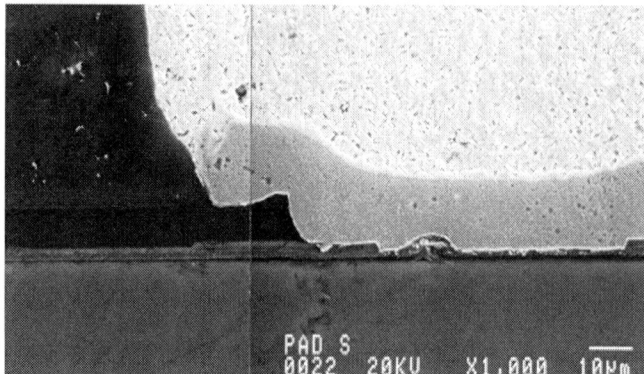

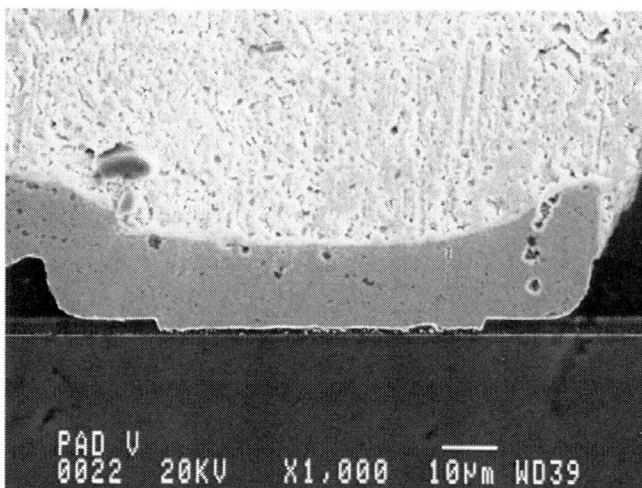

Figure 6: DPA sections corresponding to bumps S and V are shown. Bump S shows a delamination under the bond pad and extending under the adjacent passivation layers. Bump V shows a bonded bump and bond pad with no delamination of the passivation.

Air Gap Defects (voids and cracks) Within the Volume of the Solder Bump

In some cases the defect is at a deeper level within the solder joint such as a void within the solder bump or an open between the chip and the substrate rather than at either interface. Placing the electronic gate between the echoes from the chip and the substrate levels will display air gap defects as white features in the acoustic image (Figure 7).

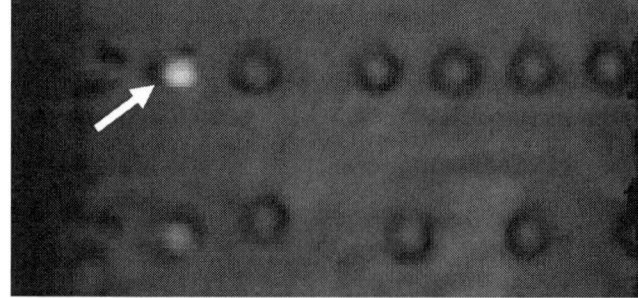

Figure 7: This 230 MHz acoustic image shows a section of a flip chip sample gated between the chip and the substrate. The mid-bump flaw appears as the bright white bump in the image.

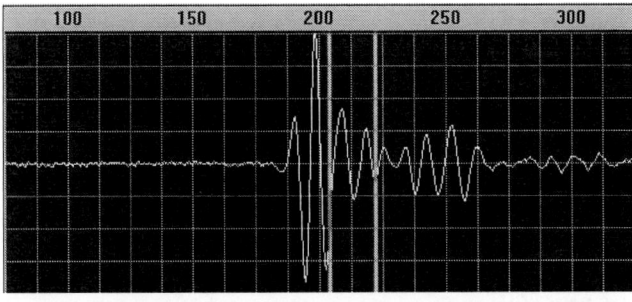

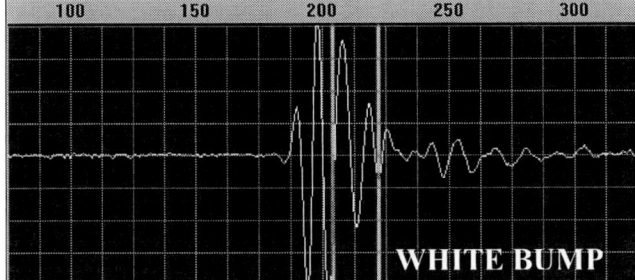

Figure 8: A-Scans showing the location of the gate (designated by the 2 vertical lines over the trace) to obtain the images of the defects between the chip and the substrate. The white bump shows significantly higher amplitude than the echo from the darker adjacent bump.

Open Solder Bump Connection at the Substrate

Disbonds of the solder bump connections can also occur at the bump/substrate interface. Similar to the previous examples by placing the electronic gate at the proper position the substrate level of the device can be imaged. In this example a device was showing an open site using electrical tests. Examination of the chip/bump level showed no anomalies of the bump bunds at the chip level. The image of the substrate level however showed a bright, white bump site indicative of an air-gap type defect (delamination) at the substrate (Figure 10). The position of the white bump correlated with the open bond site detected by electrical test.

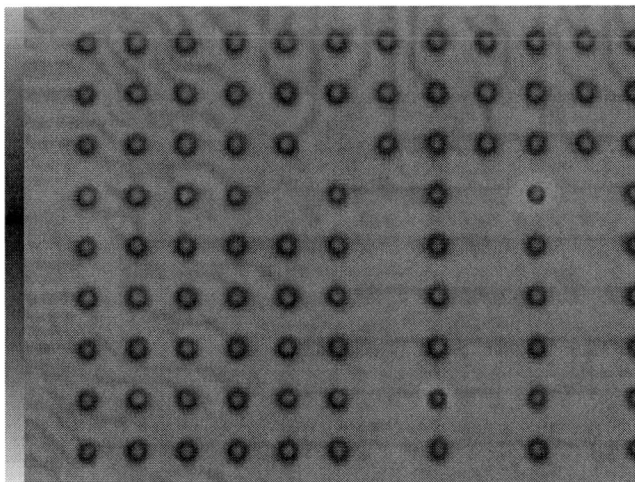

Figure 9: 230 MHz acoustic image showing no bump bond defects at the chip/bump interface.

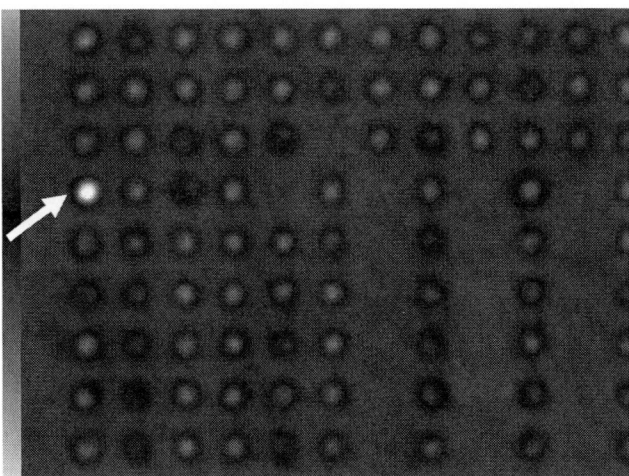

Figure 10: 230 MHz acoustic image of the same device shown in figure 9 at the substrate level. An open connection (white bump) is present at the substrate level.

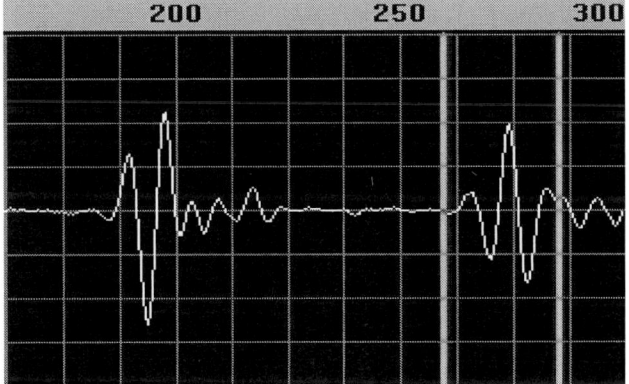

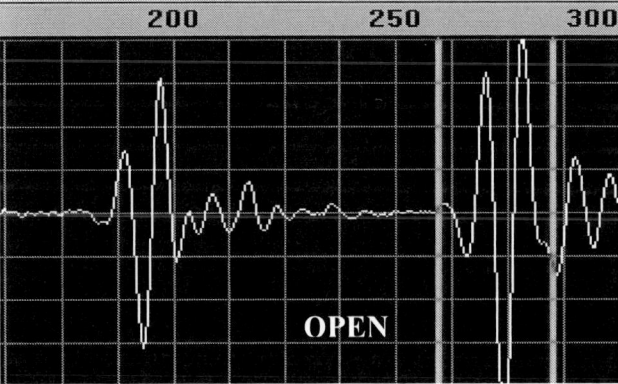

Figure 11: A-Scans showing the location of the gate (designated by the 2 vertical lines over the trace) to obtain the image of the open connection at the substrate. The white (open) bump at the substrate shows significantly higher amplitude and an inverted polarity compared to the echo from a darker (bonded) bump.

Underfill Voids

Underfill voids can also be detected using AMI at the die or substrate level or within the bulk of the underfill using similar methods to the detection of solder bump bond issues. This example illustrates an instance of where Fourier Domain Imaging was used to enhance the detection of underfill voids [1]. Voids in underfill can reduce the field life of the device particularly when the voids are in close association to the interconnect bumps. Halo voids that surround the bump totally or partially at the chip/bump level typically result from flux residue or other contamination on the chip surface. Figure 12a displays a time domain image of a flip chip that shows halo voids associated with two of the bump bond sites. Figure 12bb shows the corresponding frequency domain image. The image was made again using the higher frequency content of the pulse. In the FFT enhanced image many other halo voids become apparent.

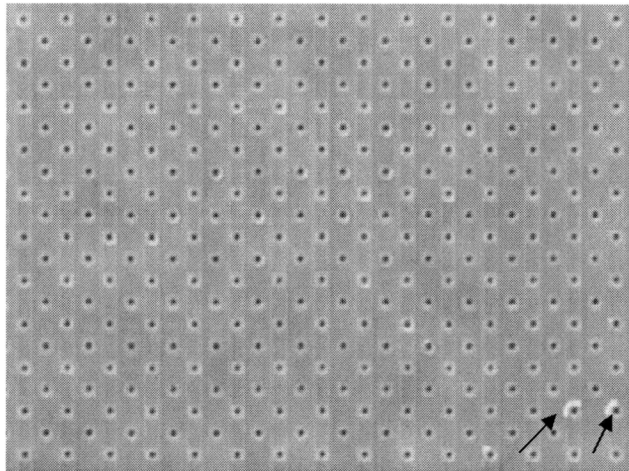

Figure 12a -Time domain image of flip chip with halo voids in the underfill

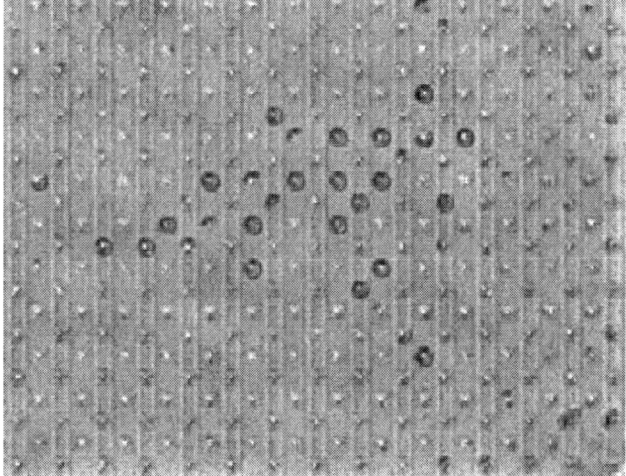

Figure 12b - frequency domain image of flip chip. Now many more bumps show associated halo voids.

CONCLUSION

The examples presented here include the main categories of flip chip failure modes encountered over years of experience using AMI. In addition to developing the acoustic methods to analyze the different kinds of defects that can be present in flip chip devices correlative analysis has been performed in many instances (including DPA, X-Ray or electrical testing) to verify the presence of the defects. This information base is valuable when evaluating future samples however flip chip technology is continually evolving in

complexity. In the future AMI developments will continue to meet the challenges presented by changes in the design and manufacturing of flip chips and the catalogue will be updated to reflect the changes.

REFERENCES

[1] Janet Semmens and Lawrence Kessler, "Acoustic Micro Imaging in the Fourier Domain for Evaluation of Advanced Packaging", Proceedings of Pan Pacific Conference, Maui, Hawaii, 2002.

[2] Sridhar Canumalla and Bryan P. Schackmuth, "Metrology of Thin Layers in IC Packages using an Acoustic Microprobe: Bondline Thickness", Proceedings of the 49th Electronic Components and Technology Conference (ECTC), San Diego, CA, June 1-4, 1999

[3] J.E. Semmens and L.W. Kessler, "Characterization of Flip Chip Interconnect Failure Modes using High Frequency Acoustic Micro Imaging with Correlative Analysis," Proceedings of IRPS 1997, Denver, CO.

MODERN 2D / 3D X-RAY INSPECTION -- EMPHASIS ON BGA, QFN, 3D PACKAGES, AND COUNTERFEIT COMPONENTS

Evstatin Krastev and David Bernard
Nordson DAGE
Fremont, CA, USA
evstatin.krastev@nordsondage.com; david.bernard@nordsondage.com

ABSTRACT

With PCB complexity and density increasing and also wider use of 3D devices, tougher requirements are now imposed on device inspection both during original manufacture and at their subsequent processing onto printed circuit boards. More complicated and dense packages have more opportunities to exhibit defects both internal to the package as well as to the PCB. As components increase in complexity their cost increases, making counterfeiting them a potentially lucrative business for unscrupulous individuals and organizations.

Recent years have brought significant improvement in the capabilities in the 2D/3D X-Ray Inspection systems. New X-ray sources, detectors, and ergonomic features improve the efficiency and productivity of the inspection process.

This paper reviews the methods of finding defects in BGAs, QFNs, and 3D packages using X-Ray inspection with real-life examples provided. Voiding, cracks, shorts, open joints, and head in pillow (HIP) will be discussed. Comparison of the relative merits of the 2D and 3D (CT) X-Ray inspection for investigating 3D packages is presented with examples. Using X-Ray inspection for detecting counterfeits is discussed at the end.

Key words: BGA, QFN, X-ray inspection, Computer Tomography, CT, 3D packages, POP, PIP, SIP, stacked die, defect

INTRODUCTION

Traditionally, the use of 2D x-ray inspection provides important and at the same time a non-destructive method for investigating all aspects of device production and subsequent PCB assembly. In recent years increased use of area array packages like BGAs and QFNs, CSPs and flip chips makes traditional optical methods of inspection ineffective as the joints to the PCB are hidden under the package. In order to use optical means to inspect the above cases the device needs to be physically removed prior to inspection, which practically destroys the assembly. In addition, during the process of physically removing the device, vital information can be lost or additional defects introduced. X-Ray inspection is also used to inspect the wire and die attach quality inside the individual package without opening the package. This technique is very handy for detecting counterfeit components while keeping the parts in their sealed, as-received packaging.

With increasing system integration, new 3D packages like package in package (PiP) and package-on-package (PoP) are replacing standard leadframe packages. These new packages incorporate multiple dies stacked on the top of each other, multi-level wire bonding and interconnection. The ultimate goal is to achieve greater circuit density resulting in better overall performance of the assembly. These new and more complex devices are bringing their own requirements and challenges to the inspection and quality control process during device assembly, test and subsequent assembly onto printed circuit boards. In some cases this increased complexity requires the use of 3D Computer Tomography (CT) technique in addition to the traditional 2D oblique angle X-Ray inspection. This is facilitated by the fast and straightforward switch between 2D and 3D CT inspection mode offered by the modern X-Ray system. Practically the range of applications of the modern X-Ray system is very large with the electronics inspection being one of the main groups. X-Ray is also a useful tool for inspecting medical, other mechanical and optical devices.

BASICS OF AN X-RAY SYSTEM

The 2D X-Ray inspection systems are essentially X-Ray microscopes (Figure 1 and Figure 2). A wide cone of X-rays is emitted by the source (X-Ray tube). The inspection is accomplished by moving the sample inside the X-Ray emission cone. All materials absorb the X-Ray radiation differently depending on their density, atomic number and thickness.

Thicker and/or denser material will absorb more of the X-Rays. The resulting image, composed of shades of grey with darker areas corresponding to higher X-Ray absorption, is registered by the X-Ray imaging device, usually a very high quality digital image intensifier or a flat panel. The closer the sample is to the X-Ray source the higher the magnification level (Figure 1). The ability to inspect at an oblique angle view is crucial for finding defects like cracks and open joints. Generally two ways are used to accomplish oblique, or angled views - tilt the sample or tilt the imaging device. As seen in Figure 2, tilting the imaging device has the huge advantage of producing maximum magnification at all tilt angles.

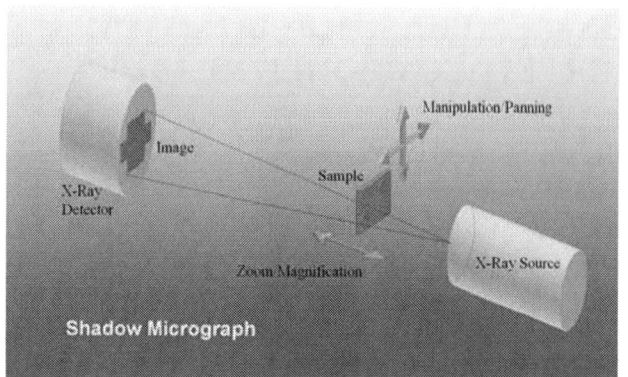

Figure 1. Basic operation principle of X-Ray inspection system

The modern X-Ray machine tilts the imaging device up to 70 degrees at the same time permitting images from 360 degrees around any inspection point. A sophisticated combination of software and hardware keeps the point of interest (for instance a defective joint) in the center of the field of view while the imaging device is rotating around and examining from all directions. Additional advantages of this method are that there is no need to secure the sample (PCB, component or another device) and no risk of dropping, damage or collision during sample manipulation.

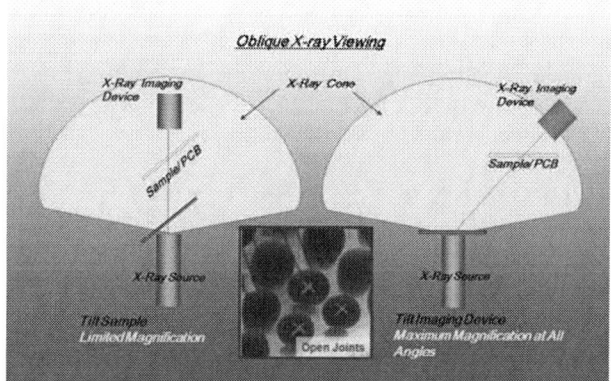

Figure 2. Oblique angle viewing

During the last several years, X-ray source and digital detector technology have significantly developed. Submicron feature recognition as fine as 100 nanometres (0.1 micron) is achievable in 2D mode as well as system magnification levels of 12,000X. These advancements allow inspection of finer detail and a corresponding increase in the detection of potential defects. Sealed transmissive, filament-free X-ray tube technology combines the highest performance levels with maintenance-free or minimal-maintenance operation, which reduces downtime in an active production environment.

The latest imaging systems provide real-time digital inspection at 2.0 mega-pixels, 30 frames per second and 65,000 greyscale levels, viewed on 24" ultra-high-definition LCD monitors. The important point here is that these advancements in image quality and enhanced feature recognition make the inspection process much faster, more effective, and highly reliable.

3D Computerized Tomography (CT)
Computerized Tomography (CT) is an X-Ray imaging method where mathematical geometric processing is used to generate a 3D virtual model of an object from a large series of individual 2-D X-Ray images taken as the object is rotated through 360 degrees in the x-ray beam.

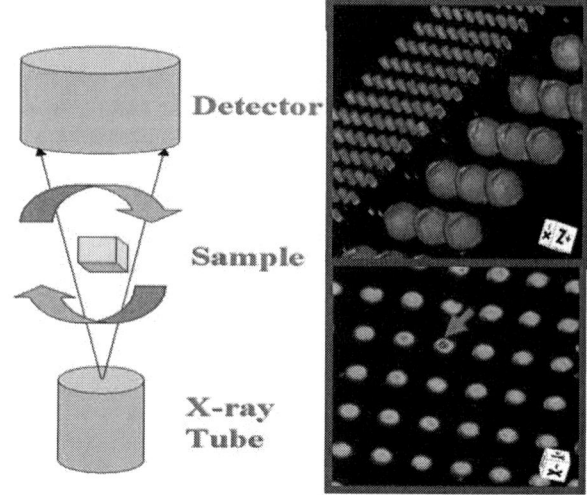

Figure 3. Typical layout of a CT system and an example of a CT model

The layout of a CT system is shown in Figure 3. Once the CT model has been produced, it enables 'virtual micro-sectioning' by allowing the operator to investigate any two-dimensional plane within the entire model as well as full, real time manipulation of the 3D model. In this way, different layers and different features within the package can be viewed whilst being isolated from other, potentially confusing, detail to enable improved analysis. This allows complete examination of features or defects within a device or package that would otherwise remain hidden by multi-level interconnections. For example, the tiny 100-micron solder bumps within the multi-layered device shown on Figure 3 would be obscured by the much larger 500-micron BGA balls.

The modern 2D/3D X-Ray inspection system can be switched from 2D to 3D CT mode in a couple of minutes, facilitating the joint use of the two techniques providing powerful and *non-destructive* analytical capabilities.

Defects in BGA and QFN Packages
Common BGA defects like voiding, shorts, cracks, and head-in-pillow (HIP) or head-on-pillow (HOP) usually occur during reflow. The oblique angle capability is critical for detecting cracks and HOP defects using viewing angles of 55 to 70 degrees. Voids and shorts are easily visible using a top-down view, but in order to determine the location of the voids oblique angle 2D viewing and 3D CT come very handy. The angled view on Figure 4 reveals

voiding concentrated on the joint interface making the particular connection less reliable and possibly more prone to failure in the field.

Figure 4. Angled/ oblique 2D X-Ray view showing interface voiding

The 3D CT technique permits virtual slicing or cross-sectioning through the devices revealing the exact location of the voids as seen on Figure 5.

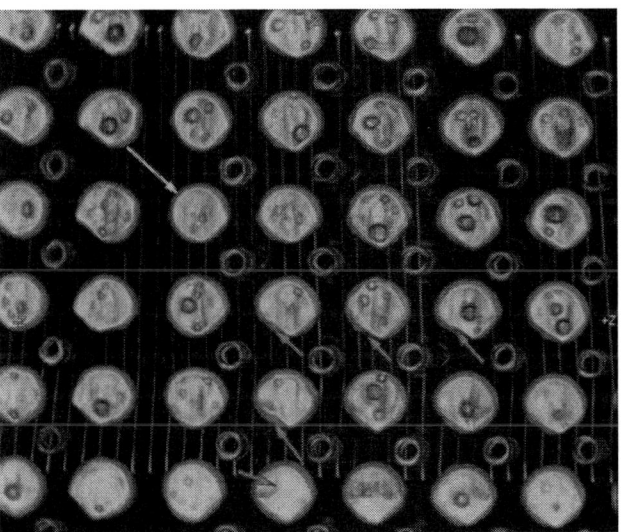

Figure 5. CT virtual section through the interface region of a BGA device.

Large amount of interfacial voiding is apparent and many BGA balls marked with red arrows are deformed/ cracked. The modern 2DX machine features automatic routines designed for precise measurement of diameter, voiding percentage, shape and area of the BGA balls. As seen in Figure 6, the BGA balls marked in red outline fail the requirement for shape or roundness, which is the first number accompanying each ball. The second number represents the total voiding percentage for each BGA ball. It is up to the operator to set up the acceptable levels and the machine will automatically flag any values which fall outside.

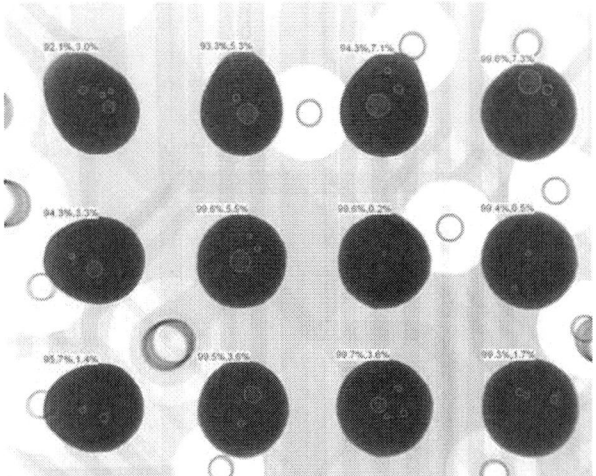

Figure 6. 2D X-Ray voiding and roundness calculation for a BGA device.

Head-in-pillow (HIP) defects, also known as head-on-pillow (HOP), occur during reflow. The solder paste wets the printed circuit board pad while not fully wetting the BGA ball.

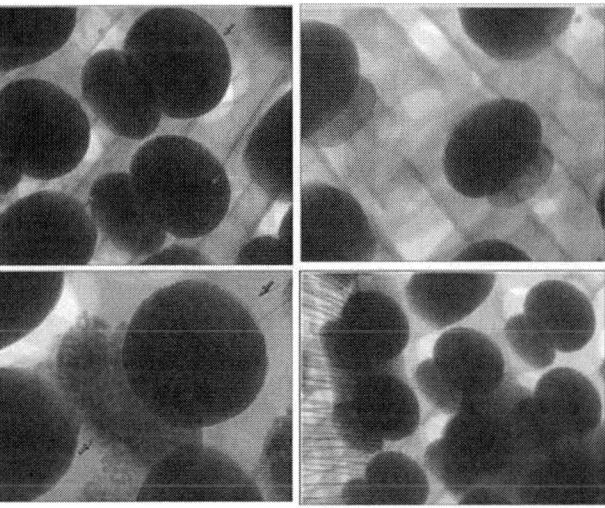

Figure 7. Various examples of HOP defects easily identifiable with off-axis 2D X-Ray inspection

Even though the HOP joint might at first exhibit electrical conductivity it lacks mechanical strength and fails in the field. In many cases the BGA device incorporating HOP defects is not functional from the very beginning. The HOP defects have become more widespread with the arrival of lead-free solder paste due to greater board warpage and solder ball lifting caused by higher reflow temperatures. Process variables impacting HOP defects include solder ball alloys, the type of reflow profile, peak reflow temperature and solder paste chemistry.

Even though HOP defects can be difficult to detect when using in-line automatic X-ray inspection (AXI) systems or lower performance 2D systems, they can be quickly and easily identified using modern high-performance 2D X-ray system with an oblique angle viewing (Figure 7). In addition the inspection process is completely *non-destructive*.

The only other practical way of confirming HOP defect residing in the middle of a BGA is by cross-section and scanning electron microscope (SEM) or optical examination; however this method is destructive resulting in damaged and unusable PCB and BGA component.

Figure 7 shows various examples of HOP defects. It is obvious why these defects are referred to as head-in-pillow, or better head-on-pillow. The defective BGA solder ball appears to be laying on the reflowed solder paste instead of forming a single joint after reflow. Figure 8 is a virtual CT micro-section through the BGA device revealing a 3D view of the HOP defect.

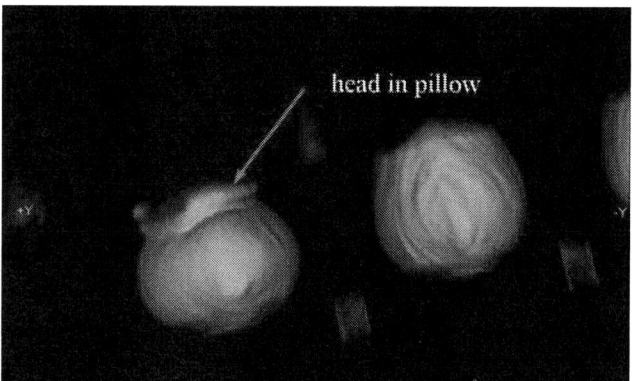

Figure 8. 3D CT section showing HOP defect in a BGA device

The traditional method for detecting HOP defects is manual inspection at an oblique angle of 55 to 70 degrees performed by the operator. This method, although being very reliable, is time consuming and labor intensive. As a further step, an automatic HOP inspection routine has been developed that can identify suspect HOP defects without requiring an operator to manually inspect each individual solder joint location within a BGA. The routine scans the entire BGA device; a sophisticated artificial intelligence algorithm analyses all BGA solder balls and highlights the ones considered to be HOP defects to the operator by displaying a color coded overlay on-sreen. This is a very important development providing additional confidence in solder joint integrity and helping to prevent HOP defect escapes.

The Quad Flat Pack No Leads (QFN) style of leadless package is also becoming more and more popular because of its low cost, low profile and excellent electrical and thermal parameters. It is widely used in the wireless, automotive telecom and many other applications because of the merits outlined above. Although most commonly known as QFNs, the same, or similar, package types are also known under other names. The most common alternative is Land Grid Array (LGA).

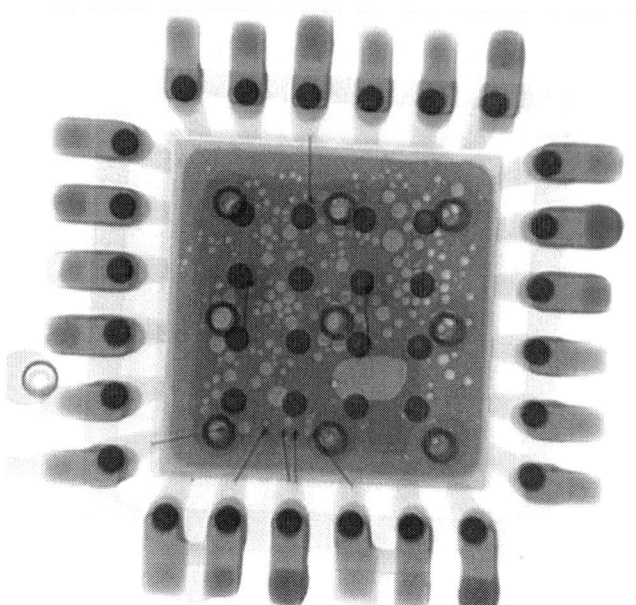

Figure 9. Top X-Ray view of QFN package

As shown on Figure 9, the package terminations are all located under the device. This makes the traditional Automatic Optical Inspection not practical. As with the BGA devices, cross section followed by SEM or optical evaluation is an alternative, but the method is destructive and permanently damages the device and PCB.

Modern 2D X-Ray systems provide a fast, effective and non-destructive method for inspecting QFNs. A relatively inexperienced operator can quickly assess and quantify the analysis within the production environment. With a lesser X-Ray inspection system that lacks good magnification, resolution, contrast sensitivity, and high angle oblique viewing the analysis could be less straightforward and reliable.

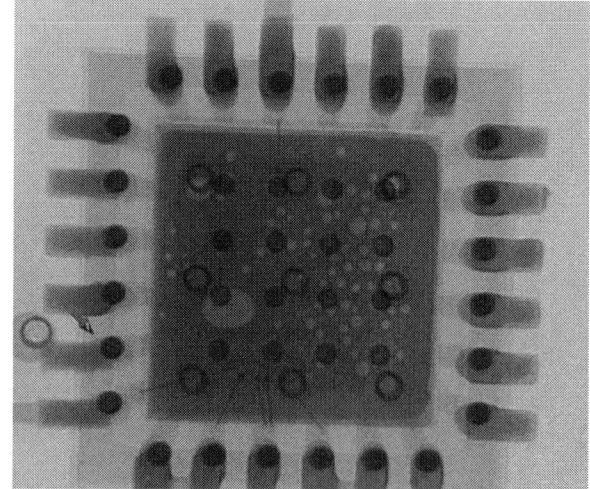

Figure 10. Top X-Ray view of LGA package. Suspected pin pointed by arrow.

Figure 10 shows a low magnification image of a QFN package with a red arrow pointing out to a suspect pin.

Because of the high resolution and contrast sensitivity of the X-Ray system, a possible defect is apparent even at this low magnification image.

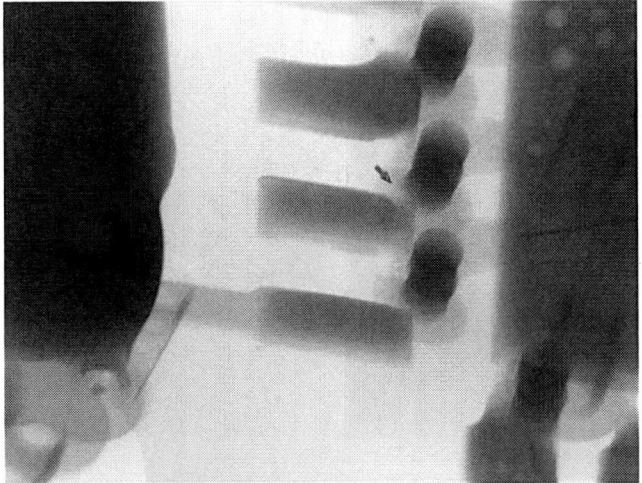

Figure 11. Higher magnification oblique angle view of defective pins from Figure 10.

Using oblique view and higher magnification one can study the open pins in detail as shown in Figure 11. However, even a low magnification oblique view is sufficient to confirm the defect as seen in Figure 12.

The fact that the defective pins can be identified at such low magnification levels makes the X-Ray inspection process fast, efficient and reliable. Figure 13 represents an oblique angle 2D x-ray view of different QFN device. The red arrows are pointing towards open pins compared to the good pins (green arrows).

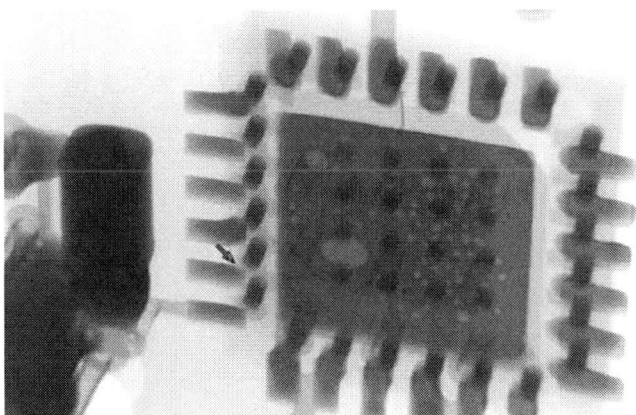

Figure 12. Low magnification oblique X-Ray image showing defective (open) pins in QFN device.

It is important to point out that in addition to the quality of the side connection, the voiding within the central large pad is also crucial in some cases when large amounts of heat need to be transferred out of the device in order to prevent overheating. The example (Figure 9) shows significant voiding in the central area which would be a concern or reason for rejection in a high-power application.

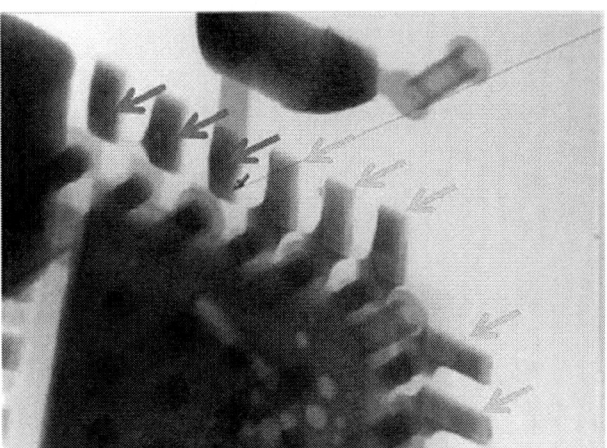

Figure 13. Oblique 2D X-Ray view of defective QFN device. Red arrows point toward open pins and the green arrows identify good pins.

3D Packages

With the continuing trend for subsystem integration, advanced 3D packages including PiP, PoP, SiP and flip-chip devices are replacing standard lead-frame packages. Many of these exotic packages meet the demand for greater circuit density and improved electrical performance, however the increased complexity generates unique challenges for the inspection and quality control process during device packaging and subsequent assembly. Traditionally, the use of 2D X-ray inspection provides a vital and non-destructive method for investigating all aspects of device production and PCB processing. However, with 3D package investigation, 2D X-ray imaging may be limited since all layers within the device are seen at the same time, projected on a plane. Analytically, this can be confusing to the operator because the multiple dies and multiple layers of wire bonds will appear to overlap each other in the x-ray image.

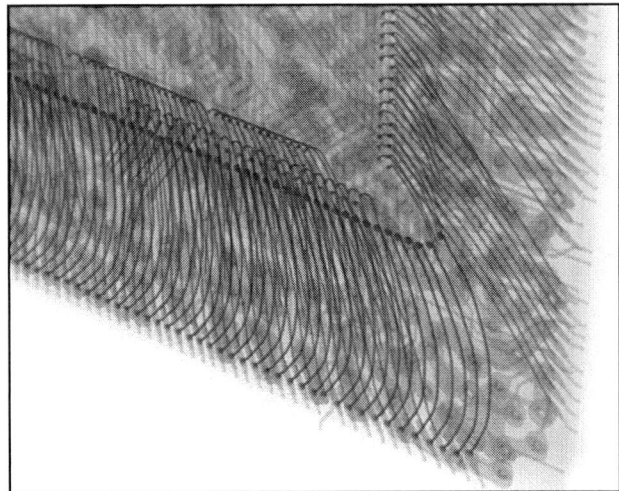

Figure 14. 2D X-ray oblique view of stacked device.

As seen on the 2D X-Ray image in Figure 14, the two wire bonding layers of this stacked device cannot be easily separated for analysis looking for shorted or open connections.

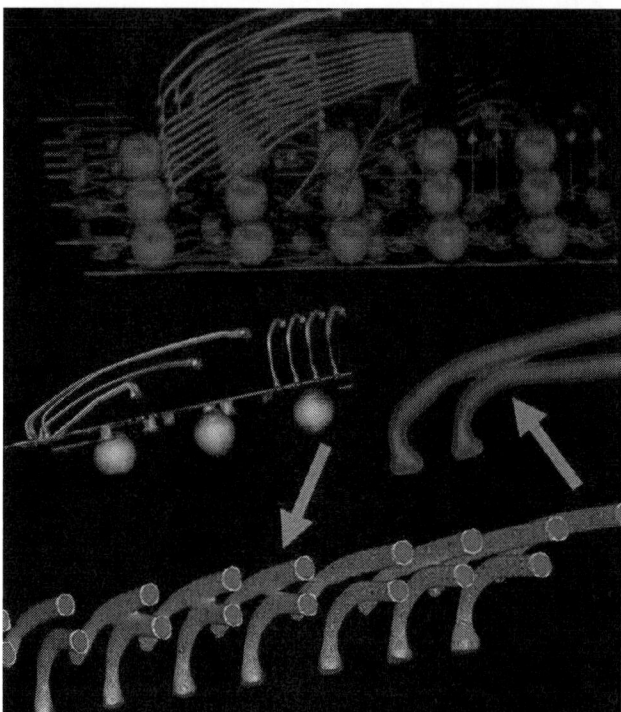

Figure 15. CT images of stacked devices showing shorted bond wires.

Figure 15 is a combination of CT images showing various stacked devices. The CT technique permits the operator to isolate the area of interest and carefully examine for potential failures using virtual micro-sectioning. As seen on Figure 15 the shorted wires are easily identified using CT and this cannot be easily and reliably accomplished using 2D methods.

Another example of the powerful CT technique is shown in Figures 16 and 17.

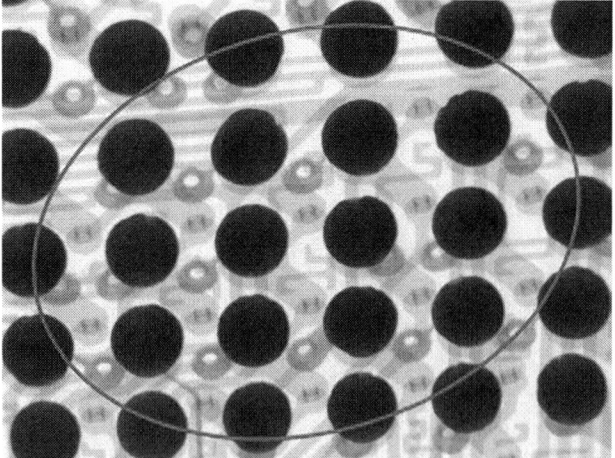

Figure 16. 2D X-Ray image of an area containing suspect shorted vias in a 3D package

The examined 3D device has shorted vias in the region covered by the 2D X-Ray image on Figure 16. Clearly identifying the defect is quite difficult due to the complexity of the 2D X-Ray image. The CT model of the same package is shown on Figure 17. The CT model is also quite complicated, but once the micro-sectioning capability is engaged the shorted vias pop up easily detected -- see inset of Figure 17 representing a virtual micro-section trough the CT model.

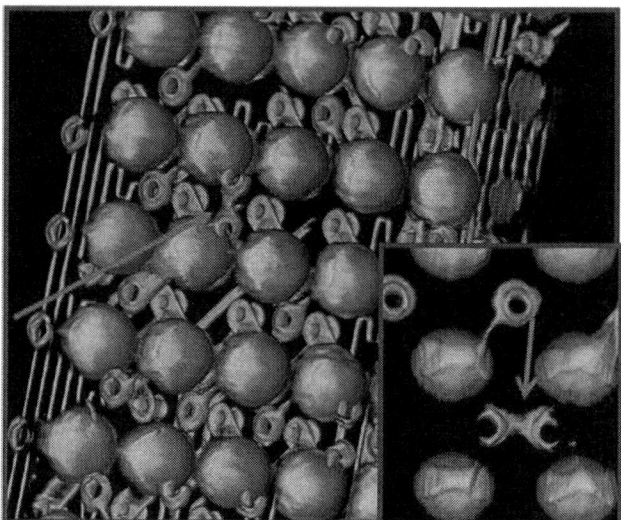

Figure 17. CT model of the 3D deice shown on Figure 17. The inset is a virtual cross-section easily detecting the two shorted vias.

Counterfeit Components
The problem of counterfeit components penetrating the supply chain has been growing during recent years, with most board assemblers admitting to having received some recently, costing them prestige and lost money. Low cost (less than $10 apiece) counterfeits tend to be devastating as it is hard to justify thorough extra tests, while counterfeiters make it even more difficult by providing some good samples at the end of the reel.

Figure 18. 2D X-Ray image revealing big differences between components looking similar optically, thus easily identifying the counterfeit.

As shown in Figure 18, the real and fake components might look very similar optically, but an X-Ray image reveals big differences. The modern 2D X-Ray inspection provides a fast and effective method to identify counterfeit components. Being *non-destructive* and offering high magnification, resolution and contrast sensitivity it makes it easy and straightforward for the operator to identify the fake components by observing die placement, internal wiring and comparing to known good samples.

CONCLUSIONS

Modern 2D/3D X-Ray inspection systems are powerful tools for finding defects in BGA, QFN and 3D packages and also provide fast and straightforward method for identifying counterfeit components. During the last several years, significant advancements in the X-Ray technology including sealed-transmissive X-Ray sources, extremely high quality digital imaging intensifiers, and ergonomic features, has brought the X-Ray inspection to a new and much more advanced level, making the X-Ray inspection much more effective, faster and reliable.

ACKNOWLEDGEMENT

We would like to acknowledge great help from Shoukai Li.

REFERENCES

[1] Modern 2D X-Ray Tackles BGA Defects facilitating process and yield improvements by identifying common BGA defects, Zhen (Jane) Feng, Jayapaul Basani and Murad Kurwa, Flextronics International, and David Bernard and Evstatin Krastev, Nordson Dage SMT 2008

[2] Investigating Defects in 3D Packages Using 2D and 3D X-ray Inspection, David Bernard and Evstatin Krastev , Nordson Dage, SMTA International, August 2008

[3] *X-ray Inspection Identifies Flip-Chip Defects, Evstatin Krastev, Nordson Dage, Advanced Packaging e-newsletter, September 17, 2008*

[4] David Bernard, Bob Willis, Common Process Defect Identification of QFN Packages Using Optical And X-Ray Inspection, SMTA 2007

[5] A Suggested Process for Detecting Counterfeit Components, Dr. David Bernard and Bob Willis, SMTA 2008

[6] Bernard D., "A Practical Guide to X-ray Inspection Criteria & Common Defect Analysis", Dage Publications 2006. Available through the SMTA bookshop.

HEAD STACK ASSEMBLY PERCUSSION SWAGE PROCESSING MORPHOLOGIES FOR HARD DISK DRIVE PRODUCTS

Jeffry S. Bennin and Michael R. Tiller

Hutchinson Technology Inc.

Hutchinson, MN, USA

jeffry.bennin@hti.htch.com and michael.tiller@hti.htch.com

ABSTRACT

Hard Disk Drive memory storage products comprise an assortment of mechanical and micro-electronic components and subassemblies attached and combed into a finished product. One such attachment involves a mechanical ball swage interference fit, providing a solid, stable and economical means of integrating the precision head suspension spring to the E-block arm of the Head Stack Assembly. Challenges arise, in optimizing component design and processing to ensure a robust attachment of multiple head suspensions to the arm stack without negatively impacting the attitude of the head as it flies mere nanometers above the rotating disk surface. Operational temperature extremes, high lateral seek and shock loads, dissimilar materials, process rework, cleaning, and chemical compatibility requirements all underscored with a driving necessity to minimize costs, has retained swage attachment as the predominant head suspension attachment process in HDD products for over 20 years.

The approach is not without its compromises, however, as an intricate balance of mechanical stiffness, minimized mass, and complex compressible features are designed into the baseplate component of the suspension assembly to attain both operational performance and processing objectives. Swage processing applies axial loads of 180 – 270N (40 – 60lbf), converted to compressive radial loads within the intricately patterned, ring-shaped boss tower, securely restraining the suspension assembly in a patterned hole in the E-block arm. High axial loads, however, can result in unfavorable deformation of the swage plate flange structure, with the potential to negatively impact head flight attitude. This paper will study the effects of impact or percussion swage processing as a means of transforming the power spectrum input to the system, reducing the axial force applied to the baseplate. Should impulse energy loading alter the static to dynamic friction states at the onset of compression swage, it may be possible significantly reduce the baseplate flange deformation and its associated impacts on flight attitude without deploying prohibitive lubricants, or significant modifications to established tooling and equipment infrastructure. Potential benefits may include future design and process modifications to minimize variation and further reduce feature complexity and resulting cost.

Keywords: Impact, Vibro-Impact, Percussion, Swage, Head Stack Assembly, HSA, Head Gimbal Assembly, HGA, Suspension Assembly, Hard Disk Drive, HDD, Baseplate, Swagebase, Dynamic Mass, Gram Load, Static Friction, Dynamic Friction.

INTRODUCTION

Mechanical ball swage of head suspension assemblies to rigid actuator arms used in hard disk drive storage products has provided an admirably stable, cost effective, and repeatable means of attachment for head stack assembly operations. In response to industry areal density and data-rate roadmaps, which consistently look to escalate the storage capacities of new drive introductions, all aspects of the electromechanical systems which comprise hard disk drive products are routinely analyzed for potential improvement opportunities. This paper seeks to explore the potential for improving upon the existing process methodology utilized in the ball swage operation for head suspension to actuator arm attach. Notably, an initial exploration of vibro-impact or percussion swage processing will be explored as a means of reducing detrimental performance impacts, improving process variation, and maintaining lowering overall processing costs. Should the exploration yield favorable results, a more complete study will be pursued.

BACKGROUND

Head suspension assemblies provide critical structural support, positioning, and stability to magnetic head sliders flying over the rotating disk substrates. The suspension assembly, typically a 3-4 piece precision spring of 300 series stainless steel, is shaped and patterned using photo-chemical, forming, laser welding, measurement and adjustments in automated, high volume production equipment. Head suspension assemblies are mounted to a rigid actuator arm patterned within an aluminum extruded and machined E-block, which when coupled with a bearing assembly, preamp circuitry, and a driving electromagnetic voice coil motor, comprises a Head Stack Assembly. Suspension assemblies are attached through the use of a swaging process which integrally bonds the baseplate or swagebase component of the suspension assembly to either one or both sides of the actuator arm of the head stack assembly.

Figure 1. Hard Disk Drive (HDD) data storage product

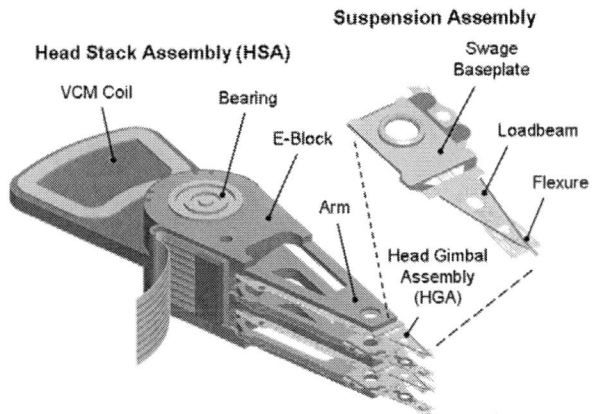

Figure 2. Head Stack Assembly and Suspension Assembly Components of a Hard Disk Drive

The attachment provides a secure mount which must not rotate or fall out during subsequent drive assembly or operational conditions. Early products utilized a threaded attach of the rigid mountplate of a suspension assembly to the actuator arm, but improvements in performance consistency and cost were realized years ago after adoption of ball swage attach of a stamped and annealed baseplate. Compressible features in the baseplate boss structure allow the annular ring of the baseplate to plastically deform, imparting an outwardly radial, and primarily elastic deformation to the rigid actuator arm, providing a stable attachment for the life of the drive product. The ability to rework or remove and replace damaged or defective components during processing also imparts additionally functionality into the baseplate and swage attach process. Gram load, or the force in grams applied by the suspension assembly to the head slider at a given vertical attitude or "Z-height" at operating conditions within the drive, ranges between 1.5 - 2.5g with a typical tolerance of ±0.075 – 0.125g. Gram load change, typically in the range of 0.1 – 0.2g, during swage operations of the head stack assembly sequence, arguably defines the single largest contributor to flyheight variation in the drive. Process and product improvements that could noticeably reduce the gram load change typically observed during head stack operations without adding significant cost and complexity would be of particular interest to the hard disk drive industry.

As a mechanical swaging process can impart significant stresses to the components involved, careful considerations in baseplate component design are taken for compressive feature management and stress isolation. The baseplate, a 300 series progressively stamped and annealed component of the suspension assembly, comprises both a flange structure typically 0.125 – 0.200mm in thickness and a shaped and formed ring or boss feature situated within the central region of the flange. The flange provides a stable attach platform for the more compliant load beam and flexure structures of the suspension assembly, looking to provide an attitude-stable and consistent platform from which to fly the head over the rotating disk surface. The boss tower, typically 2.0 – 2.75mm in O.D. and 1.5 – 2.0mm in I.D. with a height of 0.25 - 0.32mm, is sized such that it slips into a through hole patterned within the actuator arm with approx. 25µm of diametrical clearance. Features within the boss tower including O.D. and I.D., back bore, chamfers, relief ring features at the circumferential attachment to the flange, fillet radii, etc. all provide functional structure for the attach sequence.

The swaging operation itself consists first of a fixturing and positioning sequence followed by the ball swaging operation. The stack fixturing sequence involves the placement of suspension assemblies within awaiting actuator arms by inserting the baseplate boss features within the actuator arm holes, and sliding in an intervening swage key assembly which positions a precision clamp arms between each actuator arm and its paired suspension assemblies prior to final positioning and full camp. Clamp forces to 445N secure head stack for ball swaging. Two hardened 400 series stainless steel swage balls, sized at 2.019mm and 2.057mm dia. respectively in some applications, are sequentially driven through the aligned head stack assembly with a swage pin and applied forces ranging from 180 – 270N or more. The swage balls, typically oversized by 35 – 75µm on dia. in relationship to a given baseplate boss I.D., plastically compress the boss tower material radially outward to such a degree so as to provide effective attachment strength in elastic compression within the actuator arm to resist a torque-out load requirement of > 35N-m.

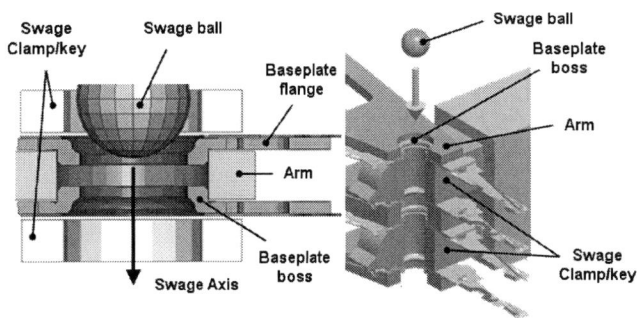

Figure 3. Section views of Head Stack Assembly during ball swage attachment operations

Of particular interest during the swage attach operation, are the potential impacts upon suspension assembly, and effective drive level performance. As suspension assemblies become more precise and important as a structural, spring, positioning, and electrical components of the high speed data transmission path between the head and the drive circuitry, the need exists for an improved method of mounting a head gimbal assembly on an actuator arm which reduces or potentially eliminates deformations due to the mounting process. Deformation potentially imparted into the baseplate flange region from stresses induced during the swaging process, can deform the flange, alter the gram load (force) applied by the suspension radius region spring to the head, and cause the heads to fly at different heights over the disk surface. Additional impacts in dynamic modal performance can also affect the servo tracking and positioning performance of the drive.

The object of this study seeks to understand if altering the loading state of the ball swage sequence from the existing near static swage process to a vibro-impact or percussion swage process might have the ability to reduce baseplate flange deformations, reduce the resulting gram load change, and improve the overall dynamic flight performance stability of the head in hard disk drive products.

SYSTEM DESCRIPTION
As outlined by Han, et.al.[1], vibro-impact systems have a long history of applications is such fields as the mining industry and civil engineering applications where in the modern sense, percussive drilling has been progressively employed since its first realization in the mid-1800's. First hints of its utilization, however, actually date back to the 4th millennia where its use in shallow water well drilling was documented in China. Extensive research and analysis in Vibro-impact system tools and methodology has occurred over the past 50 years as industry seeks to expand output, improve efficiencies, and lower effective costs. Consistent improvements have been made in hammer design, performance, and reliability in the large scale mining industries, where percussive drilling has extended well beyond empirical testing into numerical analysis methodology and coupled to experimental system correlation further enumerated by Han, et.al.[2,3]. As studies have revealed a rich system of dynamics and nonlinearities in system characteristics (Ho and Woo[4], Hinrichs et.al.[5]), it is understandable that vibro-impact system applications and technological insights are still growing. Applications, however, are observed in a broader sense with the expansion in percussion hand tools available to the common user across many fields including percussion hammers, reciprocating drills, ratchet wrenches, electromagnetic engravers, etc. Another application includes the use of vibro-impact absorbers to couple large masses with smaller mass linear system vibrations as analyzed and enumerated by Mikhlin and Reshetnikova[6], and percussive impact drilling with dry friction implications by Franca, et.al.[7].

The primary model for deploying vibro-impacting energy in this application shall be the free and constrained vibration in two-degrees-of-freedom system. Quite simply, the system defines a pair of masses m_1, and a substantially larger mass

m_2 coupled with a pair of springs K_1 and substantially smaller K_2. An applied force input F_s from the large mass will act in equal an opposite fashion upon the small mass in such a manner in a constrained case as to transfer a large contact force, F_{cont} to an external body by means of applied kinetic energy from the high rate of force application from the input F_s.

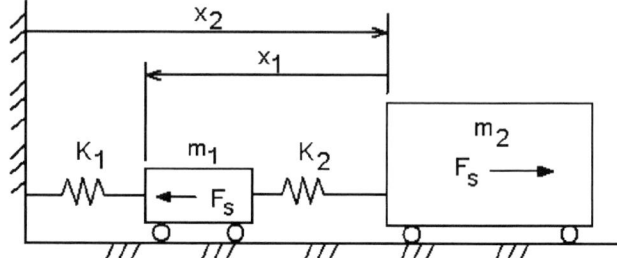

Figure 4. Two Degree of Freedom dynamic system defining the vibro-impact mechanism employed.

Referencing the 2-DOF system above, the Conservation of Energy equation provides,

$$U = \frac{1}{2} \cdot K_1 \cdot (x_1)^2 + \frac{1}{2} \cdot K_2 \cdot (x_2 - x_1)^2$$

Through differentiation with respect to x_1, x_2, this translates to the LaGrange Equation of Motion,

$$\begin{bmatrix} m_1 & -m_1 \\ -m_1 & m_1 + m_2 \end{bmatrix} \cdot \ddot{x} + \begin{bmatrix} K_2 + K_1 & -K_1 \\ -K_1 & K_1 \end{bmatrix} \cdot x = \begin{bmatrix} F_s \\ 0 \end{bmatrix}$$

This provides,

$$F_s = m_1 \cdot \ddot{x}_1 - m_1 \cdot \ddot{x}_2 + (K_1 + K_2) \cdot x_1 - K_1 \cdot x_2$$

$$0 = (m_1 + m_2) \cdot \ddot{x}_2 - m_1 \cdot \ddot{x}_1 - K_1 \cdot x_1 + K_1 \cdot x_2$$

For the first case, consider $K_1 = 0$ for no initial contact, and $m_1 \ll m_2$ providing after reformulating the equations of motion,

$$F_s = m_1 \cdot \ddot{x}_1 - m_1 \cdot \ddot{x}_2 + K_2 \cdot x_1$$

$$0 = -m_1 \cdot \ddot{x}_1 + (m_1 + m_2) \cdot \ddot{x}_2$$

This eventually simplifies down to the case where the larger mass m_2 has very little if any acceleration or displacement ($x_2 \approx 0$). This has reasonable merit for most familiar with vibro-impact systems in the sense that the larger, stabilization mass (tool handle, percussion drill body, etc.) moves to a lesser degree, while the smaller, dynamic mass vibrates or oscillates to a great degree. This effect is magnified with increasing difference between the smaller moving mass, and the larger stabilization mass as is the case with a vibration isolator. At steady state, mass m_1 undergoes simple harmonic motion prior to contact with the left hand boundary condition, and deploys a velocity amplitude which when defined by its kinetic energy gives,

40

$$E_1 = \frac{1}{2} \cdot \frac{m_1 \cdot \left(F_s\right)^2 \cdot \omega^2}{\left(K_2\right)^2}$$

The kinetic energy upon contact with the boundary condition indicates,

$$E_{cont} = \frac{1}{2} \cdot K_1 \cdot \delta^2$$

Equating the two energy terms and solving for force upon contact, F_{cont},

$$F_{cont} = \frac{\sqrt{m_1 \cdot K_1}}{K_2} \cdot \omega \cdot F_s$$

Realizing the contact stiffness K_1 upon impact as significant, the contact force can be as high as 20 – 30x the system driving force F_s, clearly explaining effectivity of the percussion hammer effect in functional applications. $F_s \approx$ 40N in the experimental device, $F_{cont} > 800N$ would be sufficient to promote successful ball swage.

Additional factors requiring further study are the potential transformations from static to dynamic friction at the ball to baseplate boss interface, the effect of successive increments of short cycle high energy impacts on baseplate boss deformation, and the post swage ball exit velocity effects upon kinematic deformations of the subsequent baseplates in full stack swage.

EXPERIMENTAL SETUP

In the deployment of vibro-impact percussive action to the hard disk drive ball swaging process, a suitable mechanism, widely available, and at comparatively low cost needed to be identified. Additionally, the device needed to apply force to at least the reasonably targeted range typical of ball swaging operations near 180 – 270N as previously identified. [As a side comment, high load piezo stack actuators such as model PI-016.000H available from Physik Instrumente GmbH & Co. of Karlsruhe, Germany were trailed initially. However, the key factor of reciprocating, dynamic mass action necessary in percussive systems was not employed. As such, no force amplification, high velocity impact, or state change from static to dynamic friction occurred, and little effectual change was noticed in comparison to initial controlled conditions.] The model 290-01 electromagnetic Dremel Electromagnetic Engraver from Dremel Inc. of Racine, WI, subsidiary of Robert Bosch LLC, Bosch U.S. of Farmington Hills, MI, however, did appear to have the characteristics necessary for further study. Dremel[8] clearly understood the beneficial, percussive nature of the engraver tool conceptualized.

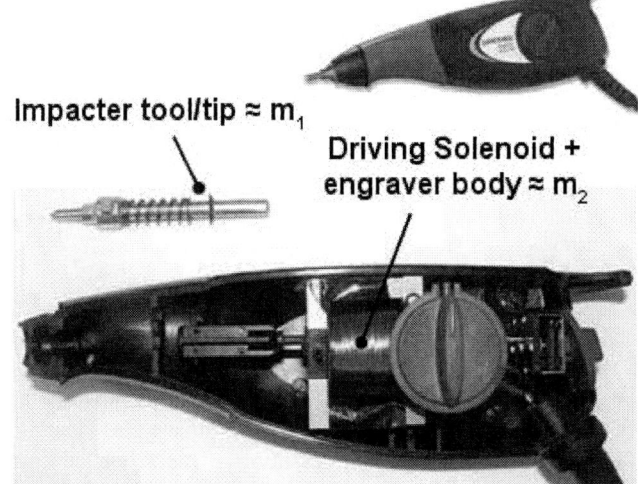

Figure 5. Dremel 290-01 Engraver with visualization of effective system masses m_1, m_2.

The Dremel 290-10 engraver specifications include:
- Current draw: 0.2 Amps
- Line frequency: 60 Hz
- Line voltage: 115 V AC @ 60Hz
- Operational freq.: 7,200 strokes per min. (120Hz)
- Variable intensity 1-5 setting
- Free air displacement amplitude: 0.5mm
- Unit functional weight: 175g
- Dynamic mass weight: 9.25g

To apply the ball swage force effectively to the head stack assembly, a close approximation of the registration and clamping mechanism employed in production swage operations in the functional sense were replicated in the lab environment. The application of sufficient (445N) clamping force to the suspension assembly was employed with simple, levered clamp assembly coupled to a 25mm dia. pneumatic cylinder to ensure sufficient clamping force throughout the swaging operation. A spherically gimbaled top clamp was also worked into the clamp mechanism to ensure a flat, full surface clamp across the swage base flange datum structure.

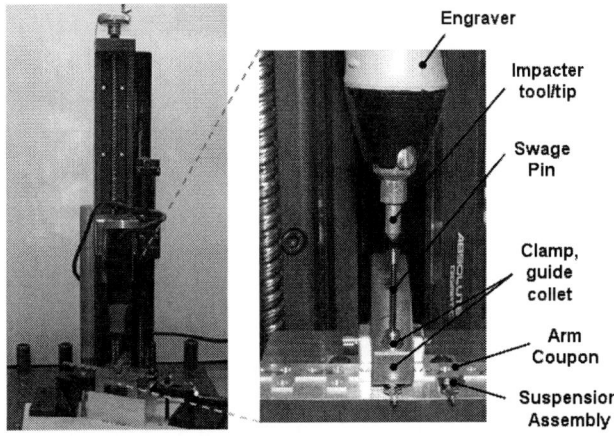

Figure 6. Full swage parts positioning, clamp setup, vibro-impacting engraver mount experimental setup.

The engraver was mounted to a right angle bracket and tooling plate of sufficient rigidity, and positioned in-line with the ball swage central axis of the setup, looking to apply percussion energy down through a swage pin positioned within a collet with an accurate sliding fit atop the ball resting within the constrained baseplate swage boss at initial state. The swage pin contained a self centering conical feature at the carbide engraver tip interface, and a spherical radius slightly smaller than the ball diameter at the work piece interface.

EXPERIMENTAL PROCEDURE
Two full testing conditions were pursued. The first, being an established control setup which deploys all of the fixturing and methods necessary to replicate the standard head stack assembly swage operation. This sequence does not use a percussion or vibro-impact swage sequence, but rather, employs a relatively static force application to the swage ball until sufficient deformation of the boss tower is attained and swage ball pass through occurs. The second setup, as described here, employs a percussive or vibro-impact swage sequence with respect to force application to the swage ball itself. All other considerations (sample preparations, clamping, analytical measurements, etc.) remain the same.

A total of 40 suspension assembly samples were prepared for both the control and the percussion swage sequences. All parts were characterized for gram load prior to the test using an off-line measurement fixture. Additionally, a smaller set of baseplates were also characterized using white light scanning interferometry as a means of accurately characterization the baseplate deformation. All characterizations were in paired sequence, looking to make one to one before and after comparisons during the swage operation for critical performance factors.

The ball swage sequence for each part involved a relatively brief sequence of events which culminated in the swage attach of the suspension assembly to the arm coupon. Upon opening the clamp fixturing, a suspension assembly is placed baseplate boss oriented down, atop a swage arm coupon strip, and positioned to ensure perpendicular orientation with respect to the arm coupon strip tooling holes. A 2.019mm dia. swage ball, first dipped in isopropyl alcohol for contaminant reduction and lubrication (typical forming or machining oils are avoided during normal processing as prohibitive contaminants in the drive system), is placed atop the baseplate boss, and the entire arm coupon strip is positioned within the clamp fixture. The suspension and swage arm are soft clamped, checked for alignment, and then hard clamped to 445N clamp force upon actuating the pneumatic cylinder. The swage pin is inserted through the clamp collet to rest upon the ball surface. With the engraver backed away from clamp setup, an engraver power level setting of "1" is selected and the voltage variac is powered evenly up from a 0% to 100% setting over a 3 second timeframe. The engraver is slowly lowered to attain tool tip to ball contact and within a brief moment, the swage ball

passes through the bottom of the fixture. Of particular note is that fact that a force equitable to the engraver body (approx. 1N), is sufficient to push the first ball through the baseplate boss.

The second, larger 2.057mm swage ball is passed through the baseplate boss in a similar manner with a few small differences. The second ball utilized a contact sequence between the swage pin and the ball throughout the voltage variance activation sequence. The second, larger ball required an engraver full power setting of "5" as well as a slightly larger down force to the engraver body; approx. 10 – 15N to complete the swage sequence. Neither ball is reused after swage. Various power level settings on the engraver device as well as variac settings ranging from 20% to full 100% line voltage were trailed and compared with the primary, rapid feedback mechanism in the testing setup; visual baseplate deformation. The larger the baseplate deformation, the larger the effective gram load change from swaging operations and the greater chance for detrimental impacts on flight and dynamic performance in the drive.

During multiple tests, a number of minor considerations were noted. Towards the goal of obtaining a flatter flange surface, the matter of contact initiation of the engraver chain with the swage ball appeared to make a difference. Initial tests with full power free-state activation followed by sudden contact with the ball surface, appeared to result in instability conditions and "large" amplitude battering of the swage pin against the ball surface. Optimization trials indicated that initiating swage of the second ball by maintaining contact with the ball through the fixturing and engraver chain, appeared to reduce the larger amplitude instability conditions upon attaining full power through the voltage variac, for a smoother and more effective pass through of the second ball during operation. Multiple settings were trialed on both the voltage variac, and the engraver, but the system setup effectively finalized on minimal power level at max. voltage variac settings for an effective pass through of the first ball, and full power level settings at max. voltage variac for an effective pass through of the second ball.

RESULTS AND DISCUSSIONS
Upon completed measurement characterizations of both the control and percussion swage sample groups for gram load change, torque retention, and baseplate deformation, there appeared to be differences between the two sample groups which may justify further exploration. Gram load change measurements show a distinct difference between the two groups when the data is visualized in both individual value and boxplot formats as shown in Fig. 7 and Fig. 8. Also, the percussion swage group does indicate roughly the same tendencies (slight gram loss for all sample parts vs. a mixture of gram rise and gram loss) as well as an apparent similarity of sample variation. A rule of thumb indicates the potential for a statistical difference in the sample group means as the mean values of each distribution lies outside

the 25th to 75th quartile ranges of the comparative distributions.

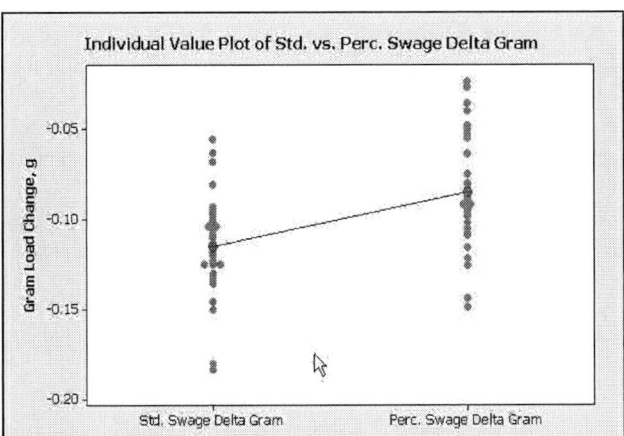

Figure 7. Individual value plot of standard vs. percussion swage gram change.

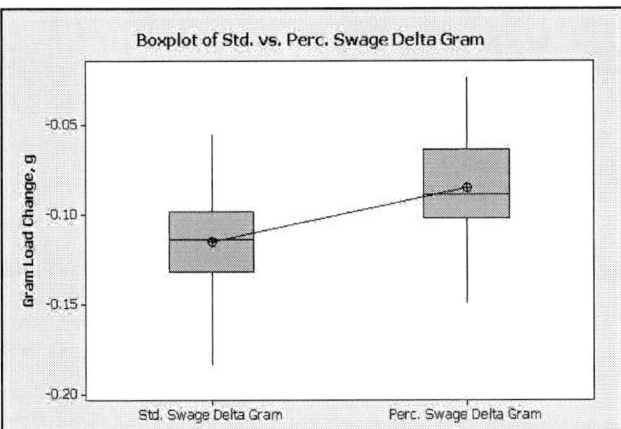

Figure 8. Boxplot showing 25th, 50th, and 75th quartiles of standard vs. percussion swage gram change.

As outlined in Ott[9], a Student's Two Sample t-Test to compare the means of the two distributions (inferences about μ, σ unknown) using a 95% confidence level (α = 0.05), does indicate a statistical difference in gram change mean values observed in the swaging operation between the two sample groups. Setting the null hypothesis H_o: μ_{std} = μ_{perc}, we calculate the test statistic of t = -4.73 and identify the tabular value of t_α = 1.668, and reject the null hypothesis H_o since t < - t_α (indicating $\mu_{std} \neq \mu_{perc}$). The sample group statistics shown below indicate a mean value gram change 25% less for percussion vs. standard swage control group.

Two-Sample T-Test and CI: Std. Swage Delta Gram, Perc. Swage Delta Gram

```
Two-sample T for Std. Swage Delta Gram vs Perc. Swage Delta Gram

                      N     Mean   StDev  SE Mean
Std. Swage Delta Gram 39  -0.1149  0.0264   0.0042
Perc. Swage Delta Gram 39 -0.0852  0.0291   0.0047


Difference = mu (Std. Swage Delta Gram) - mu (Perc. Swage Delta Gram)
Estimate for difference: -0.02977
95% CI for difference: (-0.04231, -0.01724)
T-Test of difference = 0 (vs not =): T-Value = -4.73  P-Value = 0.000  DF = 75
```

A comparison of the variances between the two sample groups using ANOVA tests for equal variances discloses a

different view of the sample sets. An F-test for checking the validity of an equal variance assumption (that is $\sigma_1^2 = \sigma_2^2$) using a 95% confidence level (α = 0.05), does not indicate a statistical difference in gram change variation observed in the swaging operation between the two sample groups. Setting the null hypothesis H_o: $\sigma_{std}^2 = \sigma_{perc}^2$, we calculate the test statistic of F = 1.22 and identify the tabular value of F_α = 1.72, and accept the null hypothesis H_o since F < F_α (indicating $\sigma_{std}^2 = \sigma_{perc}^2$). While a tighter value for variance would be desirable in the consideration of an alternative processing method, an alternative process providing equitable variance to the existing with other benefits (lower mean value for gram load change) would be acceptable as well. Follow-on testing with larger sample groups, over a longer range of time would provide a better representation of true process variation.

Test for Equal Variances: ANOVA Response versus ANOVA Factors

```
95% Bonferroni confidence intervals for standard deviations

          ANOVA Factors   N     Lower     StDev     Upper
Perc.Swage Delta Gram   39  0.0231497  0.0291266  0.0390093
Std. Swage Delta Gram   39  0.0209649  0.0263778  0.0353278


F-Test (Normal Distribution)
Test statistic = 1.22, p-value = 0.544


Levene's Test (Any Continuous Distribution)
Test statistic = 0.17, p-value = 0.679
```

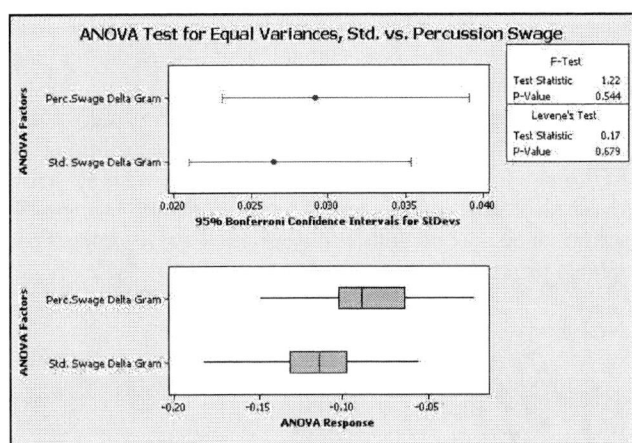

Figure 9. ANOVA plots for equal variance tests of standard vs. percussion swage gram change.

The second factor of comparative analysis involves the torque-out or retention torque value necessary to defeat the swage base to arm coupon bond for both the standard and percussion swage sample groups. With no distinct test and sample group characteristics expected to impact torque values specifically, a histogram plot completes the analysis, indicating comparative values for both sample groups with respect to mean and variation comparisons. While a slight increase in torque retention was noted for percussion swage, its sample group variance was comparatively wider than standard swage group. Again, follow-on testing may provide further insight of process variation and mean value tendencies of retention torque performance.

43

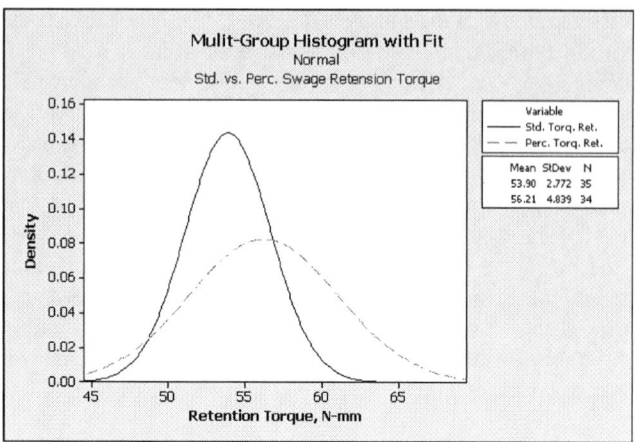

Figure 10. Multi-group histogram with fit of standard vs. percussion swage retention torque.

The final comparative analysis observes differences in baseplate flange flatness between the two sample groups. As previously indicated, white light interferometry for a full baseplate topographical, quantitative scan of both pre and post swage conditions of each sample group was captured. A small sample set of baseplates swaged to coupon arms with both swage processes, provides indication of performance implications. On the 2D X-Y profile scans, a stamped baseplate has very little deformation, ranging approx. 4µm full range X-profile, and even smaller say near 2µm of max. tip deflection (to the right of the large, circular boss region mask cutout shown below), if the plot were normalized about the boss cutout. Y-profile at the tip is an admirably small 2µm across the width of the part. As the vertical scales of each of the 3D scans are signicantly magnified, care must be taken in the perceptions of performance acceptance for each of the sample parts involved. All parts would pass existing acceptable performance specifications for hard disk drive products in volume production applications.

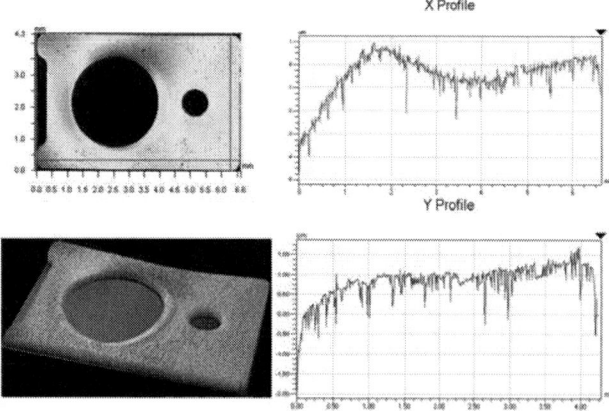

Figure 11. Representitive baseplate profile after stamping and before to swaging operations.

In comparing both swage group profiles in Fig. 12 & 13 to the initial profile of Fig. 11, it becomes evident that both groups portray a marked increase in tip deflection in both X-Y orientations as a result of the stresses incurred during the swaging process using either methodology. However, one notes a representative increase in the baseplate deformation or tip deflection on the standard swage product vs. the percussion swage. This intuitively equates to the increase in gram change noted for the standard group in comparison to the percussion group. A higher tip deflection shifts the load beam and head away from the disk surface, decreasing the gram load to a slight degree. This equates to an increase in gram loss observed in the previous data sets for the standard vs. the percussion swage group.

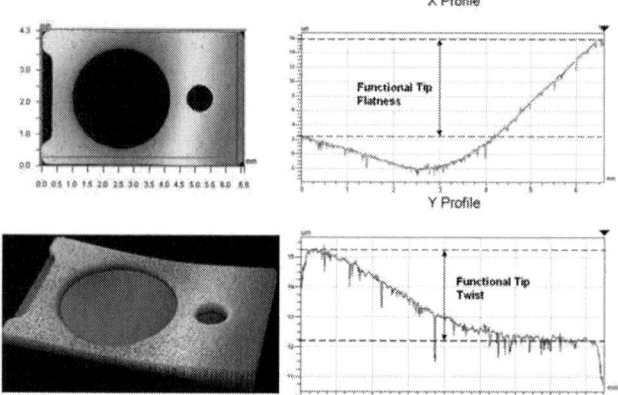

Figure 12. Representitive baseplate profile after standard swage operation

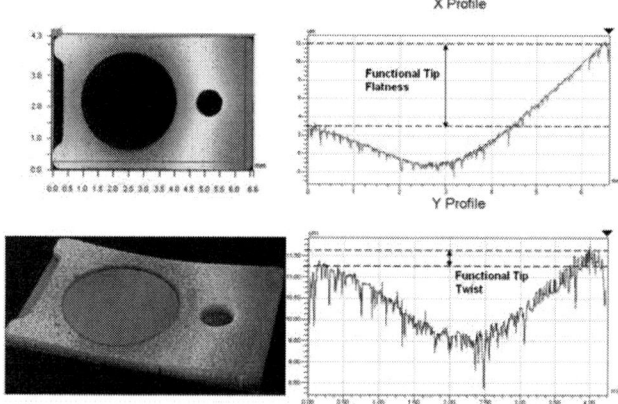

Figure 13. Representitive baseplate profile after percussion swage operation

A quick comparison indicates the standard swage sample tip deflection of 13.75µm visualized in the X-profile plot in Fig. 12, while the percussion swage sample tip deflection measures 9µm for a reduction of nearly 35%. Additionally, the difference in amplitude and the characteristic shape in baseplate tip twist as visualized in the Y-profile plots for both samples exists as well. Greater baseplate tip twist can induce off-track displacement during bending mode excitation from actuator seek and settle movements or from other conditions such as rotating disk windage or external vibrations in an operating drive. Tip twist for the standard swage group measures 3µm across the width of the baseplate in the Y-profile plot of Fig. 12, while the percussion swage sample cups symmetrically across the baseplate width indicating a tip twist in the Y-profile plot of Fig. 13 of less than 1µm.

CONCLUSIONS

Mechanical ball swage of head suspension assemblies to rigid actuator arms used in hard disk drive storage products does show a potential for improvement using a percussion or vibro-impact mechanism for swage force application. A simple, percussive hammer device augmenting existing fixturing and process methodology has the potential to offer a stable, cost effective, and repeatable means of improving the resulting performance of head stack assembly operations. Critical performance measures of gram change through swage and baseplate flatness and tip deflection were demonstrated to improve by 25% (a gram change reduction of 0.03g) and 35% (a tip flatness improvement of 4.75µm) respectively with the use of a percussive vs. conventional swage methods. While some additional improvement was noted in tip twist using percussive swage, torque retention appeared comparable for both methods. Of particular interest is the fact that little modification of tooling or processing methodology would be required for deployment to existing manufacturing sequences.

The functional performance improvements noted for percussion swage of head suspension assemblies for disk drive products appears notable enough to justify further analysis and empirical study. Future work will seek to verify early results with increasing sample sizes, expansion to multi-arm swage setups to understand the potential for reducing kinematic deformation, and further analytical studies to support empirical results.

ACKNOWLEDGEMENTS

The authors would like to thank Reid Danielson of Hutchinson Technology Inc. for support and guidance. Analytical contributions from Shane VanSloun, as well as empirical, measurement, and tooling support from Dennis Johnson, Ann Kasper, and Edward Melberg were much appreciated. A special thanks to Michael Christensen for his trouble shooting and extensive empirical testing support throughout the effort.

REFERENCES

1. Han, G., et.al., "Percussion Drilling in Oil Industry: Review and Rock Failure Modeling", American Association of Drilling Engineers, 2005, 59th National Technical Conference and Exhibition, Houston Texas, April 5-7, 2005.
2. Han, G., et.al., "Dynamically Modeling Rock Failure in Percussion Drilling", ARMA USRMS 05-819, Alaska Rocks 2006, 40th U.S. Symposium on Rock Mechanics (USRMS), Anchorage, Alaska, June 25-29, 2005.
3. Han, G., et.al., "Lab investigations of percussion drilling; From single impact to full scale fluid hammer", ARMA USRMS 09-962, Golden Rocks 2006, 41st U.S. Symposium on Rock Mechanics (USRMS), Golden, Colorado, June 17-21, 2006.
4. Ho, J., Woo, K., "Bifurcations in an Electro-vibroimpact system with friction", University of Nottingham, Malaysia, *Journal of Theoretical and Applied Mechanics*, **46**, 3, pp. 511-520, Warsaw, Poland, 2008.
5. Hinrichs N., Oestreich M., Popp K., 1997, Dynamics of Oscillators with impact and friction, *Chaos, Solutions, and Fractals*, **8**, 4, 535-558.
6. Mikhlin, Y., Reshetnikova, N., "Dynamical interaction of an elastic system and a vibro-impact absorber", Mathematical Problems in Engineering, **2006**, 37980, 1-15.
7 Franca, L.F.P, Weber, H.I., "Drilling under percussive vibro-impact with dry friction", PUC-Rio, Dept. of Mech. Eng. Rio de Janeiro, Brazil, XXI International Congress of Theoretical and Applied Mechanics, Warsaw, Poland, 2004.
8. Dremel, A.J., "Reciprocating type electric tool", US Patent 3,317,764, Patented May 2, 1967.
9. Ott, R.L., *An Introduction To Statistical Methods and Data Analysis, Fourth Edition*, Duxbury Press, Div. of Wadsworth, Inc., 1993.

NEW PACKAGING AND INTERCONNECT TECHNOLOGIES FOR ULTRA THIN CHIPS

Christine Kallmayer and Rolf Aschenbrenner
Fraunhofer IZM
Berlin, Germany
kallmayer@izm.fhg.de

Julian Haberland and Herbert Reichl
Technical University Berlin
Berlin, Germany

ABSTRACT

This paper shows different approaches to use the availability of ultrathin chips for the realization of new packages with high density and improved performance.

For several years technologies have been developed for the embedding of chips in circuit boards in order to achieve 3-D-packages using conventional processes from PCB manufacturing.

Ultrathin chips are suited to be integrated in rigid circuit boards as well as on and in multilayer flexible substrates. The use of interposers prior to embedding can facilitate the embedding of components with ultra fine pitches.

An example for a complex RFID-based product will be shown which is enabled by the integration of ultrathin dies.

INTRODUCTION

A visionary approach for the future production of printed circuit boards was the development of integration technologies for active and passive components in the different layers of the boards. This enables extremely high density of functionality and components in 3 dimensions. The goal is to achieve this progress while maintaining the processes typical for PCB manufacturing. Different approaches have been described for the embedding in FR4 [1,2]. A suitable solution for the integration of chips with very fine pitch is the iBoard technology [3]. It is based on the flip-chip assembly of the fine pitch chips on thin interposers with fan out design.

Besides the rigid organic substrates flexible circuits become increasingly important in PCB industry for applications ranging from consumer products to medical implants. Recent developments – e.g. in the project SHIFT [4,5] – allow the integration of components also in multilayer flex substrates which allows an even higher density with low overall thickness. The technologies which are the prerequisites for this solution and experimental results are shown in this paper.

Important aspects which were investigated for the different new technologies are the constraints regarding pitch and chip geometries [6]. In order to be embedded in a substrate the components have to be thin. For the integration in rigid circuit boards, 50 μm chip thickness were chosen, for the flexible substrates 20 μm are required.

TECHNOLOGIES
Ultrathin Chips
The wafer thinning is a commonplace procedure in front end wafer fabrication. Typically wafers are thinned down to 120 or 80 μm thickness before dicing into single chips and subsequent packaging. Using commercial thinning services wafer thicknesses of 50 μm are also readily available. Thinning below 50 μm, however, still is critical. This regime is therefore called ultra thin. Wafer handling and dicing becomes more subtle. Micro cracks at edges and corners of chips which may be induced by mechanical dicing are prone to propagate into the bulk of the chips and cause failure. This is especially critical for processes with high mechanical loads like flip chip bonding. For ultra thin chips it is therefore advisable to use a dicing by thinning technique, were separation grooves are etched into the wafer prior to grinding.

In the present studies chips with a thickness of 50 μm for FR4 and 20 μm for flexible multilayers have been used to develop and investigate the performance of two distinct embedding technologies.

Details of the thinning process and the encountered problems have been the topic of several publications [7,8] will not be presented and discussed in the present paper.

Assembly of Ultrathin Chips
It has been shown that ultrathin chips are suitable for flip-chip processes [9]. Especially in combination with flexible substrate it is important to achieve thin flip-chip contacts in similar dimensions as the two contact partners. For chips and substrates down to 10 μm the interconnect height should also not exceed 10 μm.

This is possible with solder as well as adhesive bonding technologies.

Thermode Bonding

For thin solder joint CuSn bump metallurgy is a very good solution. It can be deposited by electroplating with very fine pitch. The bump is formed by plating a thin layer of Sn on top of Cu sockets. Uniformity of the thin layers during electroplating is most important to ensure consistent quality of flip chip bonded assemblies. Bump height deviations of 3% have been achieved using diffuser plating rings in the cup plater. All bumps have been used in the as plated condition without prior reflow process. The reflow would consume already too much of the little solder volume by IMC formation. The topography of a CuSn-bump as plated can be seen in Figure 1.

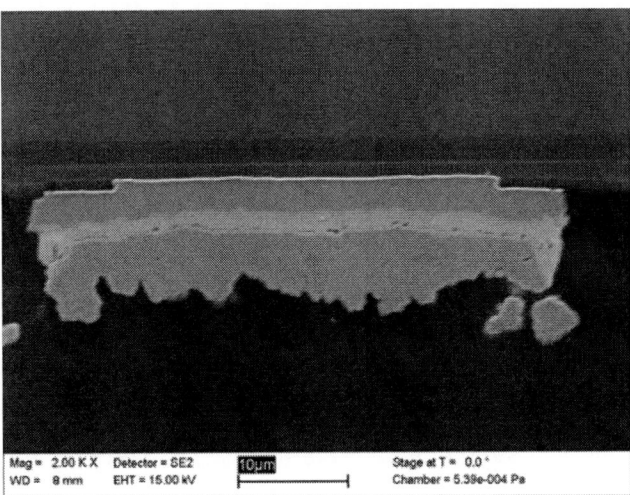

Figure 2: Ultrathin CuSn solder joint

ACA Bonding

Common bumps for ACA technology are either made by mechanical stud bump bonding [6] or various chemical deposition technologies, ranging from evaporation to plating processes. Electroless nickel deposition is often used, as it benefits from its low cost potential [7,8]. The standard electroless nickel UBM for high reliability has a thickness of 5 μm but only a minimum of 1 μm is necessary to have a closed and void free nickel layer [9]. For thin interconnects NiAu with a thickness of 3 μm is used typically. Driver or RFID chips are often available with electroplated Au bumps which are also very well suited for ACA processes. The bump thickness is typically >10μm for the commercial chips but the process works well with 5 μm.

Figure 1: Electroplated CuSn bump on test chip, (bump height 6μm, chip thickness 30μm)

The thermode bonding technology is based on fast reflow soldering by pulse heating. The fast process allows the use of low cost materials with low temperature resistance for flip chip soldering at high temperatures without damage of the flex. It is even possible to apply the underfill material before the placement of the die and perform underfilling and bonding in one step. The preapplication of noflow underfiller can either be done by stencil printing or by dispensing. The use of noflow underfiller today is limited by soldering temperatures. In case of eutectic SnCu solder noflow underfiller was successfully established. A typical thin solder joint is shown in Figure 2.

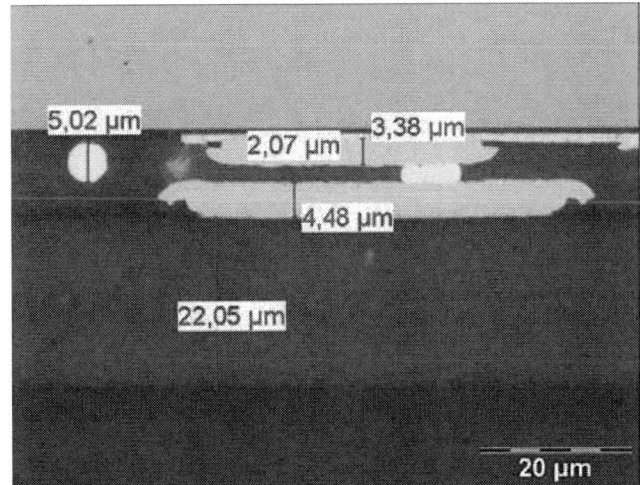

Figure 3: ultrathin ACP contact, on PI thin film substrate with CuNiAu metallization

The ACA bonding process requires pressure to trap electrically conductive particles between the chip bumps and the substrate metallization to form an electrical contact. Applied heat on the bonding tool and chuck is needed to cure the adhesive, which is usually an epoxy based material.

Adhesives are of paste or film form either. A thin contact with NiAu metallization on both sides is shown in Figure 3.

Embedded Flip Chip

In the past flip chip on flex aimed at the highest degree of miniaturization. Recent developments use flip-chip on flex as a prerequisite for embedding.

For the build up of a multilayer flex, commercial PI sheets of 25 μm polyimide with 17 μm thick copper wiring are used. Onto the substrates ultra thin flip chips are mounted. The electrical interconnect is established using either anisotropic conductive adhesive or thin solder interconnects. No difference regarding yield or performance could be found in comparison of those technologies. Multiple flex boards with assembled chips and adhesive foils are then stacked together and laminated in a commercial stage press. Subsequently the layers of the stack are interconnected by trough hole drilling and metallization. Finally the outer surfaces of the whole stack are structured.

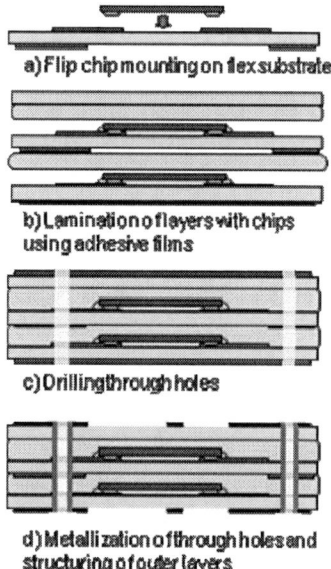

a) Flip chip mounting on flex substrate

b) Lamination of flayers with chips using adhesive films

c) Drilling through holes

d) Metallization of through holes and structuring of outer layers

Figure 4: Process flow for the Flip Chip in Flex technology

Using this process flow multilayer flexes with 4 layers of embedded dies and an overall thickness of 450 μm were realized [Figure 5]. No breaking of the chips was observed during lamination.

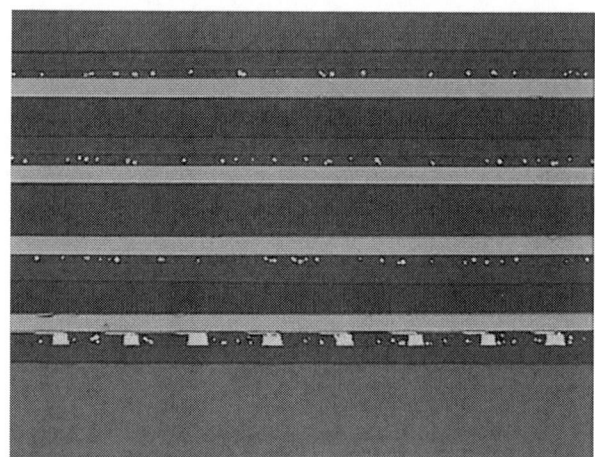

Figure 5: Multilayer flex with 4 layers of embedded dies

Embedding in FR4

Ultrathin chips can be integrated in rigid circuit boards by different technologies, e.g. Chip in Polymer [2, 10]. If the pitch of the chips is very fine (< 100 μm), it is suitable to assemble the ICs on a flexible interposer (e.g. PI) with thin Cu metallization (5μm) which enables the required fine structuring.

An example for a driver chip with 50 μm pitch on a thinfilm HiCoFlex [11,12] is shown in Figure 6. The chip thickness is relevant for the yield during the integration process. The chips are typically 50 μm or thinner to avoid damaging of the silicon during lamination. The difference between the ultrathin die and a conventional flip-chip assembly is shown in Figure 4.

For pitches > 100 μm thin FR4 interposers can be applied. The design is a fan out of the I/Os so that the resulting Pitch is uncritical for the following processes. The use of an interposer separates the chip assembly from the embedding process. This interposer can be tested before being integrated in the board.

Figure 6: Driver IC with 50 μm Pitch on HiCoFlex

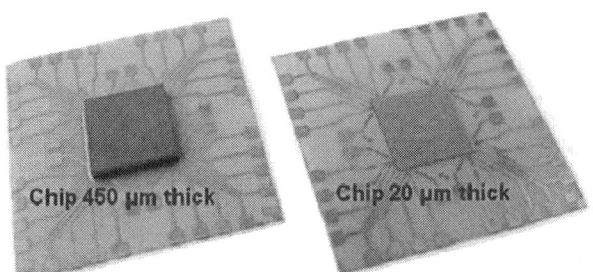

Figure 7: Interposer with thick and thin flip chip in comparison

The interposers are placed between the prepegs of a multilayer and then laminated. Due to the fan out the interposer can be connected by conventional methods which are always used in the manufacturing process of multilayer boards. The process flow is shown in Figure 8. If the interposer is in the neutral area of the build-up the holes are drilled conventionally. For the integration directly underneath the outer layer, the interconnection by laser vias is used successfully.

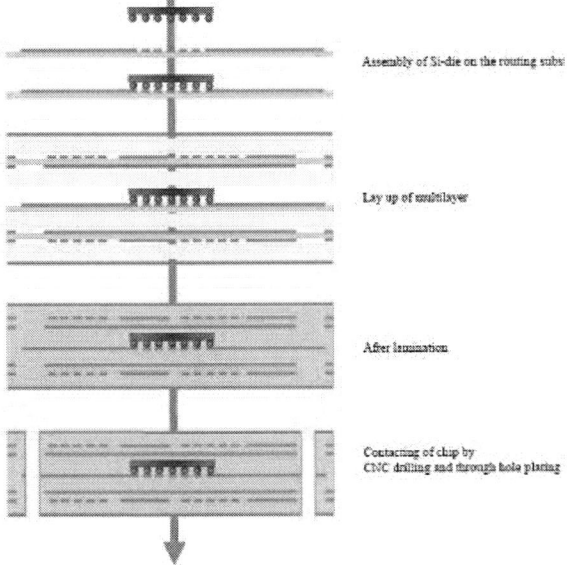

Figure 8: Process flow of iBoard integration process

APPLICATION EXAMPLE

A complex module with 6 integrated ICs is described as an example for flip-Chip assembly of ultrathin chips and the integration in PCB.

The goal was the replacement of paper labels with text and barcode on transport containers for letters. As a solution with high potential for automation a passive RFID-Label with E ink display was developed. Energy and data are transmitted according to ISO15693 to two different antennas (13,56 MHz) which are positioned outside the display area. In order to realize a display with 982 segments and symbols 4 driver ICs with 324 I/Os each were required. These commercial ICs have a pitch of 50 µm. Therefore a flip-chip integration on a conventional printed circuit board is not possible. Depending on the process flow some embedding

technologies do also not yet allow to contact 25 µm pads, e.g. directly by drilling vias to those pads. The integration by iBoard technology was chosen to solve this problem. In a first step the 50 µm thin chips with Au bumps had to be assembled in ACA technology on a thinfilm HiCoFlex interposer [11,12] with CuNiAu end metallization. The overall thickness of the assembled interposer was only 90 µm. The cross section in Figure 9 shows how critical the positioning accuracy is for such fine pitches with a bump distance of < 8 µm.

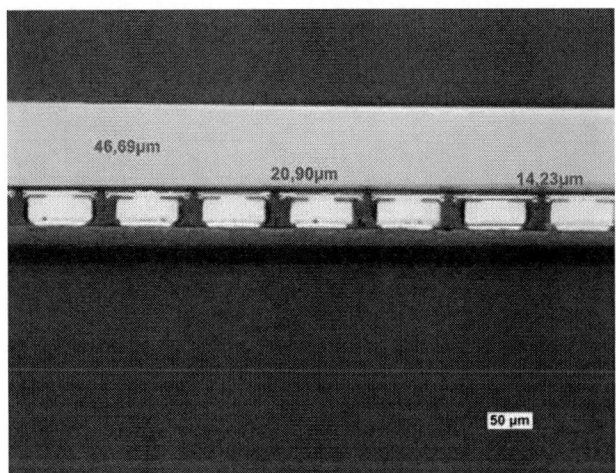

Figure 9: cross section of ACA contacts on thinfilm interposer

The other necessary ICs are a microcontroller and the RFID frontend. As these chips have only minimum pitches > 100 µm, they could be assembled on thin FR4 interposers with Cu end metallization.

The 4 driver interposers, controller interposer and RFID interposer were then integrated in a 6-layer FR4 board as shown in the process flow in figure 8. The driver chips do not break during the lamination process although their geometry is critical (Figure 12). The red mark in Figure 11 shows the position of a driver in an X-ray image of the final board.

A few necessary passive SMD components were then assembled on the back side of the board. On the front side the e-ink FPL was laminated together with protection layers.

Powered by the reader the content of memory and display can be changed even in motion. The display then remains unchanged with high contrast for months without energy supply. The final module is shown in Figure 10.

Figure 10: Display RFID demonstrator with 6 embedded chips and E ink foil (160mm X 65mm X 1mm)

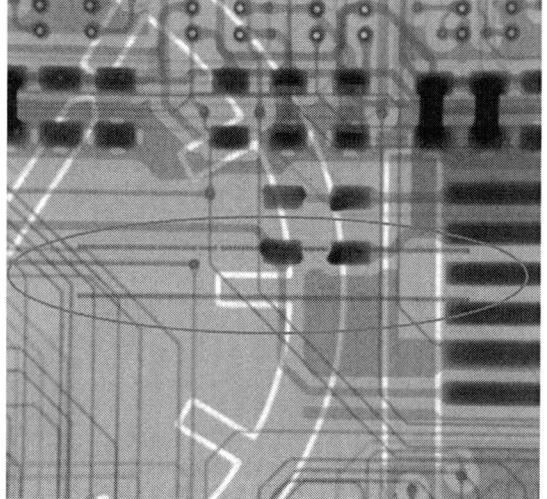

Figure 11: X-ray image of integrated driver IC

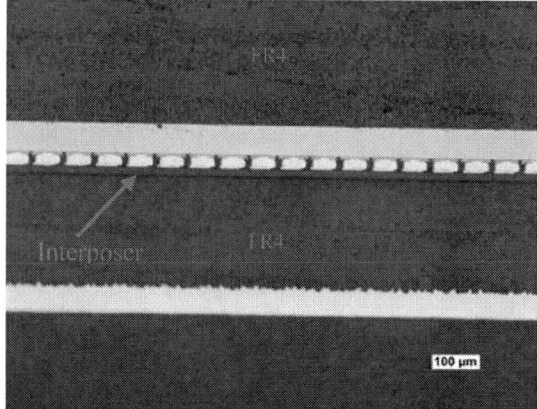

Figure 12: Cross section of integrated driver IC

CONCLUSION

In general the supply of ultra thin chips or components is the bottleneck for the new enabling embedding technologies of active components. The assembly of thin chips can be realized with good results by various flip-chip technologies. Contact thicknesses < 10 µm can be achieved.

Embedding of thin chips in flexible multilayer flexes can be used to obtain thin modules with several layers of active components. As a demonstrator 4 layers with –ICs have been realized with 450 µm thickness.

For the integration of chips with pitches < 100 µm the iBoard technology has shown very promising results. Using either FR4 or thinfilm PI interposers, a wide range of chip types with different bump metallurgies can be integrated with chip thicknesses of 50 µm or even higher.

As an example for the potential of this technology a display RFID module with 6 integrated ICs has been described.

Although there are many potential applications for such technologies the manufacturers are reluctant to use modules with embedded components processed in printed circuit board technology. Only the growing necessity to use this technology in order to overcome the gap between chip pitches and line geometries on FR4 has lead to the big interest in recent years. One prerequisite to achieve a higher degree of acceptance would be the availability of the corresponding design tools for this 3D integration to facilitate the design flow.

ACKNOWLEDGEMENTS

The work presented was partially done within funded projects: The project SHIFT was funded by the European Commission and the project Pariflex by the Federal Ministry of education and research.

The authors want to thank Thomas Gottwald and Alexander Neumann from Schweizer Electronic AG for providing the iBoard technology for the display RFID modules.

REFERENCES

[1] R. Aschenbrenner, A. Ostmann, A. Neumann, H. Reichl, Process flow and manufacturing concept for embedded active devices, Proceedings of 6th Electronics Packaging Technology Conference, EPTC 2004

[2] Chen, Yu-Hua., "Chip-in-Substrate, CiSP, Technology", Proc 6th Electronic Packaging Technology Conference 2004, Singapur, pp. 595-599.

[3] T. Gottwald, U. Ockenfuß; The iBoard Technology - A New Method for "System in Board"; IEEE Workshop 3D Integration, Munich 2007

[4] T. Löher, M. Seckel, B. Pahl, L. Böttcher, A. Ostmann, H. Reichl, Highly integrated flexible electronic Circuits and Modules; Proc. IMPACT 2008

[5] B. Holland, R. Mcpherson, T. Zhang, Z. Hou, R. Dean, R. Johnson, L. Del Castillo, and A. Moussessian, Ultra Thin, Flexible Electronics, Proc. ECTC; Orlando (2008), p. 1110-1116

[6] Ostmann, A. Neumann, J. Auersperg, C. Ghahremani, G. Sommer, Aschenbrenner R., and Reichl H., "Integration of Passive and Active Components into Build-Up Layers", EPTC 2002, Singapore, 2002

[7] C.Landesberger, C.; D. Bollmann; M. Bleier; S. Scherbaum; H.-P. Spöhrle; K. Bock; Thinning of bumped

wafers or bumping of thinned wafers, Workshop Thin Semiconductor Devices, München, 2004

[8] Gerhard Klink et al., „Innovative Packaging Concepts for Ultra Thin Integrated Circuits", Proc. 51st ECTC, Orlando, Florida, 2001

[9] J. Haberland, B. Pahl, Ch.Kallmayer, R. Aschenbrenner, H. Reichl: Super Thin Flip Chip Assemblies on Flex Substrates - Adhesive Bonding and Soldering Technology, Proc. 39th International Symposium on Microelectronics IMAPS 2006, San Diego, 2006

[10] T.Löher, A. Neumann, L. Boettcher, B. Pahl, A. Ostmann, R. Aschenbrenner, and H. Reichl, *Smart PCB manufacturing technologies*, Proc. ICEPT; Shenzen (2005)

[11] A. Fach, Y. Athanassov, U. Brunner, D. Hablützel, B. Ketterer and J. Link, Multilayer polyimide film substrate for interconnections in microsystems, Microsystem Technologies, Heidelberg: Springer Berlin Volume 5, 1999, pp. 166-168

[12] Hans Burkard, Thin Film Resistor Integration into Flex-Boards, 5th International Workshop ‚Flexible Electronic Systems', November 29, 2006, Munich

CHIP-LAST INTERCONNECTION TO THIN ORGANIC PACKAGES WITH ADVANCED DILECTRICS AND FINE PITCH I/Os

Rao R. Tummala, Ph.D., Venkatesh Sundaram, Ph.D., Nitesh Kumbhat, and Nithya Sankaran
Packaging Research Center
Georgia Institute of Technology
Atlanta, GA, USA
rao.tummala@ece.gatech.edu; venkatesh.sundaram@ece.gatech.edu;
nitesh@ece.gatech.edu; nithya.sankaran@gatech.edu

ABSTRACT

Three miniaturization developments have been underway for more than a decade. These include thin core, fine line wiring and embedded chip technologies. This paper proposes and describes the latest advances and results in all these. It describes advanced low-loss dilectrics and wiring down to 10 microns, thin organic cores down to 50 microns thickness, and chip-last interconnection as a simpler and lower cost alternative to chip-first.

Chip-First, Chip-Last, Dilectrics, EMAP, Embedded Actives, Fine Pitch, Interconnect

INTRODUCTION

Ultra miniaturized and low-profile mobile products are driving the need for embedded active and passive component integration technologies. Two such technologies are currently being pursued by the industry. Both are based on "chip-first", in two versions: 1) embedding by wafer-level fan-out with chip-first concept by US and European companies [1], and 2) embedding in organic substrates by chip-first and chip-middle by Japanese companies [2]. These technologies are based on either the chips simultaneously mounted on detachable-tape or molded by a carrier or molding compound, and then interconnected using thin film package wiring processes. An alternative to this is to mount the chips on rigid-core surfaces and then interconnected using thinfilm package processes.

Georgia Tech PRC (GT PRC) has pioneered and demonstrated a third option to overcome some of the challenges in chip-first technology. This is referred to as Embedded MEMS, Actives and Passives (EMAP) Technology with chip-last (CL) interconnection but with chip-first benefits. As seen in Figure 1 the chip-last embedded actives and passives (EMAP-CL) technology is targeted at highly integrated systems with multiple 2D or 3D ICs for RF, Digital, Analog, MEMS and passive devices, all in a single package or module. Ultra-thin ICs and passives are embedded in high-precision cavity structures on both sides of thin core organic substrates with high I/O density, vertical and horizontal interconnections.

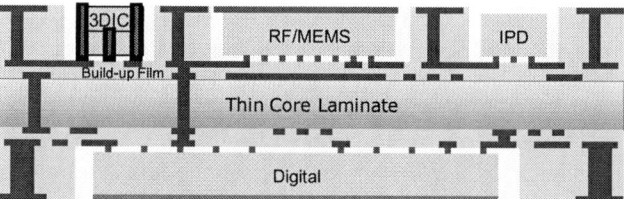

Figure 1. Cross-section of EMAP Chip-Last Embedded Actives

The GT PRC's EMAP-CL approach has many advantages that include:

1. Module thickness less than 0.5mm including 3D stacked components
2. Ultra thin dielectrics, conductors and vias leading to ultra thin substrate
3. Double-side wiring and components due to through-package-vias (TPV)
4. Known-good substrate and known-good die before assembly
5. Repairability after assembly, if necessary
6. Ultra fine-pitch and short pad-to-pad interconnection
7. Minimal change to current package manufacturing infrastructure
8. Shorter time-to-market
9. Flexible testing due to the exposed dies on surface
10. Overcome TCE mismatch reliability issues due to the use of ultra thin underfill
11. Precise placement of dies using current assembly processes and tools
12. Not limited by panel or wafer size due to the post placement of ICs
13. Allows embedded dies, IPDs and discrete passives with different thicknesses and substrates (Si, GaAs, SiGe, Glass, Ceramic, Laminate) due to the multi-depth cavities
14. Top surface cavity allows MEMS devices; ideal low temperature cavity packaging option for MEMS
15. Easier heat transfer due to the exposed die
16. Embedding of TSV chips and 3D-ICs for high I/O density
17. Low-K and ULK/ELK die embedding with low stress interconnects

18. Flexible die designs – accommodates current cell and pad designs
19. Flexible interconnect materials – Cu, Au, lead-free solder, adhesives

ELECTRICAL BENEFITS OF CHIP-FIRST AND CHIP-LAST

One of the major driving forces for embedded chip technology is its high electrical performance. Parasitics associated with the package and the interconnections, affecting the signal and power integrity of the system significantly [3]. It is well known that the parasitic inductance degrades power integrity and the parasitic capacitance degrades signal integrity.

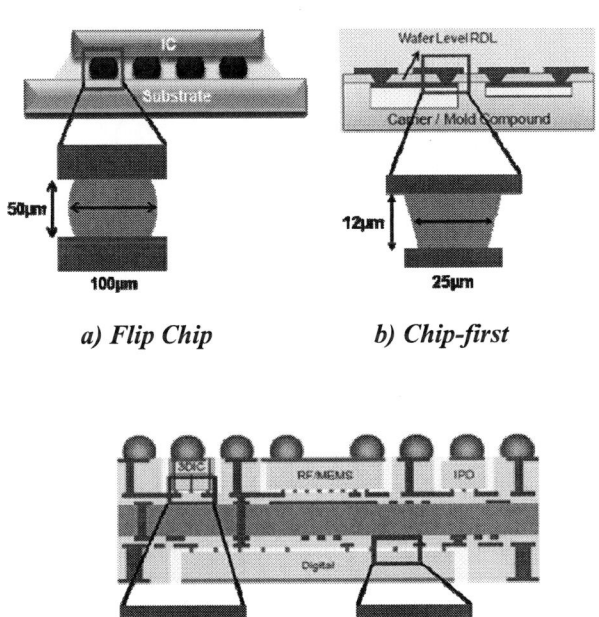

a) Flip Chip   *b) Chip-first*

c) Chip-last

Figure 2. Flip Chip, Chip-First and Chip-Last Interconnect Schemes and Differences

Power Integrity

Low-inductance packaging and low-impedance power distribution network are essential for preserving power integrity. Parasitic inductance offered by the package power supply lines and interconnections result in supply voltage variations, which can affect the system performance. As an example, in an ADC die, the voltage range over which the ADC can operate has a maximum limit called the reference voltage. In some ADCs, supply voltage is set as the reference voltage. If there are any fluctuations in the supply voltage at the input of the ADC due to interconnect parasitics, the reference voltage is destabilized, and the conversion accuracy of the ADC is affected. Hence, the

parasitic inductance offered by the package and first level interconnects should be minimized as much as possible to achieve a clean supply voltage. Also, the supply voltage variations give rise to simultaneous switching noise (SSN) in the package [4]. Mutual parasitics of first level interconnects can affect the on-chip SSN [4].

Figure 2 illustrates schematically the difference in dimensions between flip-chip, chip-first and chip-last interconnection technologies. Table 1 shows the interconnect parameters for chip-first and chip-last technologies, comparing and contrasting with the traditional flip chip interconnections. The parasitic values are estimated, based on empirical relations and experimental validation of the formulas used in the estimation of interconnection parasitic [5]. The self inductance (L) and resistance (R) of chip-first and chip-last configurations are benchmarked against flip-chip. Mutual parasitics (Lm, Cm) are calculated for a pitch of 150 μm (edge-to-edge spacing 50μm) for flip chip interconnections and for an edge-to-edge spacing of 25 μm for chip-first and chip-last interconnections. As can e seen in Table 1, the chip-to-substrate interconnections of chip-last and chip first configurations have low values of self (L) and mutual (Lm) inductances in comparison to flip-chip, as shown in Table 1, which help in preserving the overall system power integrity.

Additionally, the chip-last embedded active structure has the provision to accommodate dedicated power and ground planes, as shown in Figure 3. In this structure, the M1 – M3 are the metallization layers, the planes forming M2 and M3 (circled in red) are for power and ground while M1 can be used as a redistribution layer. The power and ground planes offer low inductance values and help reduce the magnitude of power supply variations.

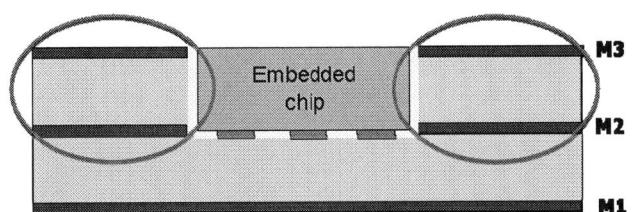

Figure 3. Cross-section of Chip-Last Embedded Actives Package with Embedded Chip

Signal Integrity

Low-loss and low-dielectric constant materials, seamless transitions, smaller discontinuities and parasitics and impedance-matching are important for achieving good signal integrity. Parasitic capacitance adversely affects signal integrity. First level interconnects of chip-last embedded actives score well in terms of reduced capacitance as shown in Table 1. Signal distribution, through multiple layers, can cause impedance-mismatch and return path discontinuities. However, with interconnections of short pitch and associated substrate technology, as in this program with fine pitch, it is possible to use only one redistribution layer for chip-last embedded chips. Ultra-thin organic substrates used for supporting chip-last embedded chips have low-loss tangent and low-dielectric permittivity that reduce signal transmission losses and achieve faster signal transmission. The all-copper chip-to-substrate interconnects in chip-last are better than solder-based flip chip interconnects in electrical conductivity and electro-migration resistance, as copper has better electrical properties, compared to solder.

Interconnect Parameter	Flip-Chip (150 μm)	Chip-First (25 μm)	Chip-Last	
			(Non-Repairable (25 μm)	(Repairable) (25 μm)
L (pH)	10.8 (100%)	3.27 (30.3%)	1.79 (16.57%)	2.25 (20.83%)
R (mΩ)	2.19 (100%)	0.85 (38.8%)	1.5 (68.49%)	1.2 (54.79%)
Lm (pH)	9.612	3.156	1.78	2.228
Cm (fF)	8.96	1.571	0.48	0.757

Table 1. R, L and C Parameters for Different First Level Interconnect Technologies

ULTRA-THIN ORGANIC SUBSTRATES FOR DC TO 110 GHZ RF/DIGITAL APPLICATIONS

The focus of EMAP is in halogen-free organic dielectrics with low-loss, low-CTE and high reliability to compete with well-established ceramic packages for RF, microwave and mm-wave applications. Additional advantages of EMAP include higher I/Os in minimum number of layers, leading to thinner substrates with acceptable warpage. A comparison of the Georgia Tech's EMAP substrate technology with other substrate approaches is shown in Table 2.

Technology	Dk	Df	RF-4 Process Compatible	Wiring Density	Thickness	Cost
FR-4 Build-up Organic	Medium	High	Standard	Medium	Medium	Low
Ceramic	High	Low	No	Low	Thicker, Heavier	High
Teflon, LCP Based	Low	Low	Limited	Low	Medium	High
Current Si Interposer	High	High	Limited	Highest	Thin	High
EMAP 2 Organic Platform	Low	Low	Yes	High	Thin	Low

Table 2. Summary of EMAP Substrate Benefits

High Frequency Performance

The integration of embedded RF passive components in the EMAP substrate places an added requirement for stable low loss dielectrics. The dielectric material used in the current EMAP substrate consists of two distinct low-loss thermo-setting polymers; (a) a thin-core glass reinforced laminate (RXP-1), and (b) a thin dry-film build-up dielectric (RXP-4). The RXP-1 core is a glass fiber reinforced organic laminate with thickness in the range of 50-125μm. The RXP-1 has a stable dielectric constant of 3.25 and loss

tangent of 0.004 from 1 to20GHz. The RXP-4 build-up dielectric film has a dielectric constant of 2.92 and dissipation factor of 0.003, stable over a wide range of 12 GHz. Both RXP-1 and RXP-4 have been recently characterized from 1 to 110GHz, with stable dielectric constants and loss tangents below 0.005 [6,7].

Low Stress and High Reliability

Low-CTE core laminates and low-surface modulus build-up films were combined to provide a low-stress interconnect, especially for ULK and ELK on-chip dielectrics. A new generation of ultra-thin halogen-free BT laminates (HL-832NS) with progressively low CTEs from 14ppm/C down to 6.5ppm/C are evaluated for warpage, stress and reliability of chip-last Cu interconnections at 30-50μm pitch. The 832NS laminates also have high Tg and higher modulus to minimize warpage at lead-free reflow temperatures. The loss tangent of this material has been reported to be around 0.006 at 1GHz. Core thicknesses of 35μm, 50μm and 100μm are integrated into chip-last embedded active test vehicles. The RXP-1B laminate has a medium CTE (10-13ppm/C) and high Tg (~300C) above the lead-free reflow temperature of 260C, which leads to excellent thermal stability during assembly processes. The intermediate CTE of both these laminates provide the optimum solution for 1st and 2nd level interconnect reliabilities.

High I/Os and Fine Pitch

The substrate design rules include a minimum copper line width and space of 15μm, blind microvia diameter of 25-40μm and through-hole diameter of 30-50μm. The through vias were designed and fabricated with 50μm diameter at 125-150μm pitch, 40μm diameter at 100-125μm pitch, and 25-30μm diameter at 75-100μm pitch. The line width and spaces of 10μm were designed for bonding Cu microbumps at 30μm pitch. Line widths below 15μm were targeted with semi-additive processes (SAP), while lower cost subtractive etching was used for >15μm lines.

FR-4 Compatible Low-Cost Processes

A major emphasis of the EMAP is to maintain process compatibility with current FR-4 substrate manufacturing infrastructure for low cost and faster time-to-market. To achieve 30-50μm through-via diameters and 25-40μm blind via diameters, the UV laser ablation was chosen as the front-up approach. The CO_2 laser via formation has also been demonstrated down to 40-50μm diameter through vias. Metallization was accomplished using state-of-the-art low-stress electroless copper seed plating, and copper-filled through and blind-via electroplating. A novel super-filling process technology based on the Inplate DI^{TM} system has been demonstrated. This process uses a proprietary set of additives to control the plating distribution within and outside the through vias. These and other improved process technologies resulted in the demonstration of copper-filled through-vias of 40-50μm diameter in the RXP-1 thin core and 25μm blind vias in RXP-4, as shown in Figure 4. Combinations of RXP-4 on BT core have also been optimized. The finest through-package vias (TPV) demonstrated were 30µm diameter at 75µm pitch.

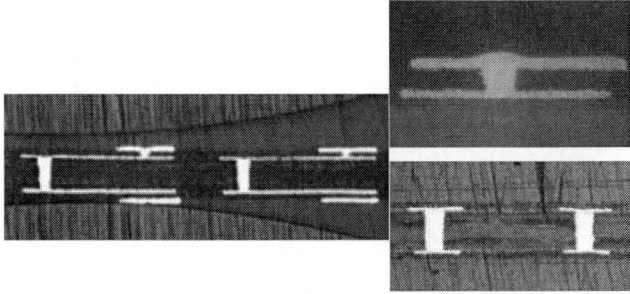

Figure 4. Micro-sections of Ultra-Thin Organic Substrate with Cu Filled Through and Blind Vias in RXP-1 and RXP-4 Dielectrics After Passing 2000 Thermal Shock Cycles.

The reliability test conditions were based on JEDEC standards and mobile product package test conditions including MSL-3 preconditioning, JEDEC/IPC JSTD020D-01, 3X lead-free solder reflow excursion, peak temperature 260°C, and thermal shock test according to JESD22-A104C, -55°C to +125°C. At substrate level, both the 2ML and 4ML RXP substrates with Cu filled through and blind microvias passed 1500-2000 cycles with cumulative via failure rates less than 1%. Failure analysis by micro-sectioning attributed all failures to microvia cracking due to thin via wall plating or residual dielectric at the via bottom, from process induced defects.

INTERCONNECT, ASSEMBLY AND RELIABILITY
Interconnections
Chip-last technology provides the flexibility in the choice of interconnection materials, processes, geometries and structures. In this project, consistent with industry standards and infrastructure, copper bumps were used as the interconnections. To enable ultra-high density I/Os and minimize interconnection parasitics, however, these bumps were fabricated at ultra-fine pitch of 30µm with bumps of 15µm diameter and bump heights between 8 and12µm. Electroless-Nickel Immersion-Gold (ENIG) was used as the surface finish on both the die bumps and the substrate pads in order to prevent surface oxidation of copper. The bonding media used includes both non-conducting film (NCF) and anisotropically-conducting film (ACF).

Multiple variations in test vehicle parameters were studied to understand their effect on reliability performance. Two die thicknesses (55µm and 550µm), two assembly loads (14N and 21N), two adhesive materials (NCF and nano-ACF with nano-silver as conducting filler) and two structures (cavity and non-cavity) were investigated for reliability performance under High Temperature Storage Test (HTS), Highly Accelerated Stress Test (HAST) and Thermal Cycling Test (TCT).

Assembly Process
The assembly was performed using FINETECH Fineplacer© Lambda assembly tool with +/-1µm alignment accuracy. The assembly process involved pre-bonding of nano-ACF or NCF to the substrate at 90°C, followed by removal of nano-ACF or NCF carrier. After alignment of the die and the substrate, thermocompression bonding was performed with 14 or 21N load at 180°C for 300 seconds. Based on the effective cross-sectional area of contact for all the bumps, the 14N and 21N loads translated to a pressure of ~200MPa and ~300MPa or a force of ~4g/bump and ~6g/bump respectively.

Reliability Test Results
The reliability test conditions were based on JEDEC standards and mobile-product package test conditions, including MSL-3 preconditioning, JEDEC/IPC JSTD020D-01, 3X lead-free solder reflow excursion, peak temperature 260°C, and thermal shock test according to JESD22-A104C, and -55°C to +125°C. At substrate level, both the 2ML and 4ML RXP substrates with Cu-filled through and blind microvias passed 1500-2000 cycles with cumulative via failure rates of less than 1%. Failure analysis by micro-sectioning attributed all failures to microvia cracking due to thin via wall plating or residual dielectric at the via bottom, from process-induced defects.

The measured average insulation resistance for the assemblies with nano-ACF was ~3.75E+11Ω. Similar measurement for NCF yielded ~7.82E+11Ω which clearly indicates that addition of 0.1 weight percent of silver nano-particles does not affect the leakage current and both nano-ACF and NCF are non-conducting in lateral conduction.

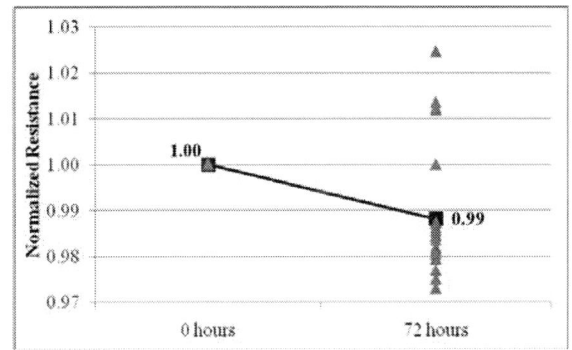

Figure 5. Normalized Resistance Change dDring HTS for 24 Daisy Chains

When subjected to HTS testing, samples exhibited no failures, as shown in Figure 5 which depicts the normalized and average resistance change for 24 daisy chains. The resistance change was within ~3% of the as-assembled resistance. These results indicate that the interconnect scheme evaluated in this project shows excellent reliability during HTS test.

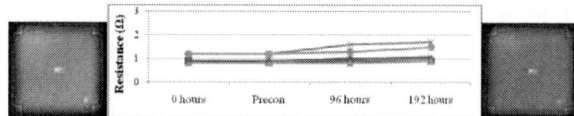

Figure 6. C-SAM Images of Assemblies Under HAST Test at 0 Hours and 192 Hours, and the Associated Changes in Daisy Chain Resistance

Moisture-induced hygroscopic swelling has been widely reported in the literature, as the primary failure mode for assemblies with adhesives [8, 9, 10]. In this project, the samples subjected to HAST test showed no moisture ingression and only marginal increase in resistance, as shown in Figure 6, emphasizing the robustness of Cu bump interconnect with nano-ACF and NCF.

The TCT performance of the assemblies was also found to be excellent. Figure 7 shows the daisy chain resistance values of a sample through more than 2000 cycles. As is evident from Figure 7, the resistance changed very slightly after 1800 cycles and the samples survived more than 2000 cycles, demonstrating thermal cycle fatigue resistance of the chip-last interconnection scheme.

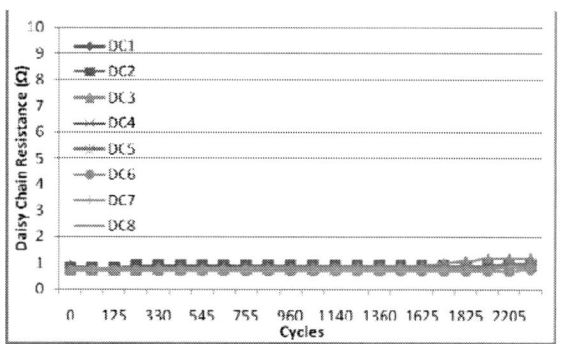

Figure 7. Daisy chain resistance values of a sample subjected to TCT

PROTOTYPE DEMONSTRATOR

The first demonstrator of chip-last, cavity-embedded IC in ultra-thin organic substrates was fabricated as shown in Figure 8 with its top view and micro-section [11, 12]. The substrate core was made of a low loss, 110µm thick RXP-1 polymer-glass laminate, while the dielectric was made of a 20µm thick RXP-4, a low loss dry film. The interconnection from IC to substrate was accomplished at 30µm pitch, Cu-to-Cu bonding using non-conductive adhesive at 160°C. The embedded ICs were 3x3mm and 7x7mm in size and in the case of 3x3mm die, a total of 360 I/Os were interconnected in a 5x5x0.3mm package.

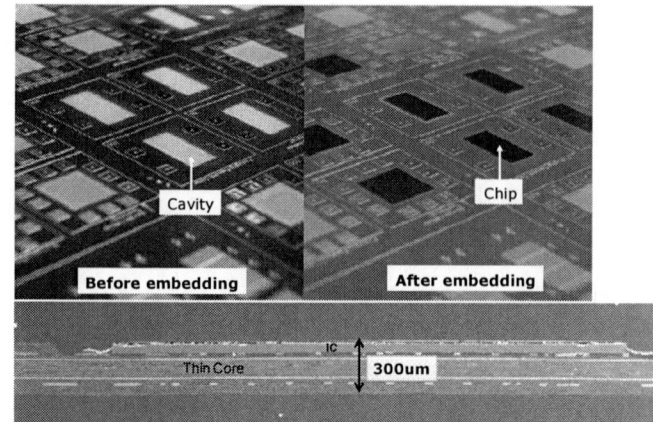

Figure 8. Top View and Cross-section of Thin IC (55µm thick) in <300µm Organic Substrate by Chip-Last Cavity Embedding

Reliability of the chip-last embedded IC substrate was evaluated using 3x3mm die, successfully passing 1000 thermal cycles (-55°C to 125°C) after MSL-3 pre-conditioning and 3x260°C reflow at 50µm I/O pitch.

The second generation of CL-EMAP technology is targeting 1) 10x improvement in throughput for the embedding cavilty processes, 2) top-side re-distribution with full design flexibility while maintaining the chip-last methodology, and 3) higher I/Os. A summary of key parameters for Gen 1 and Gen 2 CL-EMAP is shown in Table 3.

Parameter	Generation 1 (2007-2009)	Generation 2 (2009-2011)
Substrate # of Layers for Re-distribution	2 on each side	1 on each side
Substrate Wiring	15-20um L/S on build-up 30um L/S on core	10-15um L/S on build-up 15um L/S on core
Substrate Vias (Cu filled)	40-50um diameter in core 25um diameter in build-up	25-40um diameter in core 25um diameter in build-up
Cavity Layer Vias for 3D	150um diameter	40-50um diameter stacked
Substrate Core Thickness	110um	50-110um
Substrate Materials	RXP-1, BT RXP-4	RXP-1, BT, FR-4/FR-5 RXP-4, Epoxy
Peripheral Interconnect	30um pitch Cu-to-Cu	30um pitch Cu-to-Cu
Area Array Interconnect	None	75-150um pitch Cu and Lead-Free Solder
Cavity Process	Photo, Plasma etch, Laser ablation	Laser, photo or stamped with 10x higher throughput
Cavity Clearance to IC	50-100um	10-50um
Chip Size	3-7mm	3-10mm

Table 3. Georgia Tech EMAP Roadmap

ACKNOWLEDGEMENTS

The authors would like to acknowledge the contributions of a number of technical staff at the Georgia Tech Packaging Research Center, Rogers Corporation, Atotech Berlin and USA, DuPont Electronic Materials, Endicott Interconnect, Ibiden Japan and USA, and Mitsubishi Gas Chemical. This research was funded by the EMAP consortium industry members whose financial and technical support is greatly appreciated.

REFERENCES

1. Yann Guillou, "3D Integration for wireless products: An industrial perspective", *i-Micronews*, June 2009. *http://www.i-micronews.com*
2. Takaharu Yamano, Masahiro Sunohara, Hajime Izuka, Tetsuya Koyama, "Wiring board with embedded semiconductor chip, embedded reinforcing member and method of manufacturing the same", *European Patent # EP1703558*, 2006.
3. "Electrical Characterization and Design Optimization of Embedded Chip in Substrate Cavities", Sankaran, Nithya; Lee, Baik-Woo; Sundaram, Venky; Engin, Ege; Iyer, Mahadevan; Swaminathan, Madhavan; Tummala, Rao; Electronic Components and Technology Conference, 2007. ECTC '07. Proceedings. 57th, May 29 2007-June 1 2007 Page(s):992 – 999.
4. Kim, W., "Development of measurement based time domain models and its application to wafer level packaging", Ph.D Thesis, ECE, Georgia Tech, 2004
5. Trucco, G.; Boselli, G.; Liberali, V., "An approach to computer simulation of bonding and package crosstalk in mixed-signal CMOS ICs," *Integrated Circuits and Systems Design, 2004. SBCCI 2004. 17th Symposium on* , vol., no., pp. 129-134, 7-11 Sept. 2004
6. Seunghyun Hwang, Sung-Hwan Min, Madhavan Swaminathan, Venkatesan, Venkatakrishnan, Hunter Chan, Fuhan Liu, Venky Sundaram, Scott Kennedy and Dirk Baars, "Characterization of Next Generation Thin Low-K and Low-Loss Organic Dielectrics from 1 to 40GHz", Submitted to IEEE Transactions on Advanced Packaging, 2008.
7. Scott Kennedy, Al Horn, and Greg Bull, Fuhan Liu, Hunter Chan, Venky Sundaram, and Rao Tummala, "An Overview of Material Options Suitable for Today's Commercial Millimeter Wave Designs", Circuitree, May 2009.
8. de Vries, J. *et al*, "Humidity and reflow resistance of flip chip on foil assemblies with conductive adhesive joints," IEEE Trans-Components and Pkg. Tech., Vol. 26, No. 3 (2003), pp. 563-568
9. Teh, L. K. *et al*, "Moisture-induced failures of adhesive flip chip interconnects," IEEE Trans-Components and Pkg. Tech., Vol. 28, No. 3 (2005), pp. 506-516
10. Mercado, L. L. *et al*, "Failure mechanism study of anisotropic conductive film (ACF) packages," IEEE Trans-Components and Pkg. Tech., Vol. 26, No. 3 (2003), pp. 509-516
11. Sankaran, Nithya; Lee, Baik-Woo; Sundaram, Venky; Engin, Ege; Iyer, Mahadevan; Swaminathan, Madhavan; Tummala, Rao; "Electrical Characterization and Design Optimization of Embedded Chip in Substrate Cavities", Proceedings of the 57th Electronic Components and Technology Conference, May 29 2007-June 1 2007 Page(s):992 – 999.
12. Lee, Baik-Woo; Sundaram, Venky; Wiedenman, Boyd; Yoon, Chong K; Kripesh, Vaidyanathan; Iyer, Mahadevan; Tummala, Rao R; "Chip-last Embedded Active for System-On-Package (SOP)", Proceedings of the 57th Electronic Components and Technology Conference, May 29 2007 - June 1 2007 Page(s):292 – 298.

ADVANTAGES OF A NEW WAFER LEVEL INTEGRATION CONCEPT BASED ON DIRECT BONDED SILICON ON LTCC

J. Müller, M. Fischer and H. Bartsch de Torres
Institute for Micro- and Nanotechnologies MacroNano®
Ilmenau University of Technology
Ilmenau, Germany
jens.mueller@tu-ilmenau.de

B. Pawlowski and S. Barth
Hermsdorfer Institut für Technische Keramik e.V.
Hermsdorf, Germany

ABSTRACT

A new integration concept for Si-Wafers on multilayer Low Temperature Cofired Ceramic (LTCC) substrates based on a bonding technique between nano-scaled Black Silicon (BSi) and a modified LTCC-material is presented. This novel technique enables the combination of advantages from silicon and ceramic technologies, whereby a new wafer compound material (Silicon on Ceramics) becomes available. The new compound is fabricated by using a bonding procedure between a nano patterned silicon surface and a low temperature cofired ceramic (LTCC). A special LTCC tape with an adapted TCE to silicon is joined with a silicon wafer using typical LTCC manufacturing steps (lamination and pressure assisted firing). By matching the size of ceramic particles in the unfired tape to the needle dimensions on the BSi it was possible to achieve an average bonding strength above 5000 N/cm² and gas tightness at the interface between LTCC and silicon.

This "Silicon-On-Ceramic"-substrate enables a wide range of design solutions, whereby several, unfired ceramic layers are prepared with vias, wirings and fluidic channels using standard LTCC-technologies. After sintering, the ceramic acts as a carrier system with electrical and fluidic properties. The Si-side of the compound substrate can be further used to generate MEMS or NEMS-devices or for high density electrical interconnects made by thin film processes. To ensure the electrical functionality of MEMS devices, only a thin silicon layer is necessary. The separation of silicon areas can be easily accomplished by standard silicon etching processes such as DRIE or RIE, in which the ceramic works as a natural etching barrier. Consequently, the process enables wafer level system packaging.

In addition to wiring and mounting pads, the LTCC part of the compound may contain fluidic features (e.g. for active cooling) or passive integrated components. The electrical connection between the Si-wafer and the LTCC can be achieved during sintering or in a post-process. The paper addresses manufacturing and material issues as well as ongoing development strategies to improve the interconnect density of the compound substrate.

Key words: LTCC, Black Silicon, Wafer Bonding, Packaging

INTRODUCTION

LTCC is a mature substrate technology for RF-components, RF-packages, microwave and automotive modules. Typical applications for LTCC-substrates have strong requirements for reliability and durability even under harsh environmental conditions which can be found in the car engine, at the wheels or even inside the gear box [1]. LTCC technology is also known as a high density interconnect technology due to its multilayer capability along with passive integration concepts [2]. The implementation of three-dimensional structures such as cavities or channels opens up options for fluidic system packaging [3].

Beside IC manufacturing and photovoltaics, silicon is being used for micro-mechanical and micro-electro-mechanical systems, or MEMS. Further reduced and enhanced structural dimensions in the nanometer range lead to the term NEMS. MEMS and NEMS can be widely found in sensor applications or as RF-components. Figure 1 shows an example with piezoelectric micro-electro-mechanical double clamped beam resonators from epitaxially grown AlGaN/GaN layers on silicon [4]. Individual components are manufactured in large arrays on wafer level. In most cases they need to be hermetically sealed which can either be done at wafer level or by packaging after singulation and picking from the wafer. The latter process has higher design flexibility but also high cost.

Back-end processes in wafer format require wafer bonding techniques. These techniques are divided into wafer bonding without intermediate layers (e.g. direct silicon and anodic bonding) and bonding with intermediate layers using solder, adhesive, glass frits or eutectic bonding. In either case both wafers need to be matched in the coefficient of thermal expansion (CTE) to avoid bending, delamination or fracturing under thermal load. Typical bonding partners are silicon, glass [5][6] and LTCC glass-ceramics [7]. The use of LTCC would be very beneficial since it can already contain horizontal and vertical wiring. However, the typical as-fired surface roughness is insufficient for an anodic bonding procedure; expensive grinding and polishing steps are necessary. In addition, free alkaline ions need to be available in the LTCC material. The glass-ceramic material which was developed for this purpose by the Hermsdorfer Institute for Technical Ceramics (HITK) provided the basis for the development of the silicon-on-ceramics technology.

Figure 1. Scanning-electron microscope (SEM) image of beam resonators from epitaxially grown AlGaN/GaN layers The length of the resonator in the center of the image is 50 µm.

SILICON-ON-CERAMICS (SiCer)
The silicon-on-ceramics platform combines advantages of multilayer low temperature cofired technology (e.g. integrated components) and silicon microelectronic and micromachining technologies (e.g. thin film structuring or selective etching). Bonding of both materials is made at wafer level without additional interface materials by typical LTCC cofiring. Structuring on the silicon side is typically accomplished after firing.

Technology
The special LTCC tape with the matched silicon TCE (trade name BGK [7]) is processed like any other commercial LTCC. After sizing and via punching the via holes are filled with the gold paste, TC7101 from Heraeus, and dried. Alternatively, laser structuring can also be applied to create via and alignment holes. Gold conductor (TC7102) is printed on tapes with 325 mesh stainless steel screens with 15 µm emulsion. An alignment tool with pins is used for tape stacking and pre-lamination. A single sheet of release tape (sacrificial material) is added to the bottom of this tape stack to allow for a pressure assisted sintering later.

In order to provide high bond strength, the contact side of the silicon wafer contains a self-organized nano-textured surface with needles. These homogeneously distributed needles are generated by a self masking technique in reactive ion etching (RIE). Due to the light absorbing properties, the material is called Black Silicon. This typically parasitic and unwanted effect has already been described about 15 years ago [8].

For the silicon-on-ceramics process a parallel plate reactor (STS 320) with a cooled wafer electrode and a SF_6 /O_2 – plasma is utilized. Needles up to 2.5 µm in length can be achieved by this etching process. The needle diameters vary from 5 nm at the top to 100-400 nm at the bottom. Typical pitches are in the 100 to 400 nm range. Figure 2 shows a fully processed 4 inch wafer and a detailed view of the Black Silicon structure.

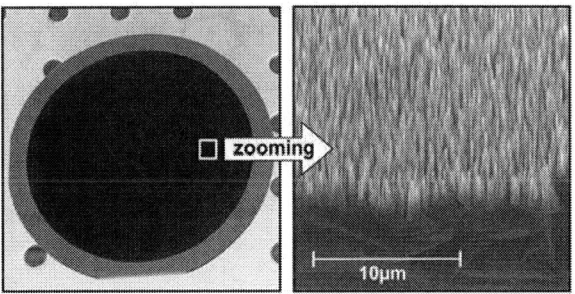

Figure 2. Black Silicon wafer and SEM image of the needle structure

The LTCC pre-laminated layer stack and the prepared silicon wafer are aligned and arranged in a way to have the needle surface in contact with the BGK-tape body (Figure 3).

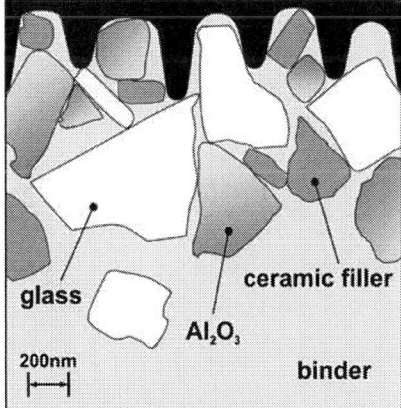

Figure 3. Schematic drawing of the interface between needles and the green BGK-body

The needles penetrate into the unfired tape by either isostatic or uniaxial lamination. The lamination parameters pressure, pressure ramp, temperature and time need to be optimized for the specific viscoelastic properties of the tape consisting of polymer binder, glass and ceramic filler in order to achieve optimum penetration and adhesion [9].

Pressure assisted firing is used to burn out organic contents, to sinter the LTCC (850°C peak temperature) and to form the bond interface between silicon and LTCC. A pressure of approximately 0.5 MPa is applied during the sintering phase. Figure 4 schematically shows the tape/silicon-compound. After firing, the release tape is removed from the BGK-LTCC by brushing.

The average bonding strength of the silicon-on-ceramics is about 900 N/cm² which is significantly higher compared to glass frit and anodic bonding (approx. 300 N/cm²) [9].

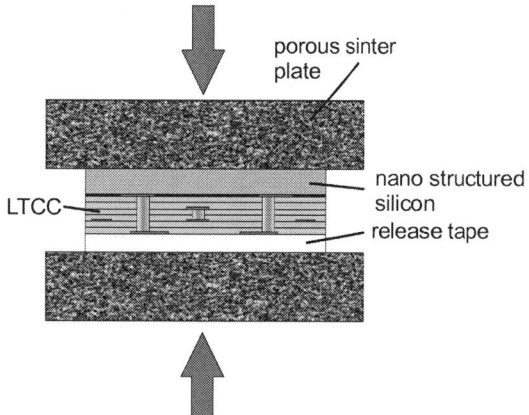

Figure 4. Layer stack of LTCC and silicon in the sintering press

Material Development

Helium leak testing and cross sectional analysis revealed the inadequate penetration of Black-Si needles into the BGK-LTCC. This is a result of the needles being too long, too flexible and the needle density too high (Figure 5).

By the use of a plasma etching post-process (inductively coupled plasma reactor = ICP) with argon, the average needle length is reduced to about 600 nm and the minimum pitch is increased to approximately 500 nm (Figure 6).

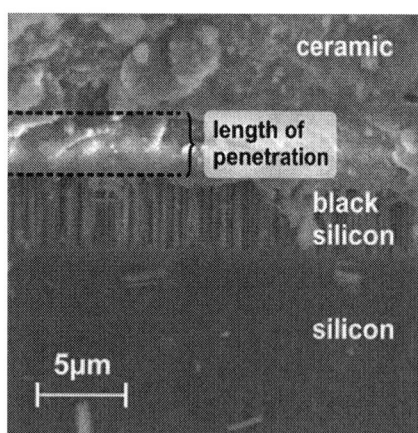

Figure 5. SEM picture of the insufficient penetration of Black-Si needles into the LTCC

Due to the changed needle geometry an average bonding strength of about 1775 N/cm² was achieved.

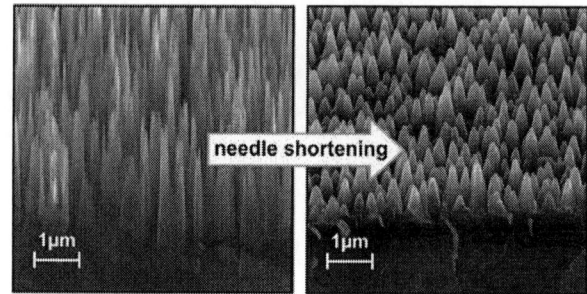

Figure 6. Black-Si needle modification by ICP-etching

The original BGK tape contains glass and ceramic particles from 50 nm up to 2.5 μm in diameter. It consists mainly of a glass, ceramic fillers and the organic matrix (binder, solvents and plasticizers). Table 1 summarizes the basic properties.

Property	Value
Material	glass ceramic composite (GCC)
Green Density	1.43 g/cm³
Free Shrinkage x,y	15-17.5%
Peak Firing Temperature	850-900°C
Dwell Time	10-30 min
Thermal Expansion (25-400°C)	3.4 ppm/K
Surface Roughness as fired	200 nm
Dielectric Constant ε_r	5.35 (25°C, 1 kHz)

Table 1. BGK material properties

Due to the size of the large particles and the low content of fine powder, the gaps between Black-Si needles cannot be tightly filled by them. In the worst case, only the organic content of tapes fills this volume during lamination. Voids are left after burnout and the bonding strength is reduced according to the limited interface area. In addition to the above mentioned modification of the needle geometries, the powder morphology was modified as well (Figure 7). As a result of the combined changes, the bonding strength was significantly improved to values above 5000 N/cm² and water tightness has been demonstrated [10].

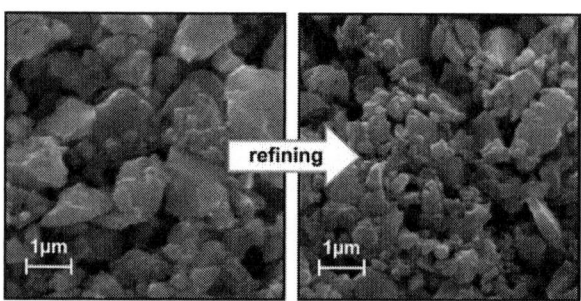

Figure 7. SEM of the original (left) and the modified (right) powder morphology

Ongoing Research

Currently, the process management for SiCer-substrate fabrication is being optimized to guarantee international wafer standards (bow, total thickness variation, cleanlyness etc.). For further applications of the silicon-on-ceramics platform, particularly for CMOS-compatibility and RF-device fabrication, it is necessary to adapt the composition of the BGK-tape on CMOS process conditions. Dielectric losses of the tape are comparable to standard LTCC at low frequencies but show an increase towards microwave frequencies (Table 2) which would limit its use for such applications.

Frequency [GHz]	tan δ [10^{-3}]	Test Method
0.1-1	~ 5.5	Agilent E4991A Impedance Analyzer
5	~ 7.5	Split-Post-Resonator
10	~ 8.5	Split-Post-Resonator
20	~ 9.5	Split-Post-Resonator

Table 2. Measured loss tangent properties of the refined BGK-LTCC at various frequencies analyzed by impedance analyzer [11] or split-post-resonator [12] methods

Another area of investigation is related to through silicon via (TSV) generation for direct vertical integration (direct connection to LTCC vias). Concepts of vias first (prior to co-firing) and vias last (post-firing) are under consideration. Additional conductive, resistive and dielectric pastes compatible to the BGK-tape will also be developed.

APPLICATIONS

MEMS-Package

The SiCer-platform is particularly of interest for manufacturing silicon-based MEMS. The LTCC is already a part of the functional package. All components are produced in an array format at wafer level. Silicon micromachining is carried out after cofiring (Figure 8).

First attempts to realize a MEMS-package are based on traditional interconnection concepts using wire bonding. However, this approach requires thick film metals in direct contact to the nano-structured surface. The surface quality of the thick film lines and pads are influenced mechanically by the silicon needles and chemically by silicid-formation. Figure 9 shows the selectively opened LTCC surface with AgPt-structures.

The final concept of an RF-MEMS package is depicted in Figure 10. In addition to the hermetically sealed MEMS device on the silicon substrate it contains vias through the silicon, a functional LTCC multilayer with integrated RF-components and an active monolithic microwave integrated circuit (MMIC).

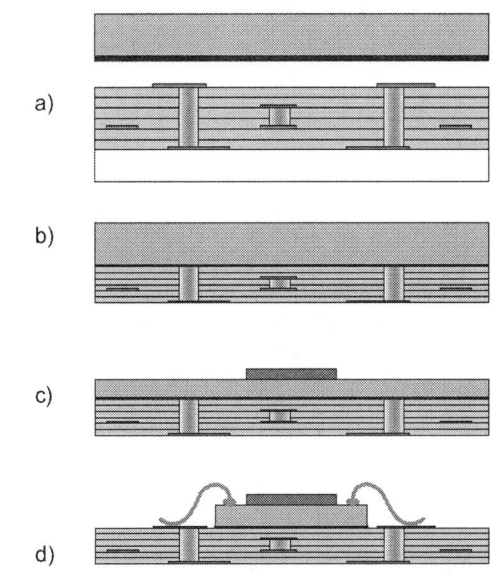

Figure 8. Process sequence for MEMS-Packages on SiCer-substrates, a) silicon nano-structuring and LTCC processing; b) lamination and pressure assisted sintering; c) MEMS-processing; d) silicon etching and wire bonding

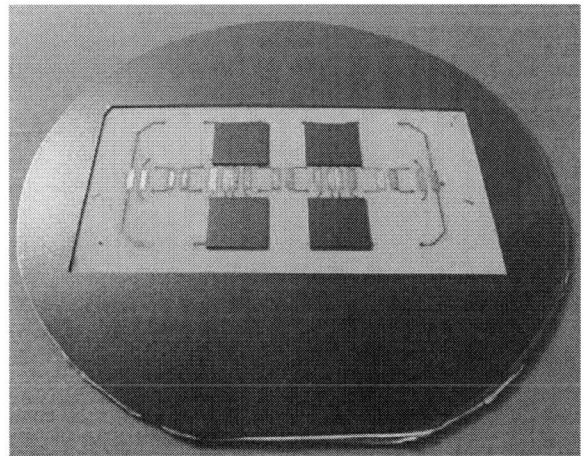

Figure 9. SiCer-substrate with selectively opened silicon area by deep reactive ion etching

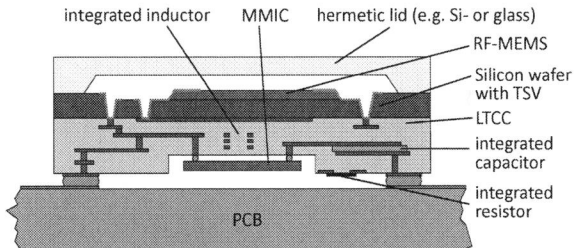

Figure 10. Concept of an RF-System-in-Package with RF-MEMS and embedded passive components

Fluidic Components

The ceramic/silicon concept allows for integrated channels and cavities on the millimeter to micrometer scale. The enlarged surface of nanostructured interfaces involves their application as a substrate for sensor layers in microfluidic devices and heat exchanger in chip cooling systems. For efficient contact between the fluid and functional layer, fine channels have to be produced which will be in direct contact with the silicon wafer. Fluidic structures in LTCC, such as open cavities or channels, are achieved prior to tape stacking by punching or laser cutting and on the laminate by laser ablation or hot embossing. Whereas punching and laser machining yields structure widths of several hundred micrometers, hot embossing can be applied for structures below 100 μm channel width. The process is shown in Fig. 11.

The implementation of such patterns is a technological challenge. Fine patterns must be formed and maintained during the pressure assisted sintering. The sublimation of carbon inlays is hindered by the dense silicon wafer. Therefore, a process without auxiliary material has been developed to avoid excessive process time.

The realisation of fine fluid channels with accurate edge forming in green ceramic tapes by embossing and lamination at reduced lamination pressure has already been demonstrated [13]. This process is adapted to form fluid channels at the silicon interface. Optimal embossing parameters are derived for DuPont's Green Tape™ 951 in [14]. Based on this work, the embossing parameters were optimized for the BGK tape to form lines and spaces with dimensions of 50 μm. Optimal moulding conditions are 85°C, a pressure of 100 MPa and a dwell time of 5 minutes. Higher temperatures would be beneficial for the embossing step, however demoulding is then hindered.

Nickel moulds with lines and spaces between 20 μm and 120 μm were embossed into a stack of 2 layers. Two different moulds were used to achieve patterns in the green ceramic tape with depths of 65 μm and 45 μm. The fired channel height was determined by cross sectioning and SEM. The result was 30 μm for the 65 μm embossed trenches and 20 μm for the 45 μm deep structures, which is close to the expected z-shrinkage of 50% due to pressure assisted sintering.

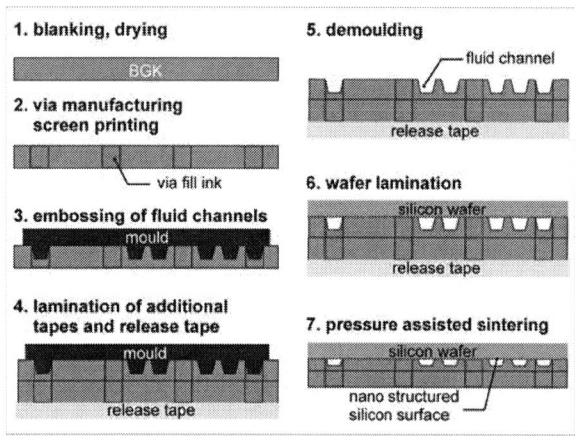

Figure 11. Manufacturing of micro channels by hot embossing

Open cavities at the LTCC surface are sealed to an embedded channel by the silicon during the SiCer process (Figure 12 and 13). These channels may be used for active cooling concepts in power packages (power semiconductor mounted on the silicon) or in the reverse mode to control the temperature of fluids by an integrated heater on the silicon substrate (e.g. in a micro reactor).

Figure 12. Cross sectional view of a silicon-on-ceramics substrate with embedded fluidic channels at the interface silicon to the LTCC.

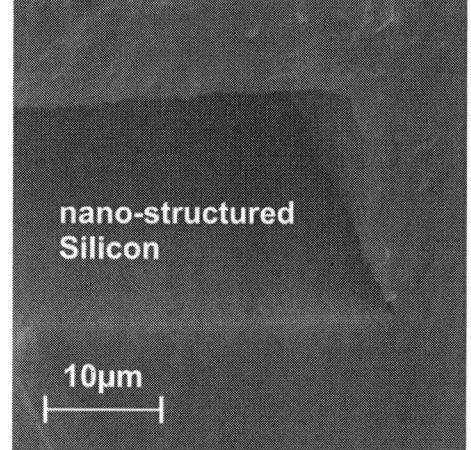

Figure 13. Detail of the channel from Figure 12 showing the nano-structured surface of the silicon substrate

MCM-C/D
Another potential application field is the use of the silicon as a very high density interconnect substrate. According to the International Technology Roadmap for Semiconductors (ITRS), the wire bond pitch is expected to shrink to 25 μm in 2015 [15]. The surface quality of silicon easily enables the deposition of thin film structures with line widths and spaces of 5 microns or below. In addition to interconnecting structures, very precise passive components might also be implemented [16].

The LTCC part of the substrate with medium wiring density (e.g. 100μm lines, spaces and vias) closes the interconnect gap to the printed circuit board.

ACKNOWLEDGEMENT
The authors gratefully acknowledge the financial support by the Thuringian Ministry of Education, Research and Culture for funding the project SiCeram (A514-09026, 2009-2012) and the Federal Ministry of Education and Research for the support within the framework Mikro-Nano-Integration (project "Nano-SilKe", 16SV3566, 2007-2008).

REFERENCE

[1] W. Röthlingshöfer, U. Goebel, Innovative Applications with LTCC for Automotive ECU's, Proc. IMAPS Conference & Exhibition on Ceramic Interconnect Technology, April 7-9, 2003, Denver/CO.

[2] R. L. Brown, A. A. Shapiro, P. W. Polinski, Integration of passive components into MCMs using advanced low-temperature cofired ceramics, International Journal of Microcircuits and Electronic Packaging. Vol. 16, no. 4, pp. 328-338. 1993.

[3] L. J. Golonka, Technology and applications of Low Temperature Cofired Ceramic (LTCC) based sensors and microsystems, Bull. Pol. Ac.: Tech. 54(2) 2006.

[4] K. Brueckner, F. Niebelschuetz, K. Tonisch, S. Michael, A. Dadgar, A. Krost, V. Cimalla, O. Ambacher, R. Stephan, and M. A. Hein, Two-dimensional electron gas based actuation of piezoelectric AlGaN/GaN microelectromechanical resonators, Appl. Phys. Lett. 93, no. 17, art. no. 173504, 2008;

[5] Zhihong Li, Yilong Hao, Dacheng Zhang, Ting Li, Guoying Wu, An SOI-MEMS technology using substrate layer and bonded glass as wafer-level package, Sensors and Actuators A 96 (2002) 34-42.

[6] Byeungleul Lee, Seonho Seok and Kukjin Chun, A study on wafer level vacuum packaging for MEMS devices, J. Micromech. Microeng. 13 (2003) 663–669.

[7] E. Müller, T. Bartnitzek, F. Bechtold, B. Pawlowski, P. Rothe, R. Ehrt, A. Heymel, E. Weiland, T. Schroeter, S. Schundau, K. Kaschlik, "Development and Processing of an Anodic Bondable LTCC Tape", European Microelectronics and Packaging, Brugge, Belgium, June 2005.

[8] H. V. Jansen, M. Boer, R. Legtenberg, M. Elwenspoek, The black silicon method: a universal method for determining the parameter setting of a fluorine based reactive ion etcher in deep silicon trench etching with profile control, Journal of Micromechanical Microengineering 5, pp. 115-120, 1995.

[9] M. Fischer, H. Bartsch de Torres, M. Stubenrauch, J. Müller, M. Hoffmann, Bonding of LTCC and silicon substrates using adapted Black Silicon, Conference on Wafer Bonding for MEMS Technologies and Wafer Level Integration, Halle/Saale December 9-11, 2007, in: Bagdan J. and Knechtel R. (Ed.); Book of Abstracts, pp. 62-64, 2007.

[10] M. Fischer, H. Bartsch de Torres, B. Pawlowski, R. Gade, S. Barth, M. Mach, M. Stubenrauch, M. Hoffmann, J. Müller; Silicon on Ceramics - A new Integration Concept for Silicon Devices to LTCC, 2008 IMAPS/ACerS 4th International Conference and Exhibition on Ceramic Interconnect and Ceramic Microsystems Technologies (CICMT), April 21 -24, 2008, Munich, Germany.

[11] Solutions for Measuring Permittivity and Permeability with LCR Meters and Impedance Analyzers, Application Note 1369-1, Agilent Technologies, October 28, 2008, 5980-2862EN.

[12] J. Krupka, R. N. Clarke, O. C. Rochardt and A. P. Gregory, Split Post Dielectric Resonator Technique for Precise Measurements of Laminar Dielectric Specimens - Measurement Uncertainties, Proc. 13th International Conference on Microwaves, Radar and Wireless Communications 2000, MIKON-2000. Volume: 1, page(s): 305-308.

[13] H. Bartsch de Torres, M. Hoffmann: Embossing of microfluidic structures in ceramic multilayers, Smart Systems Integration 2007, Paris, March 27-28, 2007 / T. Gessner (Editor). - VDE Verlag, pp. 423-425.

[14] H. Bartsch de Torres, R. Gade, A. Albrecht, M. Hoffmann: Systematic characterisation of embossing processes for LTCC-tapes, Journal of Microelectronics and Electronic Packaging, (2008) 5, pp. 142-149.

[15] International Technology Roadmap for Semiconductors, 2007 Edition, Assembly and Packaging, http://www.itrs.net.

[16] G. Posada, G. Carchon, P. Soussan, N. Pham, B. Majeed, D. Sabuncouglu, W. Ruythooren, B. Nauwelaers, W. De Raedt, Microstrip thin-film MCM-D technology on high-resistivity silicon with integrated through-substrate vias, 978-2-87487-002-6 © 2007 EuMA, Munich, Oct. 2007.

FROM THE SINGLE CHIP TO THE WAFER INTEGRATION

Gilles Poupon, Jean Charles Souriau, Hervé Boutry, Jean Brun, and Nicolas Sillon
CEA-LETI Minatec
Grenoble, France
gilles.poupon@cea.fr

ABSTRACT

System integration is clearly a driving force for innovation in packaging. The need for miniaturization has led to new System In Package (SiP) architectures, which combine a whole range of different technologies. In addition, due to the increasing complexity of systems, the introduction of new components like MEMS or RF components and the still growing pressure on size, performance and cost; a general trend is to put not one but several dies in a single package. However, cost is the critical issue in SiP Packaging as individual operations are currently necessary to complete each individual package. Taking into account all the developments that have been made to date on Wafer Level Packaging, it has been proposed to establish SiP at wafer level.

Companies that desire an "in-house solution" will prefer WLP because one of the benefits of the wafer level packaging is a simplified supply and value chain. Furthermore, for small companies (not IDM's), one of main consideration is to use generic technologies which can be applied on chips coming from different sources because they have no direct access to wafer manufacturing. To be compatible for chip multi-sourcing one of the most know examples on wafer level packaging is the fan-out wafer level structure This concept, proposed by major companies (Infineon, Freescale, ...), consists of rebuilding a wafer from heterogeneous Known Good Die (ASIC, sensor, memory, optic component etc.).

After having established the current state of the art, the objective of this paper is to identify the various technological concepts and to give a focus on Chip in Wafer in Silicon technology developed at CEA-LETI.

Key words: wafer level packaging, embedded components, rebuilt wafer, 3D integration

INTRODUCTION

Electronics components follow several tendencies related to the economic conjuncture of cost reduction and time to market, on miniaturization (smaller components, level of integration) and to functionality (increase in performances, more functions). To meet this need, a universal technique of packaging doesn't exist; for each product it is necessary to choose an adapted solution which takes account all the specifications. Thus, each technological solution must take into account the form factor, the performances to be reached, the final cost of the component but also the constraints related to the application. Several concepts from packaging coexist and make it possible to answer to integration levels (Figure 1). Each of these concepts presents specific advantages and drawbacks. For example, cost acts as a brake on SiP and further developments are necessary to reduce it.

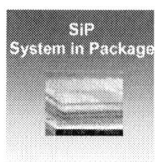  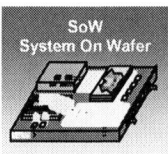

Figure 1. Packaging concepts (source LETI)

Wafer Level Packaging is one on the promising way to decrease cost and improve level of integration.

Among the emergent technological solution, 3D integration is probably one of the most promising way to improve integration level. This concept makes it possible to integrate in the vertical axis the whole of the components to be assembled which allows, for example, to decrease the size of the final assembly and to reduce the length of the interconnections. However, one of the disadvantages to this process is that it is often necessary to carry out vertical intra-connections (TSV) so that the chips can communicate between them. In general, TSV's need to be formed at somepoint before make dicing occurs.

Finally, for small companies (not IDM's), one of main consideration is have access to a limited number of chips or wafers and to use generic technologies which can be applied on individual chips coming from different sources because they have no direct access to wafer manufacturing.

BACKGROUND ON RE BUILT WAFER: FROM THE CHIP TO THE WAFER

To take into account these previous considerations, one concept which emerged recently is to rebuild a wafer from heterogeneous Know Good Die (KGD). Typically, individual dice are positioned on a temporary substrate with an accurate and high speed handler. These dice are molded in organic substances. After a backside thinning and substrate debonding, a redistributive chip layer is processed on the side to connect the pads. The main steps of process flow are explained on the next figure 2.

- **Die positioning**

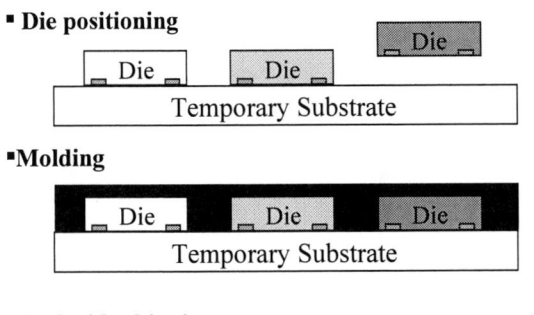

- **Molding**

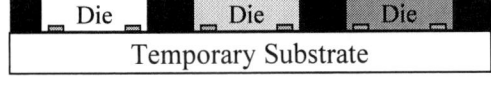

- **Back side thinning**

- **Substrate de-bonding**

Figure 2. Rebuilt wafer process

This concept was introduced by General Electric 20 years ago. In 1994, a first version, called "Neostack", had been proposed by Irvine Sensor where neo-chips for 3D stacking was formed by embedding IC chips in epoxy [1]. In the "Neostack" concept, the KGD was bumped using gold wire bonding. Now, many companies have been developing various versions of the technology.

In 2005, 3DPlus and CEA-LETI [2] in European project (Walpack) showed a new concept based on rebuilt wafers for embedded chips (figure 3) and compatible with stacked module (WDoD).

Figure 3. WDoD rebuilt wafer process (source LETI/3Plus)

The RCP (Redistributive Chip Package) has been developed by Freescale [3]. It is based on a bath process that features a build-up and metallization constructed on an embedded die. The application targets are DSP, processors, power management devices, etc...

Infineon proposes a fan out WLP structure with the embedded Wafer Level Ball grid array (eWLB)

technology [4] and "Molded Reconfigured 200mm Wafer" [5]. This technology uses a combination of front and back end manufacturing techniques with parallel processing of all the chips on a wafer.

Recently concepts of wafer level technologies form major institutes and companies (EMWLP from ITRI and IME, EM-eSiP from NEPES) using conventional die placement, molding and RDL process has been publishing [6 to 8].

With the SMAFTI concept (SMArt chip connection with Feed Through Interposer) [9], NEC demonstrated in 2006 the feasibility of a new inter-chip connection structure semiconductor package for broadband data transfer and low latency electrical communication.

A NEW CONCEPT

Generally, this process enables coplanar chip active face and Wafer Level Packaging such as pad redistribution, bumping and testing. However the consistency of these rebuilt wafers is limited by the thermal properties of the embedded resin. Processes at temperatures higher than glass transition are hazardous. Wafer deformation is also a critical issue on a rebuilt wafer with polymer. This is due to the excess polymer left by potting or injection on die. A few tens of microns lead to a deformation of several millimeters. Simulation of a 700µm silicon wafer with 50µm of polymer (polymer CTE=20ppm) deposited on it and cured at 200°C leads to a wafer bow of approximately 600µm. To get round this difficulty, we propose to use a silicon wafer as a frame. Known Good Dies are fitted into through cavities and sealed with polymer, this is the CIWIS concept (**C**hip **I**n **W**afer for **I**ntegrated **S**ystem) [10] (figure 4)

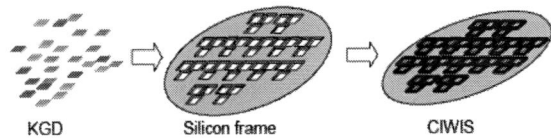

Figure 4. CIWIS concept

The CIWIS technology can be seen as an evolution of that concept : dice are inserted in a silicon or glass frame and polymer is used only to fill the gap between the frame and the dies. This configuration ensures a better mechanical behavior of the rebuilt wafer and allows the realization of higher temperature and higher resolution processes.

The process developed to make such a wafer test vehicle includes:

✓ Cavity etching in silicon wafer by laser
✓ Silicon wafer bonding on temporary substrate
✓ Die placement face down into through cavities
✓ Trench filling with polymer

- ✓ Polymer curing
- ✓ Temporary substrate de-bonding
- ✓ Back grinding

The next figure shows CIWIS process flow (figure 5)

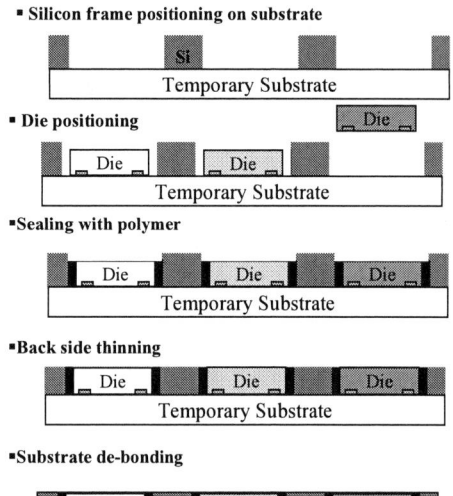

Figure 5. CIWIS process (**C**hip **I**n **W**afer for **I**ntegrated **S**ystem)

In the process, polymer is simply deposited inside the trenches. The test vehicle wafer consists of a 200mm silicon wafer with 82 dies. The die size is 8 x 8 mm², the pitch between dies is 15mm and the sealed cordon is 500µm wide. The figure 6 shows the test vehicle wafer.

Figure 6. 200mm Ciwis wafer (82 silicon dies)

During trench filling, the wafer and dies are held on the temporary substrate by an adhesive tape. This step is crucial because at the end of the process the placement accuracy of the dies must be compatible with lithography on die pads. To control the relative positioning of the dies after polymer curing, a specific photoresist pattern was made on the wafer by photolithography. Gaps between these photoresist patterns and the die pattern were measured using a microscope. The polymer curing process induces a die displacement of less than 5µm

inside the cavity. This value is compliant with pad redistribution of standard ASIC in which pads are 50x50µm².

Bows have been measured on 8 wafers and the average value is 23µm. This value is compared to the 12µm of a blank wafer. Thermal Deformation Measurement was performed to evaluate wafer deformation. The results are presented in the figure 7. It can be noted from this figure that wafer deformation has increased with temperature but is less after heat treatment than before. It is probably due to strength relaxing.

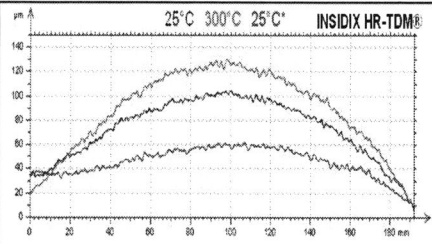

Figure 7. Thermal Deformation Measurement on CIWIS wafer

This process enables coplanar chip active face with the host silicon wafer. Chip co-planarity was observed by SEM observation (figure 8) and profilometer measurement. The typical gap between die face and wafer frame face is around 3µm which is a good value. Moreover the SEM photo shows a good continuity of polymer and silicon on the front side.

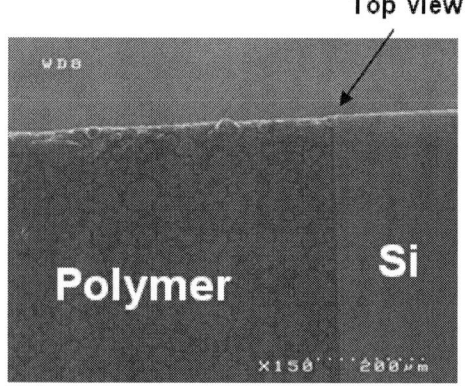

Figure 8. SEM observation on CIWIS

To make sure that the process to integrate chips in wafer does not damage the integrated circuit, a die with a CMOS was blended in a dedicated cavity of the test vehicle wafer. This die made by the CEA-LETI has CMOS with 80nm gate width. Id(Vd) was measured before and after the process. As shown in the graph (figure 9).

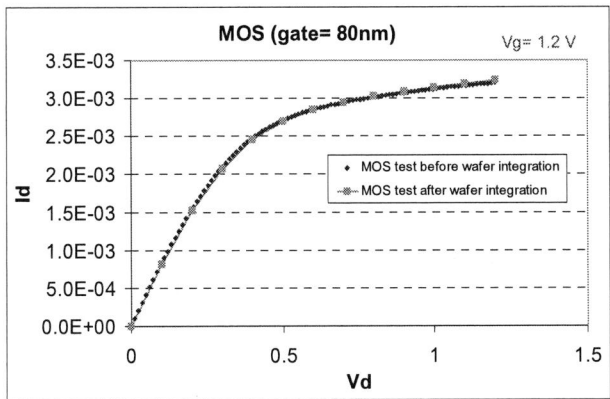

Figure 9. Id(Vd) measured before and after CIWIS process

Globally, the wafer characteristics are summarized below:
- ✓ Low warpage ~ 20 µm
- ✓ Chip placement ~ 5µm
- ✓ Good chip sealing in cavities
- ✓ Topology on wafer < 3µm

For evaluating if CIWIS technology has a potential to adopt as mass production, reliability test is necessary. There are currently in progress and further evaluation will be carried on the next months but the first results are encouraging.

Finally, the process flow is not complex and allows a short time to market. This approach can naturally address the market for all portable equipment (e.g. iPod, mobile phone, MP3, micro cards, hearing aids,...) which is extremely extensive. The CIWIS process is also compatible with other wafer frames such as glass wafer (figure 10)

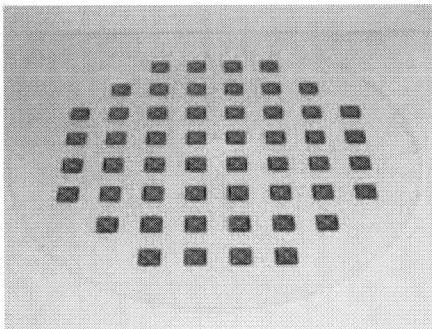

Figure 10. Ciwis on 200mm glass wafer

The next step is to work on implementation of an Above IC process on Chip In Wafer such as passive integration, rerouting, bumping and chip bonding (figure 11).

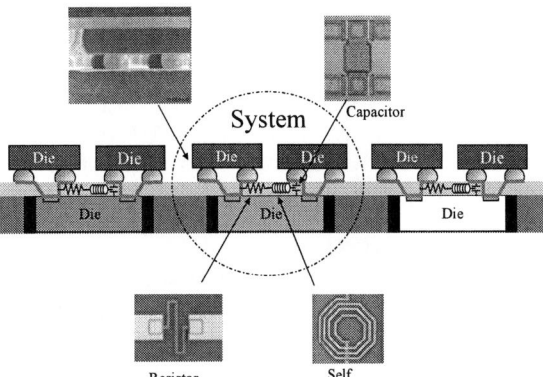

Figure 11. Embedded passive component on CIWIS

FROM THE WAFER LEVEL TO THE 3D INTEGRATION
In 2009 Sillon [11] from CEA LETI explains that we can distinguish two different approaches for Wafer Level 3D Integration, corresponding to two different business models. On one hand IDMs or their packaging subcontractors which have a direct access to design and can adapt front-end processes, will use the possibility to perform Through Silicon Vias (TSV) to achieve ultra compact /low cost assemblies. On the other hand, some end-users companies will look for generic technologies, which can be applied on heterogeneous chips coming from different sources. 3D stack will be in that case achieved thanks to vertical interconnections outside the silicon.

Using rebuilt wafers, it is now possible to process heterogeneous individual chips at the wafer level. Although these advanced technologies are competitive and, certainly improvable, it is difficult to process the embedded components to prepare the stacking or the 3D integration (particularly TSV).

HOW TO DO WITHOUT THE TSV ?
Consequently, a specific integration schemes allowing for chip stacking and chip-to-chip interconnectivity outside the die (without TSV) should be identified. The approach is based on embedding and contact redistribution of dice in film interconnect layers. Once embedded, these structures can be stacked using, e.g., solder microbump technology. This concept allows for the highest system level flexibility where thin dies with varying dimensions and shapes can be stacked and interconnected together with thin-film passive components. Different 3D architectures using plated Cu pillars have emerged on the past.

In 2002, Jung from Fraunhofer IZM [12] proposed an embedded chip in polymer. To connect the chip on the PCB, copper vias are made on thick BCB polymer and filled by electroless and electrodeposition processes. The same year, Fujitsu shows a similar approach with a chip scale module based on wafer thinning, chip stacking and re distribution technologies. IMEC, in collaboration with CNRS and University of Barcelona proposes ultra-thin-chip stacking (UTCS) [13] to form 3D compact structures and using plated copper pillars. UTCS 3D was routing in thin film dielectric. Recently CEA-LETI proposed a concept called Via Belt Technology [14].

VIA BELT TECHNOLOGY

With the objective to reduce the packaging size without TSV, an original 3D architecture has been proposed to interconnect a base wafer with standard dice. Firstly, the dice are connected to the substrate face-down by low pitch interconnects (called µ-insert). In order to obtain a new surface, the wafer is embedded in a polymer and the mold-side is thinning by grinding, enabling copper pillars to be connected. A new rerouting and interconnection system enables a second die hybridization. The "Via belt" concept is presented on figure 12.

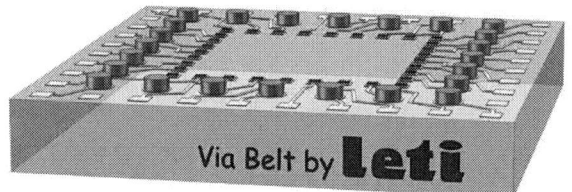

Figure 12. Via belt concept

The process starts on a base wafer (figure 13). This wafer can be just a blank wafer, an active wafer (for instance a memory wafer, memory being often the bigger die of an heterogeneous stack), or a rebuilt wafer, as described previously (CIWIS). The second step in our process is to perform flip chip, keeping the capability to integrate standard dice without any specific process on it.

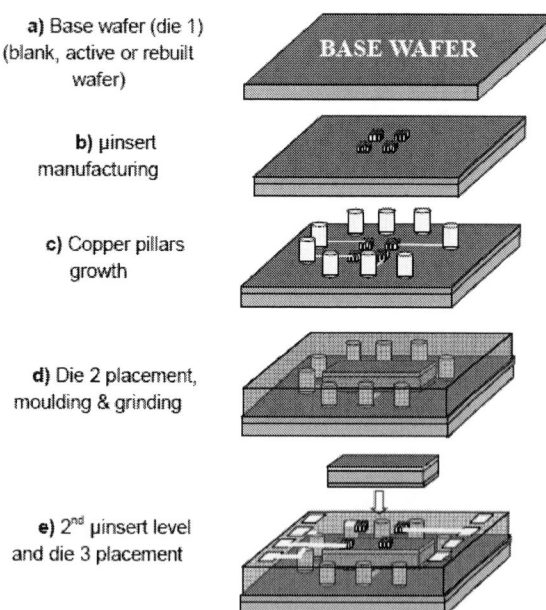

Figure 13 . Process flow for Via belt technology

The solution proposed by LETI is called µ-inserts interconnects (figure 13b). This process is based on nickel µbumps [15]. Several Ni pillars are electroplated on each pad. Each one has a diameter of 5 µm, for a thickness around 10 µm. Thanks to the pressure applied during flip-chip, Ni pillars are inserted in the aluminium pads of the top die, with no need for specific metallization (figure 14).

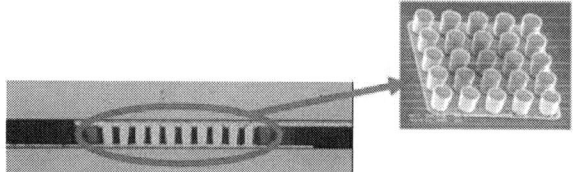

Figure 14. Cross section of a die hybridized face to down on a rerouting wafer – details on Ni µinserts (diameter 5µm)

In order to lift electrical contact to the 3rd level, a "belt" of copper pillars are grown around the footprint of die 2 as presented in figure 13c. A photo-pattern dry thick film is used for electroplating these pillars, which have a diameter of 100 µm for an initial thickness of 200 µm.

Die 2 (figure 13d) is then flip-chipped by thermo-compression on µinserts previously described, a specific polymer is spun to embed both dice and pillars, and then polymer is grinded until the copper pillars appear (figure 15).

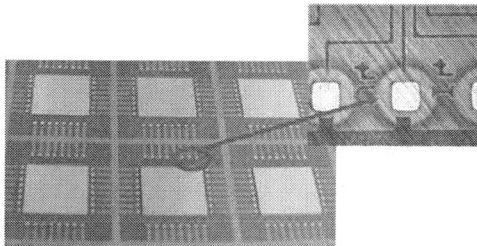

Figure 15. Top view' after epoxy glue embedding and grinding

A second level of µinserts is patterned on the polymer layer and the third level of dice is placed face-down (figure 13e).

The figure 16 shows a cross section for a 3D stack on a demonstrator. The two levels of µinserts are clearly seen. The very thin interface is a good advantage when a thin stack is targeted. The thickness of mid die is 65µm after grinding.

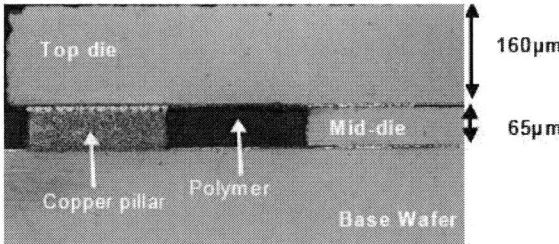

Figure 16 . Via belt cross section

Figure 17 shows a final assembly when the top die is placed face-down to be connected to the copper pillars.

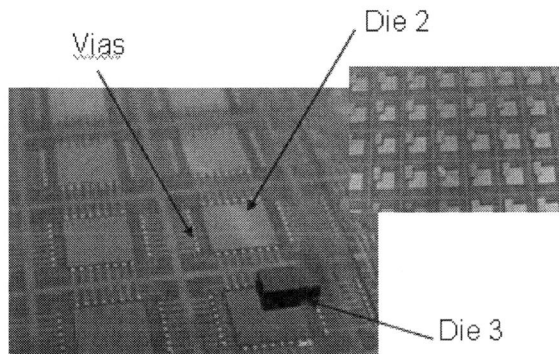

Figure 17. Via belt – final assembly

Electrical tests show an effective electrical contact between the three levels of dice. The first demonstrator has been achieved on a classical wafer base. However, rebuilt wafer coupled on via belt technology is an very interesting way to give access to wafer level technologies and 3D integration compatible with heterogeneous chips for end-users companies.

The CIWIS concept, previously presented, has been initially developed for fan out packaging or embedded wafer level packaging. Coupled to the Via Belt, we are able to propose a complete approach from a single chip to a 3D integration process including all the technical value chain.

CONCLUSION
As discussed initially, companies that want to develop "in-house" solutions prefer wafer level packaging because one of benefits is to simplify supply and value chain. Consequently with smaller companies, the main challenge is to process chips coming from different sources an without access to the wafer manufacturing

For niche markets which require only small quantities of chips or heterogeneous components of various manufacturers, wafer level technologies remain interesting but difficult to implement due to a limited wafer availability. The neo-wafer concept is really interesting because it makes it possible to select known good die, to integrate in a dedicated environment and to

interconnect at a conventional redistributive chip level. One of the main drawbacks is the susceptibility to deformation due to thermo-mechanical constraints. The technical approach, proposed by CEA LETI, consisting in using a silicon matrix and a positioning of the chips in cavities makes it possible to solve this difficulty.

It is possible to consider a 3D integration starting from reconstituted substrates comprising of heterogeneous chips. In this case, the simplest solution consists of building the intra-connections outside the chips. The process via belt developed is particularly adapted.

All these developments in embedded wafer level packaging and 3D integration can be seen as an anticipation of future heterogeneous integration systems. Today, demand is increasing for a highly miniaturised system added to a low cost solution. The wafer level approach is probably one of the most promising solutions, the challenge is to be compatible with consumer market applications.

ACKNOWLEDGEMENTS
All the technical results for CEA LETI presented on this article have been obtained in the frame of programs funded by European Community or French authorities and industrial partnership with Gemalto.

REFERENCES
[1] US patent 5953588, September, 14, 1994 from Irvine Sensor

[2] G Poupon, JC Souriau, O Lignier, M Charrier, International Symposium on WLP , 1st IWLPC conference, San Jose, 2004

[3] B.Keser, C Amrine, T Duong, O Fay, S Hayes, G Leal, W Lytle, D Mitchell, R Wenzel, Proceedings of Electronic Components and Technology Conference, 2007, pp286

[4] M Brunnbauer, T Meyer, Proceedings of 3rd Annual Device Packaging Conference, IMAPS 2008

[5] E Furgut, G Beer, M Brunnbauer, T Meyer, Advanced Packaging Conference, SEMICON Europe 2006, Munich

[6] C Ko, S Chen, C.W Chiang, T.Y. Kuo, Y.C.Shih, Y.H.Chen, Proceedings of Electronic Components and Technology Conference, 2006, pp 322

[7] A Kumar, V Sekhar, S Lim, C Keng, G Sharma, S Vempati, V Kripesh, J Lau, D L Wong, , Proceedings of Electronic Components and Technology Conference, 2009, pp 1289

[8] I.S. Kang, G.J. Jung, B.Y. Jeon, Proceedings of 2009 International Symposium on Microelectronics, IMAPS, pp 482

[9] Y Kurita, K Soejima, K Kikuchi, M Takahashi, M Tago, M Koike, K Shibuya, S Yamamichi, M Kawano, Proceedings of Electronic Components and Technology Conference, 2006, pp 289

[10] JC Souriau, ME Faivre, N Sillon, , Proceedings of 11th Electronics Packaging Technology Conference, Singapour, 2009

[11] N Sillon, D Henry, JC Souriau, J Brun, H Boutry, S Cheramy, Proceedings of 2009 IEEE-IITC

[12] E Jung et al, IEEE / CPMT / SEMI annual international electronics manufacturing technology symposium No27, 2002 , pp. 46

[13] S Pinel et al IEEE Transaction CPMT, 2002, vol 25 , pp. 244

[14] J Brun, H Boutry, R Franiatte, T Hilt, N Sillon, Proceedings of Electronic Components and Technology Conference, 2009, pp 1670

[15] A Mathewson et al, Proceedings of Electronic System and Technology Conference, 2006

MOLDED INTERCONNECT DEVICES –
PROGRESSIVE APPROACH FOR MECHATRONIC PRODUCTS AND EFFICIENT MANUFACTURING PROCESSES

Christian Goth
Institute for Manufacturing Automation and Production Systems (FAPS)
University of Erlangen-Nuremberg
Erlangen-Nuremberg, Germany
goth@faps.uni-erlangen.de

Jörg Franke
Institute for Manufacturing Automation and Production Systems (FAPS)
University of Erlangen-Nuremberg
franke@faps.uni-erlangen.de

Klaus Feldmann
Institute for Manufacturing Automation and Production Systems (FAPS)
University of Erlangen-Nuremberg
feldmann@faps.uni-erlangen.de

ABSTRACT

Growing demands for complexity and functionality arising from steady advances in technology are driving the development of new integration technologies in electronics. The increased use of electronics and the required miniaturization are forcing the pace of demand for complex mechatronic systems. In particular Molded Interconnect Devices (MID) are injected molded thermoplastic parts with integrated conductive pattern structure and allow highly integrated systems by combining electrical and mechanical functions (switches, shielding, fastening elements, etc.) that can not be achieved in this way by using conventional assembly and connection technologies.

The enormous design freedom for three-dimensional subassemblies affords scope for innovation and hence possibility of opening up new markets. Technological advances in the fields of structuring, metallization, assembly and especially in plastic materials have been accelerated and so, the more stringent demands from different sectors can increasingly be met.

Key words: Molded Interconnect Devices, MID, Mechatronic Systems.

INTRODUCTION

The use of high temperature thermoplastics and their structured metallization opens up a new dimension of circuit carrier design to the electronic industry: Molded Interconnect Devices (MID). MID are injected molded thermoplastic parts with integrated circuit traces. The MID-technology is one of the key innovations to

manufacture mechatronic systems, which finds in recent years more and more widespread use. They provide enormous technical and economic potential and offer a remarkable improved ecological behaviour in comparison to conventional printed circuit boards, they will however not replace but complement.

INNOVATION WITH MID – IMPACT FOR NEW PRODUCT SOLUTIONS
General Requirements for Electronic and Mechatronic Systems

Examination of the demands imposed on electronic devices shows that the conditions in which they are used and also in which they are manufactured have changed enormously. New installation locations (e.g. engine compartment), rising demands for reliability and an increased integration of mechatronic products leads to higher and higher, especially thermal, stresses.

General requirements for electronic and mechatronic systems are cost reduction, miniaturization and the enhancement of quality and reliability. In this context, economic benefits can be achieved with reduced interfaces and components, optimized manufacturing processes and modular technologies. Function integration and high integration on component level leads to miniaturized systems. Improved manufacturing concepts, shorter process chains and the prevention against environmental influences leads to enhanced quality and reliability.

Concurrent to the increase of the requirements there is an ongoing trend to increase the amount of electronics, especially in automotive industry. There will be a great demand for a large holistic approach of system integration. Both, the requirements and the need of highly integrated electronic systems can highly be achieved with MID.

Potentials of MID and Challenges of Mechatronic Systems

The integration of mechanical, electrical, fluidic and even optical functions in one subassembly provides high rationalization potential for the product itself and also for the manufacturing process. MID enable because of the high design flexibility the integration of many functions like housing, switches, shielding, fastening elements, etc. Thus it is possible to reduce the weight and size of the assembly and even of the whole system. Besides this, the substrate materials are inherently flame retardant and easy to recycle. [1]

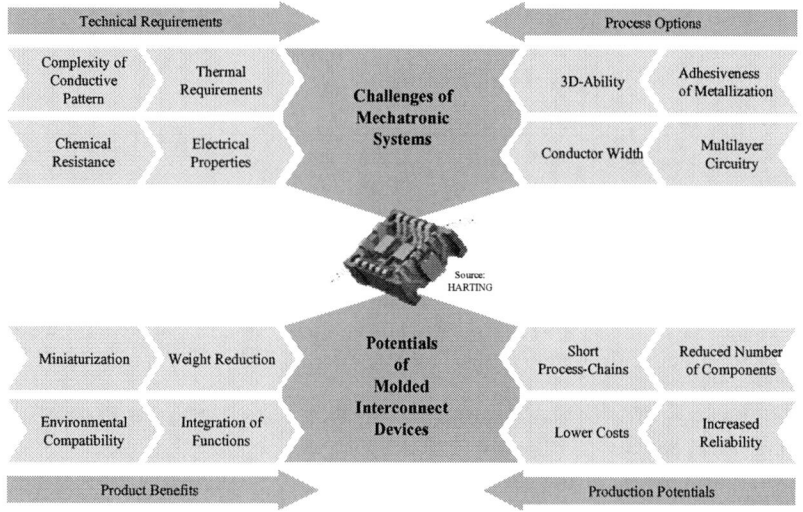

Figure 1. Challenges and potentials for mechatronic systems.

By eliminating mechanical components the amount of system components is reduced. MID offers the potential to increase the reliability of the products and the robustness of the production process because of less electrical interfaces, simplified assembly and shorter process chains. Together, all these factors lead to lower cost, higher functionality and better quality of mechatronic products.

Mechatronic systems are complex assemblies with varied preconditions determined by required functions, quality needs and costs. For this reason, it is necessary to consider the options of different manufacturing processes. They vary for example in 3D-ability, in the adhesiveness of the metallization, the conductor width and the possibility to realize vias or undercuts. Besides this, there are technical needs like complexity of the conductive pattern, thermal requirements, chemical resistance and electrical properties. [2]

DEVELOPMENT AND MANUFACTURING OF MID REQUIRES PRODUCT AND PROCESS INNOVATIONS

Key factors of electronic systems based on injection molded circuit devices are the choice of the material and a structuring procedure in line with the requirements. In principal the substrate material must be in keeping with the application conditions and the used assembly and connection technology. Further factors are a peel-resistant metallization layer since during use this has a major impact on the long-term reliability and a reliable assembly of the electrical components.

Development and Design of Three-dimensional Assemblies

The realization of innovative product ideas requires extensive development know-how and the corresponding support by software. For conventional electronic systems (e.g. ceramics or PCB) exists a complete description of the design and manufacturing processes. For the wider use of MID and a fast time-to-market are an appropriate knowledge base, an interdisciplinary understanding and spatial development tools necessary. The classical design software for circuit design does not meet the requirements of MID – the assignment is the combination of mechanical and electronic information. [3]

Consequently, the efficient design of complex MID-products requires a special design tool that considers the mechanical structure and electronic function of MID in a 3D environment at the same time. It has to provide design functions especially fitted to the needs of MID by combining functions of ECAD and MCAD systems. Existing tools only provide a rather limited support for an efficient design of complex MID-applications. In addition to the functionality provided by ECAD and MCAD-systems, a MID-CAD-system should offer at least the following functions: [4]

- 3D-component library
- MID-specific feature library
- 3D-placement and –routing
- MID-specific design- and manufacturing rule checks
- Interface to CAx-tools
- Generation of manufacturing-related information
- Integration of simulation, e.g. injection molding or electrical behaviour

This helps to design MID-products according to the concepts of design for manufacturing and design for assembly.

Significant Manufacturing Processes: Laser Structuring, Two-Component Injection Molding and Hot Embossing

There are several ways to manufacture MID-components. The most important manufacturing processes for serial applications are two-shot injection molding (two-component injection molding), laser structuring (additive and subtractive) and hot embossing.

Laser structuring can be run as an additive or subtractive process. Laser direct structuring (additive process) is based on thermoplastics that contain a specific organometallic additive as active component. The laser beam vaporizes the upper layer of polymer and activates the underlying metallization nuclei of the active component. Followed by a chemical metallization of the activated surface. There are different approaches for the subtractive process. Subtractive laser structuring presupposes a full-surface metallization with electroless copper or with a PVD process. This is followed by structuring and surface finishing processes.

Laser Direct Structuring

1 – Injection molded part

2 – Structuring and surface activation

3 – Metallization of the activated areas

① ② ③

Two-Component Injection Molding

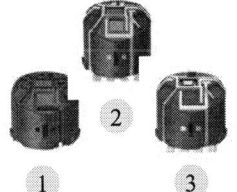

1 – First shot: plastics molding

2 – Second shot: circuitry

3 – Metallization of the second shot

Hot Embossing

1   2   3

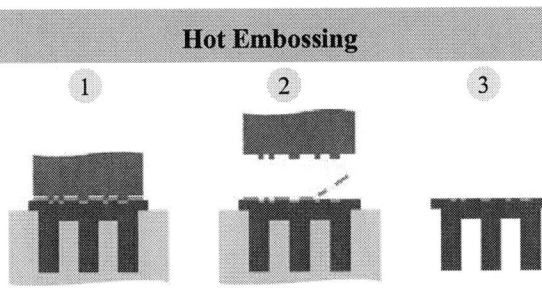

1 – Hot Embossing   2 – Removing foil   3 – MID

Figure 2. MID manufacturing processes: laser direct structuring, two-component injection molding and hot embossing.

Two-component injection molding is a process by which assemblies are molded with two different plastic types. One of them can be metallized and forms the conductor lines and the other one serves as insulation material.

Hot embossing is a very efficient and less expensive manufacturing process. After injection molding, only one job step is necessary to form the conductor layout with a heated die. Important for the process hot embossing is the material of the foil, an electrolytically obtained copper with easily shearable crystalline structure. [5]

Materials for MID

The key material requirements are processing and usage temperatures, flammability rating, mechanical and electrical properties as well as costs. There are several materials available to produce injection molded parts. As a matter of fact, the materials and processes have to be evaluated for each application and manufacturing process specifically. [6]

Characteristic	Unit	Silicon	LTCC-du Pont 951	FR4
CTE x/y	ppm/K	2,5	6	15-18
Tg	°C	n.a.	n.a.	130-170
max. T	°C	> 260 °C	> 260 °C	> 260 °C
Flammability		n.a.	n.a.	UL 94 VO

Characteristic	Unit	Pocan DPT 7140 LDS	Ultramid T 4381 LDS	Vectra E840 LDS
CTE x/y	ppm/K	36/56	30/50	12/30
Melting Point	°C	225	295	335
max. T	°C	< 255 °C	> 260 °C	> 260 °C
Flammability		UL 94 HB	UL 94 VO	UL 94 VO

UL 94 VO: Flame time after exposure to flame < 10 sec. (self-extinguishing)
UL 94 HB: not self-extinguishing

Table 1. Physical-chemical material properties of MID-materials in comparison with other circuit carriers.

Table 1 shows the different properties of silicon, LTCC-du Pont 951, FR4 and three different material types developed especially for laser direct structuring. The coefficient of thermal expansion (CTE) of plastics is usually very big, anisotropic and depends also on the shape of the MID. The mentioned LDS-materials are developed for vapour phase soldering processes. [7][8][9]

LDS-materials are expensive in comparison to conventional thermoplastics because of the specific organometallic additive. An example for a two-component injection molding material combination is LCP Vectra E820i PD as metallizable component and LCP Vectra E130i as insulation material. Besides many advantages of LCP (CTE comparable to FR4, high melting point, low water absorption), the materials don't cohesively connect and because of that cleavage can not be excluded. Therefore the design should be adapted (e.g. undercuts) to avoid this. [10]

Challenges of 3D-Assembly for MID

The lack of flexible automatic assembly machines is one of the main barriers to the wider dissemination of MID. The normally used systems in electronics manufacturing are designed for the processing of printed circuit boards. For simple 2D-MID the standard processes of dispensing a joining medium, placement of components and reflow soldering/conductive bonding can be employed. As complexity rises, however, it is necessary to modify or adapt the assembly processes, primarily the placement process. The effects of the injection molded circuit devices on the rest of the SMT (surface mounted technology) process chain depend largely on the three-dimensional complexity of the subassembly.

Spatial circuit carriers in general lead to three-dimensional placement positions and component orientations in three directions. Because of this, six different moving axes are necessary for the assembly process of 3D-MID. An assembly machine is optimized for planar placement surfaces and offers only four degrees of freedom. The Institute of Manufacturing Automation and Production Systems (FAPS, University of Erlangen-Nuremberg) developed two different approaches for the complex kinematics.

One possibility is to use a modified standard assembly machine with a permanent integrated rotatable module holder and an extended z-axis. The other solution is realized with a 6 axes industrial robot. This flexible and open programmable operation system provides high geometry flexibility for the assembly, but there are restrictions in performance and precision compared to the modified standard assembly machine. [11][12]

A new promising approach is the design of an automated, multiple MID-holder to move all subassemblies concurrently in three degrees of freedom for the orientation of the process surfaces normal to the placement direction. High potential offers the flexible integration in standard assembly machines. The mounting is moved in the PCB conveyor. The MID-adapter can be moved in height and two rotating axes to adjust the process surfaces of the convex MID circuit carriers.

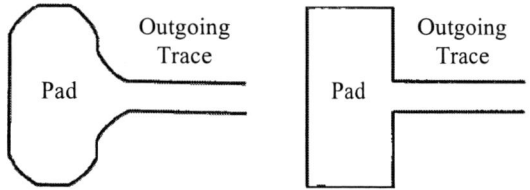

Figure 3. Long-range wetting and optimized design-rules for MID (left side: teardrop-design).

A special challenge is the synchronization of the adjustment of the MID with the assembly process itself. At the moment there is no interface for individual functions at the assembly machines. Therefore the objective is to develop a standardized interface that interrupts the given manufacturing sequence, enables the function of an external subsystem and afterwards proceed with the manufacturing process.

Beyond the manufacturing technique, there are further challenges for the process itself. In principle, the surface metallization with Cu-Ni-Au is suitable for soldering and adhesiveness processes. For the bonding process is a special rework necessary to polish the surface. Because the shape is not planar, there is no solder resist and consequently the result is a long-range wetting of the solder paste. This leads to a run-off of the solder paste of the bonding zone with a negative impact on long term reliability and the risk of short circuits. Similarly, there are sometimes cracks of the conductors in the transition zone of the pad and the outgoing trace probably because of the shrinkage stress during solidification of the solder. [13]

To avoid these problems there are specific design rules for 3D-assemblies necessary. One example is the conception of the pad design: The corners of the pads should be rounded and the transition from pad to the outgoing trace very smooth (teardrop-design to avoid stiffness jumps and notch stresses). (Figure 3)

MARKET DEVELOPMENT OF MOLDED INTERCONECT DEVICES
Technology Push vs. Market Pull
The spread of a technology can be mapped by the interplay of market- and technology-driven factors. A focused market or technology orientation leads to the misallocation of resources. The realization of the successful implementation of MID can only be settled on a successful combination of an integrative "market pull" and "technology push".

Especially in MID-technology exists an intensive "technology push". Particularly in the early stages almost only technology-driven factors played an important role. The "market pull" in the field of MID-technology is still limited. But due to the rising number of serial products references and the economics of MID-projects are demonstrated.

Key Markets for MID and Innovative Serial Applications
Developments applying MID have been accelerated as a result of technological advances in the fields of structuring, metallization, assembly, connection technologies and especially in plastic materials. More stringent demands from different sectors can increasingly be met by means of modified plastics and suitable adapted processes. Currently, numerous innovative serial applications are running in very diverse product fields.

MID is the right solution when the potentials of thermoplastic circuit carriers can highly meet the product requirements. Given that, the injection-molded parts with integrated conductors can be used in very different industry sectors. Exemplary applications are explained in the following. [14]

Industry Automation: Camera Module
An example for an innovative product in the industry automation is a MID that integrates LED as lighting for a camera module. The LED illuminate in different

74

directions and enable an exact adjustment of positioning modules in assembly machines. (Figure 4)

Medical Industry: MID Main Board
There are more and more MID-applications in the medical industry. An example is the MID main board for dentist tools. The MID integrates one LED for the illumination of the mouth, two switches to control the exhaust of water/air and also many other electrical components.

Logistics: RFID Transponder
RFID transponder (RFID: radio frequency identification) help the warehousing industry to simplify and accelerate logistical processes. The use of MID-technology allows a three-dimensional antenna structure having a range more than 5 m. The airtight and watertight welded housing meets the requirements of the protection categories up to IP67 for extreme environmental conditions and so it is particulary suitable for use in proximity of metals and liquids. The integrated chip stores all the relevant data, e. g. nature of contents, state, origin and destination.

Automotive Industry: 3D Circuit Carrier
The 3D circuit carrier is a MID for motorbikes designed to control numerous functions such as lighting, horn, bright and turn signals. The solution integrates 14 switches, many plug connectors and other electronic components. (Figure 5) Other realized automotive applications are a sun sensor for the regulation of the climate in the interior or a switch for seat adjustment. [15]

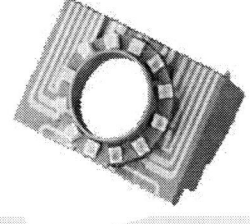

Camera Module with spatial arrangement of LEDs

Material:	LCP Vectra E820i
Process:	Laser Direct Structuring
Sector:	Industry Automation
Manufacturer:	HARTING AG/SICK AG

Figure 4. Camera module – MID serial application for the industry automation.

Telecommunication: Antenna Structures
Mobile phones with more and more functions (e.g. navigation-system with GPS, WLAN, camera system, touch screen) are highly integrated products. A further miniaturization is possible by laser structured antennas directly applied on the plastic housing. An innovative solution thereby is the combination of two MID manufacturing processes. Two-component injection molding for the basic structure and laser direct structuring for fine pitches and lots of varieties.

Consumer Electronics: Clutch System
Even in consumer electronics MID-applications are already realized. One example is a clutch system for toy-trains. It allows the automatic disconnection of trains – a former manual action.

3D Circuit Carrier for Electronic Motorcycle Handle

Material:	Ultramid® T4381 LDS
Process:	Laser Direct Structuring
Sector:	Automotive Industry
Manufacturer:	Kromberg & Schubert

Figure 5. 3D Circuit Carrier – MID serial application for the automotive industry.

MID-Development in Different Global Regions
The main MID-activities take place in Europe, Asia and the USA with different focuses:

In Europe, there exists a good knowledge base because of a great research environment with strong research activities between industry and institutes. Almost all technologies for MID are available. Preferences in the manufacturing processes are in LDS, 2K and hot embossing. But, there are still only a few companies with competences in the whole process chain. In the USA, there was in the early 1990s a transparent market with many applications from different companies (Circuit Wise, Mitsui Pathtek, UFE). In the meantime, there are only two MID-providers (Tyco and Molex) with preferences in 2K and LDS. In Japan is a high market penetration of a few local businesses with a strong product and system orientation with the processes laser subtractive structuring and 2K. In the rest of Asia, especially in China and Taiwan is high volume production (> 5 million units per month) of MID in the field of mobile phones antenna structures. There is an intensive technology transfer of the global player from USA or Europe, which are shifting their manufacturing capacities to Asia. The preferred manufacturing processes are 2K and LDS. (Figure 6)

There is no actual study about the distribution of the MID market volume. But, an indicative is the worldwide installation of the LPKF-LDS® laser systems MicroLine 3D and FUSION 3D. More than 80 % of all machines are located in Asia. But, the market segment is almost

75

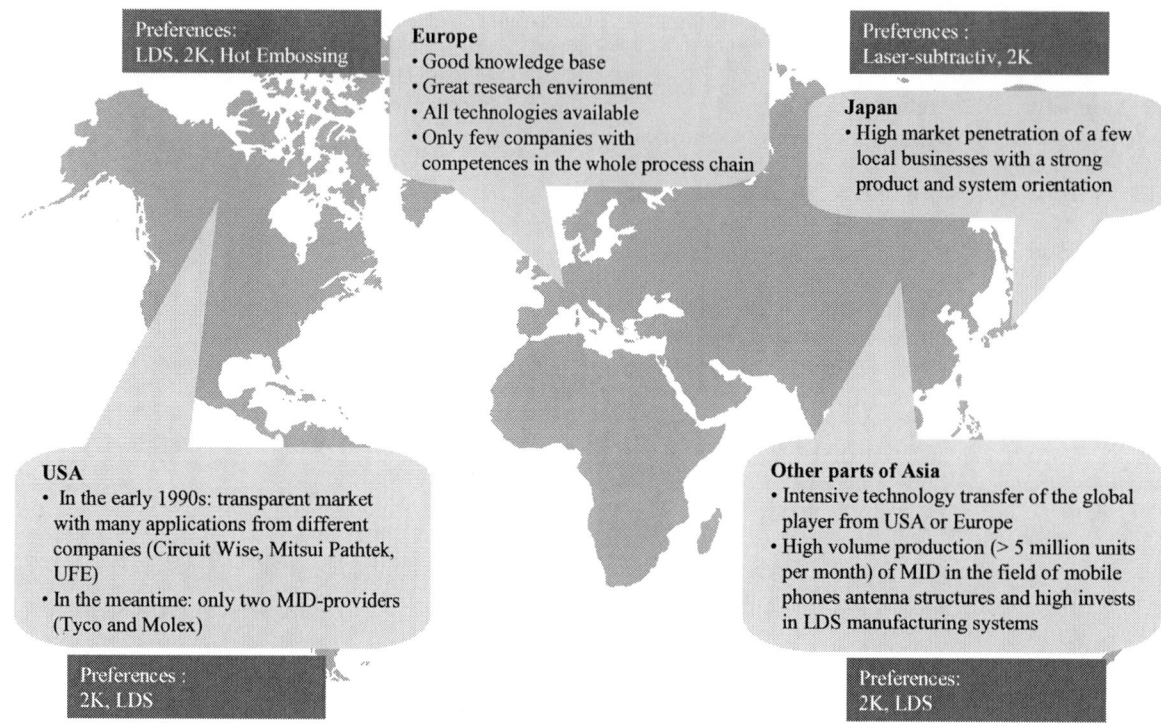

Figure 6. Development of MID-technology in different global regions.

exclusively communication with simple antenna applications. Other applications for automotive, medical engineering and other industry sectors with complex assembly and connection technology are produced mainly in Europe and Japan. [16]

Innovation Success Through Technology-Networking
Particularly in MID-technology is the formation of interdisciplinary project teams an important factor. The knowledge of all requirements of the entire process (material selection, mechanical and electrical design, injection molding, patterning, metallization and assembly and packaging) from the first conceptual idea is an essential prerequisite for the project success.

A specific initiative in this context is the collaboration between enterprises and research institutes along the process chain, which led to the research association Molded Interconnect Devices 3-D MID e. V. at a very early stage. The involved companies get valuable access to external technological know-how through the membership in this network. The objective of the research association is to support and develop the MID-technology. This includes joint research projects, promotion of experience exchange among the members and initiation of the realization of new technical opportunities by suitable publications and activities like fair participation and conference organization.

NEW DEVELOPMENTS FOR MID
The MID-technology has established in the market and further developments will support the growing importance of MID. This means the manufacturing processes for an efficient production as well as characteristics of the products to open up new fields of application.

Manufacturing Technologies: Additive Manufacturing and Printing Technologies
One problem of MID-applications is still the long time-to-market at ever-shorter development times and product life cycles. The relatively new MID-technology without sophisticated development tools and manufacturing capacities has to compete with established technologies with mature systems.

Additive manufacturing (AM) – a manufacturing technique to build up structures or features layer by layer – can be used to face these challenges. AM refers to a group of technologies used for building parts, all from 3D computer aided design (CAD) data, medical scans, or data from 3D scanning systems.

For the production of the circuit carrier are typical rapid prototyping/rapid manufacturing procedures like laser sintering or fused deposition modeling possible. The challenge is due to the layered structure the roughness of the surface. This is followed by the metallization process.

An innovative additive manufacturing technique designed for electronics industry with significant potentials for MID is aerosol jet printing – a maskless deposition process with a wide range of materials compatibility. An aerosol stream containing metal, polymer, ceramic, adhesive or even biological material is focused, deposited and patterned onto a planar or non-planar material. One great advantage of printing technologies for MID is the possibility to manufacture functional circuits without soldering processes. Thus it is possible to fully print circuits, including resistors, capacitors, transistors and interconnects with a minimum resolution of 10 microns. [17]

CONCLUSION AND OUTLOOK

An increase of cost pressure combined with a simultaneous demand for high product reliability obliges manufacturers to further system integration and miniaturization. The exclusive examples of products show that there is enormous potential to integrate mechanical and electronic functions in one application. Beyond this fact, the ongoing trend to increase the amount of electronic functions in automotives supports the demand for MID-assemblies. Powered by the pressure of market requirements for high quality and good value products as well as the technical requirements for miniaturization, ecologically friendly and multi-functional products, the MID-technology is a necessary alternative to conventional printed circuit boards.

The realization of successful MID-projects demands interdisciplinary cooperation along the whole process chain. The hedging of process-related risks must take place throughout the entire process, because standards are missing in comparison to the IPC for flexible printed circuit boards. Especially spatial design tools for 3D circuit carriers and efficient assembly technologies for 3D-MID will support the continuing success of this innovative technology. Additive manufacturing for building up prototypes (rapid prototyping) or small unit applications (rapid manufacturing) can also promote the MID-development.

In order to successfully turn the vision of new series products into reality it is necessary to have detailed process knowledge so that the growing requirements of the various sectors and the more difficult conditions of new installation locations can be met. The Institute FAPS in collaboration with the Research Association for Molded Interconnect Devices (3-D MID e. V.) keeps the track to support and develop the MID-technology further.

With regard to the increasing use of electronic with extended functionality it is necessary to exploit the opportunities provided by complex MID and by this means force the pace of technological advance. Above all the required functionality and complexity can be achieved by the systematic use of mechatronic systems.

REFERENCES

[1] Forschungsvereinigung Räumliche Elektronische Baugruppen 3-D MID e.V., *3D-MID Technologie: Räumliche Elektronische Baugruppen; Herstellungsverfahren, Gebrauchsanforderungen, Materialkennwerte.* München: Hanser Verlag, 2004.

[2] M. Immle, „Anti-Corrosion 3D-MID"in *Proceedings of the 8th International Congress Molded Interconnect Devices*, Fuerth, Germany, Sept. 2008.

[3] J. Franke.: Integrierte Entwicklung neuer Produkt- und Produktionstechnologien für räumliche spritzgegossene Schaltungsträger (3D-MID), München: Hanser Verlag, 1996.

[4] K. Feldmann, Y. Zhuo, C. Alvarez: *Horizontal and Vertical Integration of Product Data for the Design of Molded Interconnect Devices.* 5th International Conference on Digital Enterprise Technology, Bath, Oct. 2007.

[5] F. Schüßler, K. Feldmann, H. Kück, H. Richter, T. Osswald, A. Gardocki, "Molded Interconnect Devices For Applications with Advanced Thermal Requirements" in *Proceedings of the 8th International Congress Molded Interconnect Devices*, Fuerth, Germany, Sept. 2008.

[6] K. Feldmann, C. Goth, F. Schüßler, "Prospects for Micromechatronic Systems" in Kunststoffe International, vol. 6, München: Hanser, 2008, pp. 70-73.

[7] N.N.: Datasheet E 840i LDS Preliminary. www.ticona.com. Effective 07/2007.

[8] N.N.: Datasheet Pocan DP T 7140 LDS. www.lanxess.com. Effective 10/2009.

[9] N.N.: Datasheet Ultramid T 4381 LDS. www.basf.com. Effective 11/2009.

[10] H. Kueck, W. Eberhardt: Multifunktionale 3D MID Packages für die Mikrosystemtechnik, Potenziale und Trends der Mikro- und Nanotechnik, GMM – VDE/VDI-Gesellschaft Mikroelektronik, Mikro- und Feinwerktechnik. VDE-Verlag, Berlin, 2006.

[11] A. Brand, *Prozesse und Systeme zur Bestückung räumlicher elektronischer Baugruppen (3D-MID).* Bamberg: Meisenbach Verlag, 1997.

[12] S. Krimi, *Analyse und Optimierung von Montagesystemen in der Elektronikproduktion.* Bamberg: Meisenbach Verlag, 2001.

[13] Zimmermann, M.: „Aufbau- und Verbindungstechnik auf LDS-MID" in PLUS, vol. 3, Saulgau: Leuze Verlag, 2009, p. 622-632.

[14] Feldmann, K.; Franke, J.; Goth, C.: "Advanced Mechatronic Systems using MID (Molded Interconnect Devices) and FPC (Flexible Printed Circuits)" in *Proceedings of the 10th International Conference on Automation Technology*, Institute of Manufacturing Engineering, National Cheng Kung University Tainan, Taiwan, 2009, pp.24-29.

[15] D. Moser, J. Krause, "3D-MID – Multifunctional Packages for Sensors in Automotive Applications" in *Advanced Microsystems for Automotive Applications 2006.* Berlin: Springer Verlag, 2006.

[16] T. Niino, "MID Processed Electrostatic Motor – Prototyping and Development of Fabrication-Technique" in *Proceedings of the 8th International Congress Molded Interconnect Devices*, Fuerth, Germany, Sept. 2008.

[17] M. Hedges, „3D direct writing via M³D™ "in *Proceedings of the 7th International Congress Molded Interconnect Devices*, Fuerth, Germany, Sept. 2006.

MICRON LEVEL PLACEMENT ACCURACY FOR HIGH ASPECT RATIO DIE IN PRINTING PRODUCTS

Zeger Bok and Daniel D. Evans, Jr.
Palomar Technologies, Inc.
Carlsbad, CA, USA
info@bonders.com

ABSTRACT

Applications requiring ultra high placement accuracies of 1um to 5um are required in several applications: Arrayed Laser Print Heads, Arrayed Ink Jet Print Heads, P-Side-Down Laser Applications, and multi-channel optical communication products.

Arrayed print head technology, laser or ink jet, utilizes arrays of lasers or ink jets on individual die. The lithography processes used to create the arrays on each die are more than adequate to produce high quality images. Conceptually, a single long die would allow a single-pass printing on the print medium. However, there is a practical limit to the length and width of die without affecting yield and handling damage. Additionally, the mixing of monotone and color printing requires that multiple rows or columns of print pixels be precisely aligned with respect to each other. Although some correction can be accomplished in electronics by modifying the print timing of each pixel, pixels must be aligned within a few microns for high quality printed images.

Print heads typically use high aspect ratio die to maximize the number of pixels per die and to minimize the amount of silicon. Micron level placement accuracy of 1um to 5um is presented in this paper for large die with lengths greater than 30mm and high aspect ratios of nearly 10:1. Both "side by side" and "end to end" die arrays are presented along with their corresponding results.

Production equipment and process features are explored for practical automation of micron level placement accuracy. Control of adhesive is a critical process step required for ultra high placement accuracy of die. Results of fine line dispensing are presented as in input prior to die bonding. Material feeding, identification tracking, and vibration isolation relating to ultra fine placement accuracy are also discussed in this paper.

Key words: Laser Print Head, Ink Jet Print Head, Die Attach, Micron Placement, Ultra High Placement Accuracy, Die Bonding, Pick and Place Machines, Epoxy Dispense, Epoxy Die Attach, MEMS Packaging

INTRODUCTION

The application breakout given in Table 1 is useful to explore general application, attachment, and accuracy requirements for high accuracy die attach. The main breakout is by application product or technology. For the purposes of this paper, each of the applications explored is for specific attachment technologies (Epoxy and/or Eutectic) and general ranges of required placement accuracy. Only the first three applications will be studied in this paper.

Application	Attach	Accuracy
LED Laser Print Head	Epoxy	±2-5µm
Ink Jet Print Head	Epoxy	±3-5µm
Laser Marker Head (VCSEL)	Epoxy	±3-5µm
Active Optical Cables	Epoxy Eutectic	±2-3µm
P-Side Down Laser	Eutectic	±1.5-3µm
Lithography/Screen Interconnect	Epoxy	±3-5µm
Thru Via Die Stacking	Misc	±3-5µm
3D MEMS Stacking	Epoxy	±5-10µm

Table 1: High Accuracy Applications

A scale perspective is explored by using a 75µm diameter human hair as the starting point, illustrated in Figure 1. Typical SMT equipment can achieve ±40µm placement accuracy, while die attach equipment can generally achieve ±25µm placement accuracy. Another class of pick and place machines breaks the ±5µm barrier, placing to accuracies smaller than a red blood cell. Ultra high placement accuracy pick and place machines achieve accuracies of ±1.5µm, which is smaller than a typical bacteria cell.

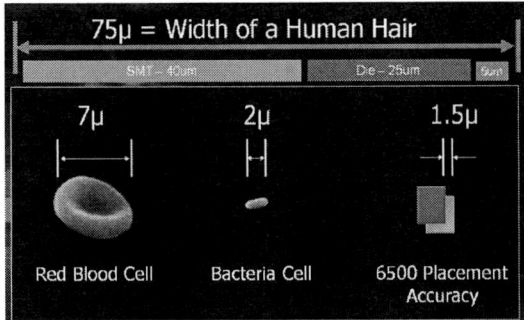

Figure 1: Accuracy Scale Reference

Terms and Definitions

The pick and place process outputs are composed of geometric accuracy and interconnect method. To completely define placement accuracy would require specifying all six (6) degrees of freedom, as shown in Figure 2. Most pick and place accuracy applications specify Z as a bond line and placement accuracy as X error, Y error, and Theta-Z error. For the purposes of this paper, the specific requirements will be listed for each of the cases studied.

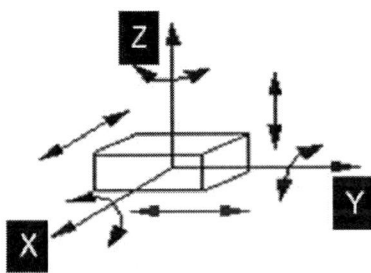

Figure 2: Geometric Six Degrees of Freedom

Interconnects are comprised primarily of two methods for optoelectronic assemblies: adhesive (epoxy) or metallurgical (eutectic solder).

These attachment options can be either in-situ (serial) or batch (parallel). In-situ attachment is completed during the placement operation for each component and will have lower throughput since the attachment time for each component is added to the pick and place time. Batch attachment is completed after all components are placed so the actual attachment can be completed as a parallel process. Batch attachment methods typically have higher throughput compared to in-situ.

Another consideration for the attachment method is its effect on placement accuracy. To understand the effect on batch attachment methods, it is important to measure the pre-cure and post-cure accuracy of the components. In-situ attachment methods can have higher placement accuracy, especially if the design does not include self-centering.

Material and Process Considerations

When approaching micron level placement accuracies, there are several important factors that need careful management and control:

- Substrate flatness, cleanliness, and fiducial clarity
- Die flatness, cleanliness, and fiducial clarity
- Attach material uniformity, shrinkage, symmetric application, and curing stability

One example of the dramatic impact of contamination on placement accuracy is given in Figure 3. Although not completely to scale, a 400μm wide by 100μm tall component can be shifted by 1.3μm if a 5μm particle is under one edge. It is generally understood that particles must be controlled, and this example shows the magnitude of the effect of a thin particle. It may be that contamination is controlled well, but that the substrate or die have burrs as a result of the dicing process, or that there is an unbalanced particle loading in the attachment material. All factors need to be controlled to achieve high accuracy placement and attach.

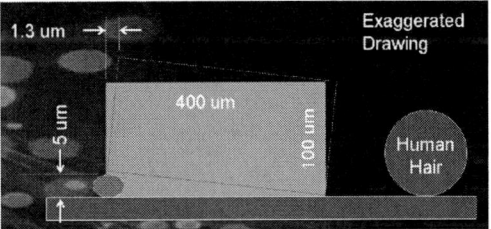

Figure 3: Particle Effect on Placement Accuracy

The clarity of fiducials and dimensional control of parts are important for accurate alignment. However, where part dimensions or fiducials cannot be controlled well in all dimensions, they should be controlled in the primary axis of alignment. Case studies will show specific examples.

The alignment tool should have the ability to diagnose machine capability and to process sub-element capabilities (X move, Y move, Z move, image find errors, etc.) so that alignment issues can be resolved down to operator, material, process, or machine issues. Figure 4 illustrates one example of diagnostics which includes single axis motion, XY axis motion and any cross coupling effects, as well as process capabilities for placement accuracy (as measured by the machine).

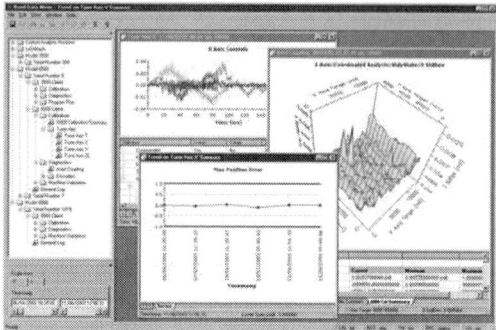

Figure 4: Process Diagnostic Tools

The measurement equipment and methods are also critical for high accuracy placement qualification. Typical measurement tools required to measure high placement accuracy include:

- Automated Optical Inspection Equipment (AOI) with ~0.5μm resolution
- Built-in part features such as vernier scales, which allow direct reading of placement accuracy with resolution of ~0.5μm
- SEM systems with ~0.1μm resolution

An example of measuring the edge to edge distance using a SEM with built-in measurement tool is shown in Figure 5. Without automated measurement tools, it would be extremely difficult to achieve a gage repeatability and reproducibility (Gage R&R), required to measure to 0.1μm.

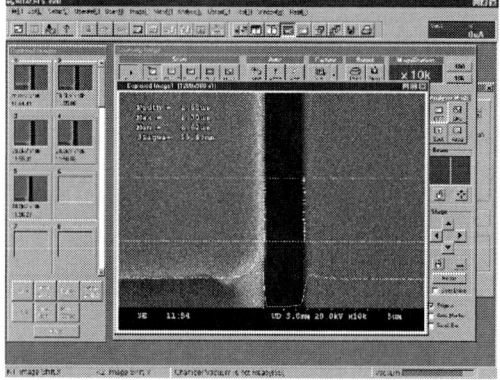

Figure 5: SEM Measurement Tool Example

PRINT HEAD CASE STUDIES
Three separate case studies are now presented to provide an overview of the various applications, key requirements, and results for micron level placement.

Case 1: LED Laser Print Head Array (End to End)
Arrayed LED print head technology uses a series (arrayed line) of LED arrays to illuminate the entire length of a photo resist drum at once rather than using a single laser and scanning wheel, as shown in Figure 6. The arrayed LED laser based system reportedly occupies 1/40th of the volume space required by a raster optical scan (ROS) system.

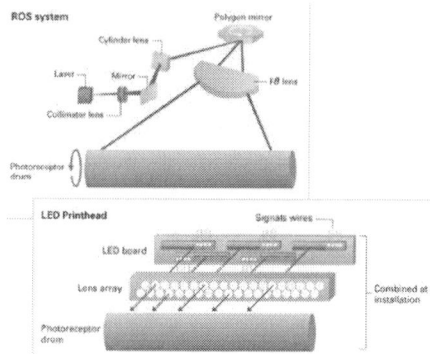

Figure 6: Laser Printer Technology Comparison - Single Laser with Scan Wheel versus Arrayed LED (Source Fuji Xerox) (1)

The primary alignment requirement for the series of LED arrays is lateral spacing uniformity (along a line), as shown in Figure 7 (top) with a series of LED arrayed die aligned End To End. Each LED Array die is essentially a singulated die, with LEDs fabricated into a specified number of LEDs at a known fixed spacing, which is determined by the lithography process. The blue circle and square on each die represent the first and last LED in the LED array on each die.

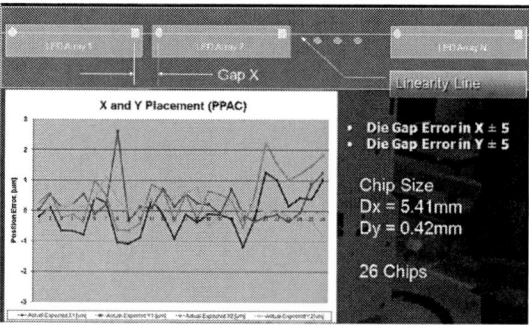

Figure 7: LED Print Head Requirements and Results in μm

Individual LED array die are lined up to mimic a continuous array of LEDs. The pick and place requirement is based on placing the outside two LEDs on each die (small blue circle and square on each die) in a uniform spacing within ±5μm in both X and Y from the theoretical line.

Dimensions of the die cut for both ends, Xa and Xc, must be controlled (shown in Figure 8) to maintain correct LED spacing as well as to avoid die colliding with each other.

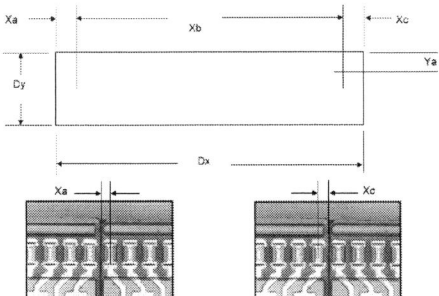

Figure 8: LED Array Die and Critical Dimensions

The attachment method specified is electrically conductive epoxy (Epotek H20E), which is cured at 150°C for 90 minutes (typical) and performed as batch cure after all LED array die have been placed.

Figure 9 illustrates specified epoxy coverage after each die is placed and cured.

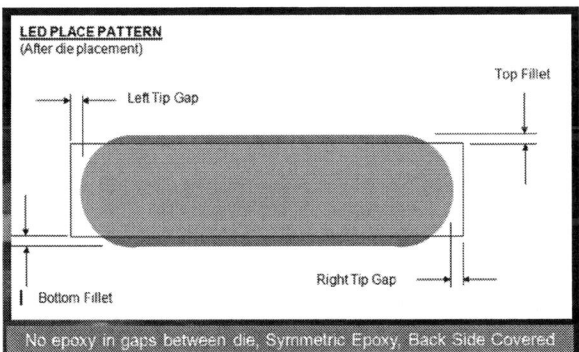

Figure 9: LED Array Die Placed in Epoxy

The assembly process consists of the following process steps:

- Dispense Epoxy on Substrate for all die locations
- Place all Die onto Substrate/Epoxy
- Measure Pre-Cure Placement
- Oven Cure Offline
- Measure Post-Cure Placement

Placement accuracy results of the measured gap between LED array die (pre- and post-cure) were well within the ±5µm placement tolerance. The results of a particular assembly with 26 LED die is shown in Figure 7.

Case 2: Ink Jet Print Head (Side by Side)
Ink Jet printers are available from several suppliers. Although details vary from supplier to supplier, they all use an array of nozzles to jet ink dots as the print head scans over the print medium. Starting in the upper left corner of figure 10, a multi-color print head composed of a single die is shown from HP (2). The top and bottom of the print cartridge are shown in the first two images. The cartridge contains the ink reservoirs for each color and the die with ink nozzles. The third image from HP shows a magnified view of the nozzle die. The single die configuration has the advantage that all of the nozzle array spacing is determined by lithographic processes. Further to the right in Figure 10, the multi-die example (3) shows

a more complex print head containing five ink jet nozzle array die. Multi-die Ink Jet print heads require each of the nozzles from each of the die to be placed relative to each other. The tolerance of placement is determined by the amount of electronic fire timing possible. The right corner of Figure 10 shows a close-up image of a three ink well print die.(4) The lower right corner of Figure 10 shows greater detail of one ink well with arrays of ink nozzles surrounding the wells.(5) The details of the nozzles and how they expel ink from the well through the nozzle varies from supplier to supplier and is not covered in this paper. Adhesive is used to attach the nozzle die to the ink source chamber which feeds the ink well. The adhesive acts as a mechanical constraint and a liquid (ink) seal. The lower left image in Figure 10 shows a multi-color print head example using multiple ink wells for different colors of ink nozzle arrays. (6)

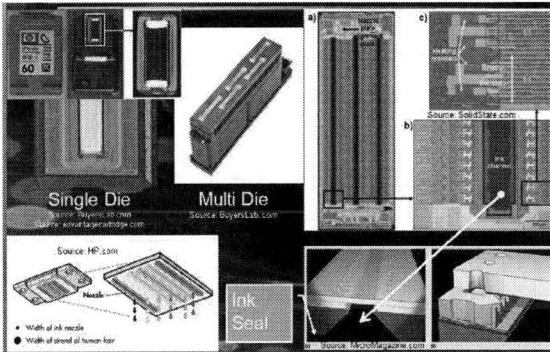

Figure 10: Ink Jet Print Head / Die Overview

The specific print head example for Case 2 is shown in Figure 11. The configuration consists of two nozzle die placed side by side. The nozzle die size is 3.5mm in X, 33.0mm in Y for a 9.4 aspect ratio of Y/X. The allowable gap error in X and Y between JetArray1 and JetArray2 is ±3µm.

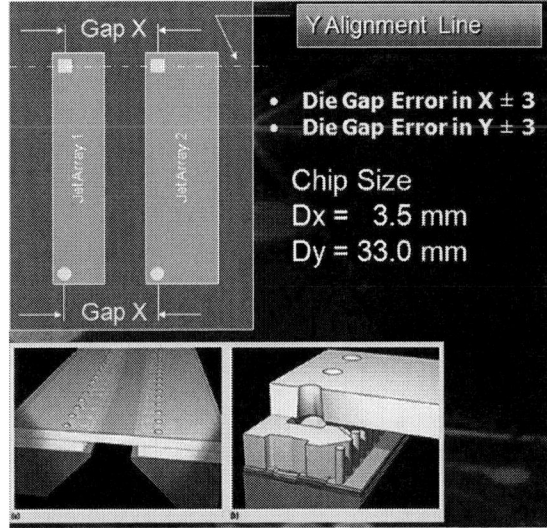

Figure 11: Ink Jet Array Die (Side by Side)

The nozzle die contains three ink wells which requires an adhesive pattern as shown in Figure 11. The epoxy must be placed symmetrically around the die and ink well locations to minimize placement errors in wet epoxy and to minimize XY shifts curing adhesive curing. The narrow geometries of the ink wells require epoxy adhesive line widths of 0.55mm and continuous lines to ensure a good seal for each ink well connection to the die. Additionally, no adhesive epoxy can flow into the ink wells.

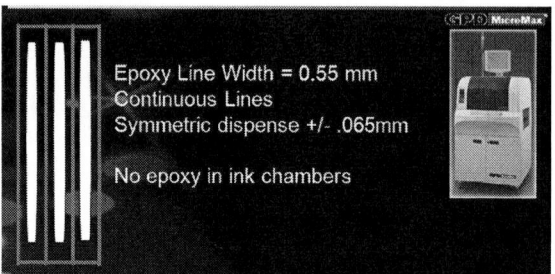

Figure 12: Ink Jet Array Die in Epoxy – Epoxy Dispense Requirements

The assembly process consists of the following process steps:

- Dispense Epoxy on Substrate
  (To Hold Position and Ink Seal)
- Place JetArray1 Die onto Substrate/Epoxy
- Place JetArray2 Die onto the Substrate/Epoxy relative to the JetArray1 placed location.
- Measure Pre-Cure Placement
- Oven Cure Offline
- Measure Post-Cure Placement

The summary results are given in Figure 13 as the relative die centroid error for X and Y by measuring the JetArray1 and JetArray2 locations and the relative error between the two die for each assembly. The data shows all Y errors within specification and all X errors within specification except for one data point.

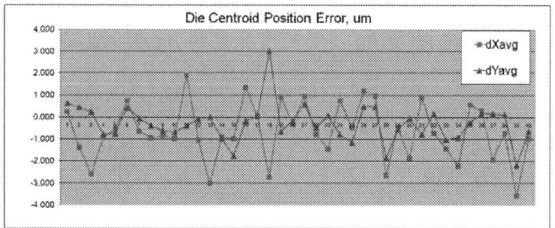

Figure 13: Ink Jet Array Die in Epoxy – Placement Errors

Further builds have shown similar results. Work is underway to optimize the adhesive pattern and placement process to improve relative accuracy.

Case 3: Laser Marker Head (VCSEL Laser)
Labeling is used for product identification and traceability and combines letters, bars, and matrix codes, as shown in Figure 14. Today, nearly everything has a label, including shipped parcels. As security becomes a larger issue, new systems that mark the inside of the paper rather than the surface are gaining acceptance. (8)

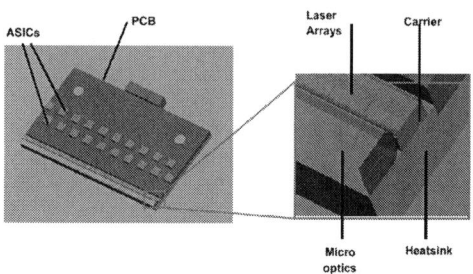

Figure 14: Laser Label Maker Output and Basic Construction (Source: Intense Photonics) (7)

One configuration of a label print head, shown in Figure 15, uses groups of arrayed lasers that are evenly spaced to mark thermal paper. These heads require die to be equally spaced with a controlled gap of ±5µm in X and Y on the left and right LEDs of each die, after epoxy cure.

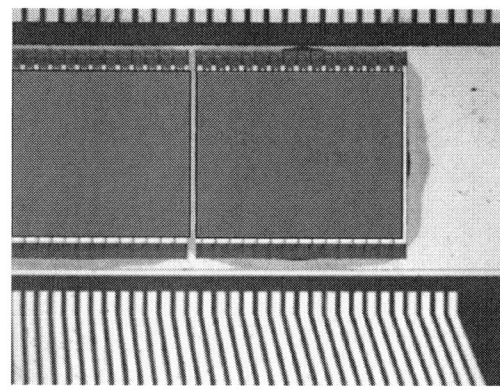

Figure 15: Laser Label Maker Close-up

The placement accuracy results shown in Figure 16 confirm that placement results are well within ±5µm of the X and Y placement accuracy requirement, with a maximum error of 3.2µm in X and -2.9µm in Y. Average placement gap errors of -0.5µm in X and -0.5µm in Y, and 3σ variation of 3.8µm in X and 3.6µm in Y, show acceptable control for ±5µm accuracy requirements.

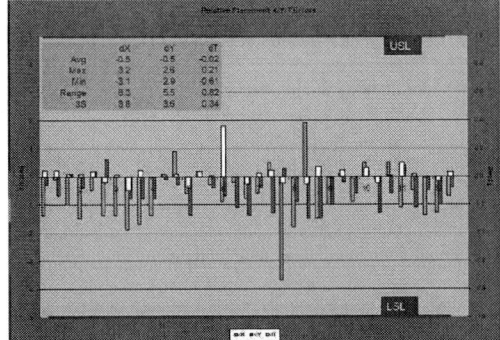

Figure 16: Laser Mark Head Placement Accuracy Results

SUMMARY AND CONCLUSIONS

Three separate optoelectronic application case studies have been reviewed in this paper. All three cases required ±5μm placement accuracy or better, using adhesive or metallurgical attachment, as shown in Table 2.

CASE	Attach	Accuracy
LED Laser Print Head	Epoxy	±2-5μm
Ink Jet Print Head	Epoxy	±3-5μm
Laser Marker Head	Epoxy	±3-5μm

Table 2: High Accuracy Applications

An overview of considerations required to reach high accuracy (±5μm or better) attachment has been presented.

Material considerations include:

- Substrate flatness, cleanliness, and fiducial clarity
- Die flatness, cleanliness, and fiducial clarity

Process considerations include:

- Die presentation format in waffle or gel pack and effect on throughput
- In-situ attachment time or batch attachment time and its effect on throughput
- Symmetry of adhesive and the effect on post-cure accuracy

An example production line is shown in figure 17. This line includes magazine input of substrates, epoxy dispense, epoxy inspection station, die ejection feeder, high accuracy die placement, and magazine output. Not shown is a line material data management system used for component traceability at the assembly level.

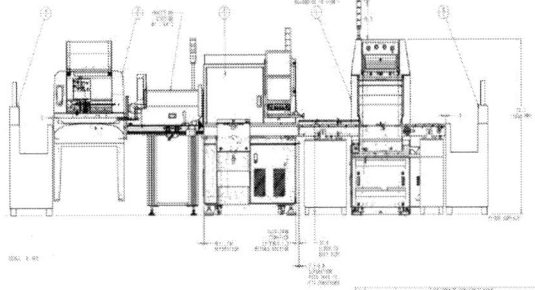

Figure 17: Production line example (Source: Palomar Technologies)

The line is currently used in production although materials and processes are further optimized to increase yield.

REFERENCES

(1) LED Print Head Technologies, Fuji Xerox, http://www.fujixerox.com/eng/company/technology/led_phead/

(2) Single Die examples Source BuyersLab.com, Source advantagecartridge.com (http://www.advantagecartridge.com/Image/16Printhead_1.jpg)

(3) Multi Die example Source BuyersLab.com (http://www.buyerslab.com/images/email/2007/edgeline-printhead.jpg)

(4) Nozzle plate detail Source: SolidState.com

(5) Nozzle plate detail Source: SolidState.com

(6) Nozzle and model representation Source MicroMagazine.com (http://www.micromagazine.com/archive/06/04/mckay.html)

(7) Multi-color print head example Source HP.com, http://h71036.www7.hp.com/hho/images/HP_INK_printhead_noCopy_370.jpg

(8) Guide to Customer High Power Laser Diodes, Intense Photonics, www.intenseco.com

(9) New Advances in Laser Arrays Hit the Spot for Coding and Marking, Professor John H. Marsh, www.intenseco.com

IMPORTANT CONSIDERATIONS IN PACKAGING A SMALL HALL-EFFECT SENSOR

Ron Molnar
AZ Tech Direct, LLC
Phoenix, AZ, USA
rmolnar@aztechdirect.com

Jeff Wise
Independent Design Consultant
Saratoga, CA, USA
wisejef@gmail.com

ABSTRACT

A Hall-effect (magnetic) sensor detects small AC or DC magnetic fields. A new design of a Hall-effect sensor has increased sensitivity to magnetic fields, even before amplification, allowing the sensor to be used for new applications, such as small-motion detection, trace substance detection, and other novel medical devices. In the case of this new sensor, small physical size enhances or enables performance in certain physical applications. The packaging objective, as with most integrated circuits, was to assemble the Hall-effect sensor chip in the smallest, high-volume package available without sacrificing reliability and performance at the lowest total cost solution. Major cost factors were die prep, package type/materials, assembly labor and overhead, final test, and packing methods. More precisely, the challenge was to specify and qualify one or more popular 4-lead to 6-lead "standard" plastic packages for a small ~250 µm x ~250 µm GaAs die that could be assembled and tested by multiple contract manufacturers in high volumes for a "die-free" cost of less than 2 cents each. The desire was to specify a RoHS-compliant package that met Level 1 moisture sensitivity without severely impacting the cost of subsequent board assembly. An overriding constraint was to select a package that contained no ferro-magnetic materials.

The search for an appropriate package and assembly facility, the issues encountered during assembly of qualification lots, the assembly process control data, and the chip/package reliability test results are described in detail. The individual constraints on the packaging, in and of themselves, are achievable with today's technologies, however in combination they limited the choice of packages, direct materials, and suppliers. The trade-offs made, the merits and disadvantages of each approach, and the ultimate decisions and results are described. Details of the final package construction, die preparation issues, the assembly qualification build, and environmental reliability test results are revealed. A total cost analysis of wafer fab and sort, die prep, assembly and final test, and packing is presented.

Initial work in developing this new Hall-effect sensor showed that the die size could be shrunk significantly, thereby reducing the current path width to as little as 0.5 µm without sacrificing performance. Thus, the size of the future sensor chip designs will be limited only by the minimum contact (wire bond or flip chip) pad sizes and their minimum pitch dictated by assembly design rules. The drive for ever smaller Hall-effect sensors will push the limits of conventional small-scale package manufacturing and may result in wafer-level packages if the cost targets can be met. Ongoing research and development is focused on low-cost, flip chip implementations in which the cost of die rework vs. the cost of sub-assembly scrap will be considered.

Key words: Hall-effect, GaAs, SC70, package, assembly

INTRODUCTION

The Hall-effect, discovered by Edwin Hall in 1879, is the production of a voltage difference across an electrical conductor, transverse to an electric current in the conductor, and magnetic field perpendicular to the current.

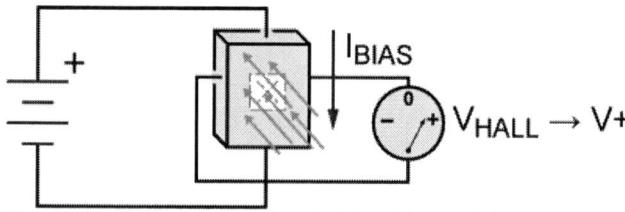

Figure 1. Hall-effect, induced VHALL, resulting from significant magnetic flux (green arrows), perpendicular to the bias current flow. (Source: Allegro MicroSystems)

Most Hall-effect devices produce a very low signal level and thus require amplification. Many devices now sold as Hall-effect sensors in fact contain both a sensor and a high gain integrated circuit (IC) amplifier in a single package. Hall sensors can detect stray magnetic fields easily, including that of Earth, so they work well as electronic compasses. However, this also means that such stray fields can hinder accurate measurements of small magnetic fields.

Hall-effect sensors may be used in various applications such as rotating speed sensors, fluid flow sensors, current sensors, and pressure sensors. Hall probes are often used to measure magnetic fields or inspect materials (such as tubing or pipelines) using the principles of magnetic flux leakage.

BACKGROUND

A new Hall-effect sensor design features a Gallium Arsenide (GaAs) substrate that optimizes sensitivity with cost and overall device size. Patented and patent-pending design features further enhance the sensitivity while reducing error due to magnetic interference and noise. GaAs provides much higher electron mobility (ie. sensitivity) than Silicon and is more resistant to temperature drift due to a higher energy band gap. There are other materials, such as Indium Antimonide (InSb) or Indium Arsenide (InAs), which provide greater electron mobility than GaAs but they either exhibit unwanted characteristics such as temperature drift or are prohibitively expensive.

Table 1. Semiconductor material properties

Material	Electron Mobility (cm^2/v-sec)	Energy Band Gap (eV)
Silicon (Si)	1900	1.12
Gallium Arsenide (GaAs)	8800	1.43
Indium Antimonide (InSb)	78000	0.17
Indium Arsenide (InAs)	33000	0.35

The vast majority of integrated circuits (ICs) manufactured today are fabricated on large 200 mm or 300 mm diameter Silicon wafers and assembled into plastic packages. An ever increasing share is manufactured by offshore wafer foundries and assembly subcontractors that take advantage of readily available labor and relatively low wages in developing countries.

Circuit Design

The final design of the Hall-effect sensor resulted in dice laid out on a 262 μm pitch with 74 μm wide saw/scribe streets. It had four (4) terminals or bonding pads with 55 x 55 μm openings in the silicon nitride passivation spaced 50 μm apart. The bond pad metallization is 1 μm thick Gold over a thin adhesion/barrier metal layer. A domestic wafer foundry fabricated the design on 100 mm diameter GaAs wafers. This design resulted in about 90,000 gross dice per wafer.

The device is fabricated as two crossed resistors in high-mobility III-V compound material using standard IC processing methods. Although the resistance values directly correlate to the Hall sensitivity (ie. coefficient), the resistance itself is the most sensitive measure of device stability.

Figure 2. Hall-effect Sensor Die Layout

The device typically dissipates less than 0.1 mW power. Each orthogonal resistor arm is on the order of 1,000 ohms in resistance value. A Hall bias current operating range of 10 – 100 μA is all that is required for adequate signal output.

Package Requirements

The package objective of the project was to select the most suitable, low cost, dual-sourced, 4-pin or 5-pin, production, surface-mount, RoHS-compliant, IC package constructed of non-magnetic materials for a single, GaAs die measuring approximately 250 x 250 x 100 μm.

Some consideration was given to the possibility of a further die size reduction by shrinking the bond pad sizes and narrowing the saw/scribe streets to reduce the die cost. Also, there was some discussion of adding a small amplifier chip and creating a two-chip package which might require up to six (6) terminals.

Assembly Requirements

The assembly objective of the project was to select and qualify one or two, low-cost, high-volume sources of the selected package using similar, if not the same, bills of material (BOM) and capable of supporting an annual demand of 120 Million units.

IMPORTANT CONSIDERATIONS

As with most IC packaging projects, the goal was to specify a "smaller, cheaper, faster, and better" package. In other words, optimize the physical size (ie. footprint) and price of the total solution without sacrificing device performance (ie. magnetic sensitivity) and reliability. The package selection had to take into consideration the price of materials, die prep, assembly, final test, packing, and downstream circuit board assembly. For example, package lead pitch often determines the minimum trace/space width required to route the design of a circuit board which in turn dictates the price of the board. Board assemblers usually specify tape-and-reel packing of the components for their high-speed pick and place equipment.

It is important to avoid the use of magnetic packaging materials, because any magnetic material, such as iron or nickel, in the immediate vicinity of the Hall sensor will distort an incident magnetic field to either increase or decrease the input to the Hall sensor, depending on the actual geometry. Thus, for precise Hall measurement applications, a completely non-magnetic package is desirable.

The selection of a high-volume package and an assembly source are intimately tied together. Selecting the assembly supplier restricts one's choice of packages to those that the supplier has tooled. Conversely, specifying a specific package restricts the choice of assembly suppliers.

Package Selection

The need for low-cost, non-magnetic package materials immediately ruled out ceramic packages, because they are relatively expensive and have magnetic Kovar leads. Thus, the choice centered quickly on plastic-encapsulated packages without ferrous Alloy 42 leads and/or Nickel-plated lead finish.

Many older plastic packages, such as PDIP and SOIC, are still assembled today with ferrous Alloy 42 leadframes coupled with older formulations of CTE-matched adhesives and encapsulants – many of which don't meet today's RoHS requirements. To meet the industry need for packages with higher power dissipation, assemblers have been converting from Alloy 42 to Copper leadframes. In recent years, they've also begun to offer RoHS-compliant formulations of Copper-compatible adhesives and encapsulants.

The need for a low leadcount, small footprint, surface-mount, package with non-magnetic materials eliminated other plastic package choices, such as PLCC, PQFP, PBGA, and the smaller Chip Scale Package (CSP). None of them offered a small enough footprint. The smallest CSPs have no less than eight terminals on 1.0 mm pitch and occupy 4 x 4 mm of board space.

The package choice was quickly narrowed to plastic Small Outline Transistor (SOT) and Quad Fine-pitch No-lead (QFN) packages. Some SOT-style packages, such as SOT23, SOT143, and SC70 (Figure 3), are now available with Copper leadframes from multiple assembly suppliers. The small SOTs have gull-wing shaped leads on two sides of the package, much like the larger TSOP and SOIC packages. The QFN-style packages are becoming quite popular, because they are lower profile and cheaper than equivalent CSP leadcounts.

Nearly all of the assembly suppliers have tooled, or are tooling, various QFN package families. They are assembled in small arrays using etched, Cu-alloy leadframes and then sawn apart after encapsulation. A variant of the QFN package is the Dual Fine-pitch No-lead (DFN) package. Instead of terminals on four sides, the DFN has terminals arranged on just two sides of the package. Although DFNs as small as 1x1 mm are available, the larger 2x2 mm DFN package with 4 to 6 terminals was considered, because its lead pitch of 0.625 mm would not require more expensive, fine line/space circuit boards and assembly.

A major concern with the popular new DFN package was that its terminals are located under the component body and difficult to inspect after soldering to printed wiring boards (PWBs). Unlike CSPs and BGAs with well-characterized solder sphere terminals and solder joints, the leadless chip carrier style DFNs are relatively new to PWB assembly lines and represent a greater risk of solder joint rework and reliability. By comparison, the SOT gull-wing leads are compliant and their solder joints can be easily inspected. On the other hand, a major concern with the SOT package was that its fragile leads are susceptible to handling damage. However, the industry has successfully dealt with this lead coplanarity issue for many years.

Figure 3. 5-Lead SC70 Package

It turns out that a 2x2 mm DFN package can be designed such that its terminals match the location of those of a small SOT, namely the SC70 package. This common footprint may allow customers to substitute one package for the other should they both be offered and a future shortage in supply arise. Dual sourcing is an important consideration and initial indications were that both of these packages were available from multiple assembly sources.

Assembler Selection

Past experience working for and with Outsourced Semiconductor Assembly and Test (OSAT) suppliers, told us that the top-tier, larger suppliers, such as ASE, Amkor, SPIL, and STATS/ChipPac would have little interest in engaging with a small, fabless, semiconductor customer – even if production volume was projected to be 10 million units per month. In fact, many of the large assemblers have discontinued production of the older generation and lower leadcount package families in favor of higher margin packages.

Therefore, we turned our attention to smaller, second-tier, OSAT suppliers, such as ASAT, Carsem, Hana, Unisem, UTAC, et al. Smaller suppliers are more willing to accept young, but growing, customers with the expectation that their revenue will grow as their new customer's product demand grows. Nevertheless, it takes a lot of effort to qualify with new customers and a few of the second-tier OSAT suppliers were still reluctant to engage with us, so the list of supplier candidates was trimmed further.

There are a lot of considerations that go into selecting an OSAT supplier, such as facility location(s), quality procedures, process controls, assembly yields, delivery time, engineering support, experience, breadth of package offering, installed assembly and test capacity, technology roadmaps, and customer support. These factors would all be considered, but our initial screening of suppliers focused on unit price.

A comparison of "budgetary" prices from three of our four OSAT contenders showed the SC70 was 30% to 45% cheaper than the 2x2 mm DFN. It's apparent that assemblers have manufactured a larger cumulative volume of SC70s than DFNs and have progressed further down the manufacturing cost learning curve. Also, there has been sufficient demand for certain SC70 leadframe designs to justify investment in expensive, open-tooled, progressive die sets to manufacture stamped leadframes which offer a lower unit cost than etched leadframes. DFN and QFN packages, on the other hand, rely on double-sided etching to create unique features in the leadframe to anchor the terminals reliably in the package body. Such features can't be achieved with low-cost, stamped leadframes.

Based on "budgetary" assembly pricing estimates, as well as the other factors described, it was clear we should proceed to qualify the SC70 package which meets the standard package outline dimensions of JEDEC MO-203. Three of the four OSAT finalists met our initial assembly cost target with the SC70 package.

We decided to proceed with qualifying the mid-priced OSAT supplier of 5-lead SC70 using an open-tooled, stamped, Copper leadframe. Although another OSAT supplier offered a lower assembly price, we were uncomfortable with their size and experience, as well as the level of their U.S. customer support. In time, they will be qualified as our second source.

Assembly Materials and Process
Lowest package and assembly pricing is achieved by specifying the OSAT's "standard" bill of materials (BOM) and "standard" assembly process flow. Further savings can be achieved by relaxing non-critical inspection criteria and eliminating unnecessary process steps. Although there were close similarities in the "standard" process flows, a comparison of the "standard" BOMs from the two OSAT finalists showed that they were not identical.

Table 3. SC70 bill of materials

MATERIAL	OSAT – A	OSAT – B
Leadframe	Copper Alloy 194	Copper Alloy 194
Die Attach	84-1 LMISR4	8390A
Bonding Wire	Gold, 0.8 mil dia.	Gold, 1.0 mil dia.
Encapsulant	EME-G600	EME-G600
Lead Finish	Matte Tin	Matte Tin

The major difference in the "standard" material sets was the conductive die attach adhesive from Henkel. One was a conventional cured adhesive and the other was a snap cure adhesive for higher throughput. Both were suitable for small die sizes. There was a minor difference in the diameter of the Gold wire, but both were suitable for our short 25 – 30 mil wire lengths. The "standard" lead finish for the OSAT suppliers was a pre-plated NiPdAu finish. An optional, lead finish of matte Tin was specified to avoid the slightly magnetic properties of Nickel. All of the direct materials chosen, including the Sumitomo molding compound, were RoHS-compliant.

Gallium Arsenide
Past experience with assembling GaAs Infrared (IR) Light Emitting Diode (LED) dice cautioned us to expect some challenges with regard to handling small GaAs dice. We knew that GaAs dice were more brittle than Silicon dice and prone to chipping and handling damage. We also knew from our LED experience that a small die size, on the order of 250 x 250 μm, was pushing the minimum die size limit for most automatic die bonders to pick and accurately place the die. Historically, GaAs LED wafers have been singulated by a diamond scribe-and-break process to minimize the width of the dicing street and achieve more gross dice per wafer.

Die Preparation
Both of the OSAT finalists were experienced with assembling GaAs dice as small as 250 x 250 μm, however neither of them had installed capacity to thin GaAs wafers and dice them to size. Instead, they insisted that their customers consign pre-thinned and sawn wafers on metal rings and dicing tape that could be expanded and placed directly on automated die bonders. There is much less demand for backgrinding and sawing or scribing GaAs wafers than there is for Silicon wafers. OSAT suppliers are concerned about contaminating the Silicon wafers from the majority of their customers if they were to thin and dice GaAs wafers using the same equipment set. Lacking sufficient GaAs assembly demand to justify the investment, OSAT supplers have been reluctant to invest in filtration and reclaim systems to prevent waste containing Arsenic from entering their local water treatment facilities. Thus, we were forced to expand the project and locate a supplier who could thin/dice our GaAs wafers prior to assembly.

Fortunately, a domestic OSAT supplier that offered GaAs wafer thinning and sawing was located in California. Although the supplier provided high quality and good service, the die prep cost per wafer without 2nd Optical inspection was double that of fabricating the entire wafer! Needless to say, efforts are underway to source a lower cost die prep process.

RELIABILITY QUALIFICATION
The simplicity of the Hall-effect sensor design and wafer fabrication process suggested that a preliminary qualification of the sensor die in a hermetic package was unnecessary. Therefore, a combined qualification of the sensor die, the production package, and the OSAT supplier was conducted to save considerable time and expense.

Qualification Build

A single wafer containing four versions of the Hall-effect sensor was thinned to 100 μm by the wafer foundry and then sawn by a domestic OSAT supplier with a 25 μm wide dicing blade to singulate the 100 mm diameter GaAs wafer into approximately 90,000 dice. The yield exceeded 99% when inspected for scratches, cracks and chip outs. The sawn wafer was left on the dicing tape and a metal film frame to be shipped to the offshore OSAT supplier.

For a minimum lot charge, our preferred offshore OSAT supplier assembled a relatively large quantity of the desired Hall-effect sensor dice in SC70 packages for reliability testing. The OSAT supplier built 9,964 units with a 99.67% assembly yield. All of the yield loss was taken at 3rd Optical inspection where 1 unit failed for a damaged bond wire and 32 units failed for one or more bond wires not sticking to a die bonding pad. This slight yield loss resulted from the optimization of wire bond parameters to prevent ball bond lifts during the wire bonder set-up.

Process monitor data was gathered from samples at each assembly operation in the process flow and is summarized in Table 4. All samples passed the monitor criteria.

Table 4. Summary of assembly process monitor data

Process Monitor	S/S	Average	Result
Bond Line Thickness (mils)	30	0.56	Pass
Die Shear Strength (kg)	30	0.53	Pass
Wire Pull Strength (grams)	30	7.04	Pass
Loop Height (mils)	10	4.44	Pass
Ball Shear Strength (grams)	30	30.79	Pass
Wire Sweep (%)	30	1.10	Pass
Plating Thickness –Top (μ-In)	30	497.82	Pass
Plating Thick – Bottom (μ-In)	30	491.95	Pass
Lead Coplanarity (mils)	30	0.00	Pass
Delamination (%)	22	0.00	Pass

Manual screening for electrical functional test of a small representative sample of the assembled SC70 units showed >98.5% final test yield.

Reliability Plan

Random samples were selected for reliability testing as a combined qualification of the Hall-effect sensor chip, the SC70 package materials, and the offshore OSAT supplier's assembly processes and facility. The testing was contracted to a respected, independent, reliability test lab in California. Test boards were designed and manufactured for HTOL and Biased HAST, and testing began in late June, 2009.

Table 5. Summary of reliability test plan

STRESS TEST	TEST CONDITIONS
Moisture Sensitivity L1 JESD22-A113 S/S = 25	24 hrs. Bake @ 125°C 168 hrs @ 85C / 85% RH 3 cycles IR reflow @ 260°C
Temp Cycle – Cond. B JESD22-A104 S/S = 25	Precondition @ MSL-1 1K cycles @ -55°C / 125°C No bias
High Temp Storage JESD22-A103 S/S = 25	Precondition @ MSL-1 1,000 hrs. @ 150°C No bias
Low Temp Op Life JESD22-A108 S/S = 25	Precondition @ MSL-1 1,000 hrs. @ -20°C Biased at 200μA
Biased HAST JESD22-A110B S/S = 25	96 hrs. @ 130°C, 85% RH, 33.5 psi, 5V from backside to topside contacts
High Temp Op Life JESD22-A108 S/S = 80	Precondition @ MSL-1 1,000 hrs. @ 125°C Biased at 200μA
Solderability MIL-STD-883 S/S = 15	8 hrs. steam age @ 93°C 1 hr. bake @ 100°C Flux w/ Kester 2335 SAC305 solder @ 245°C Clean & Inspect Leads

The HTOL, LTOL, HTS, and T/C reliability test samples, were preconditioned to Moisture Sensitivity Level 1 (MSL-1) conditions prior to environmental stress. The Solderability samples were both visually inspected and electrically tested.

Electrical Testing and Failure Criteria

A Hall-effect sensor device is fundamentally two crossed conductors (ie. resistors), each having a pair of contacts. In the SC70 package the crossed resistors are located diagonally between pins 1 and 4 and between pins 3 and 5. By other characterization work, it is known that the Hall magnetic sensitivity correlates directly with the device characteristic resistance. Thus, simple resistance was used as the metric for reliability stress measurement.

At every read point, both of the resistors on each device were measured for electrical resistance using a 2-point (non-Kelvin) test method by forcing a current of 10 μA and measuring the voltage to four (4) significant figures. This kept it simple while avoiding self-heating and measurement contact resistance effects. The sample resistance values and distributions were recorded at each read point for each stress test.

A reliability test failure was defined as follows:
1. Any obvious individual electrical open (>10K ohm) or short (<100 ohm)
2. Any change in the resistance sample mean of more than two standard deviations after a stress test
3. Any increase in standard deviation (ie. widening of a sample distribution) of 1.5x or more after a stress test

The temperature coefficient of resistance for these devices was on the order of 0.7% per °C. Since room temperature over the course of a couple of months of reliability testing was not controlled to better than about +/- 3°C, changes in the sample means that could be attributed to ambient temperature differences during testing sessions were not considered as an indication of "drift" due to reliability stressing.

Reliability Test Results
All tests passed with no failures at each read point. Electrical testing found no opens or shorts, and there was no significant shift or "drift" in the sample mean or the standard deviation of the test populations.

This is one of the first significant reliability studies on a die/package combination of this family of Hall devices, and it's quite likely that these devices are capable of passing even greater levels of environmental stress.

CONCLUSIONS / FINDINGS
The ultimate project goal of identifying and qualifying a suitable package and production OSAT supplier for a small, GaAs Hall sensor was successfully accomplished. The 5-lead SC70 satisfied all of the package requirements and met the desired, high volume, assembly price target.

It was accomplished by sourcing a popular, JEDEC-standard, plastic package constructed with a Copper alloy leadframe. Lower pricing was attained by specifying a "stamped" leadframe in lieu of an "etched" leadframe. The more popular (ie. higher volume) plastic packages tend to have open-tooled stamped leadframes available.

Although leadframes of ferrous Alloy 42 were avoided in favor of copper Alloy 194, we were surprised to discover the lightweight, 5.5 mg, tin-plated, SC70 packages were weakly attracted to a strong magnet. A quick investigation found that we had overlooked the fact that copper Alloy 194 nominally contains 2.4% Iron. For some sensitive applications, it may be necessary to specify an Iron- and/or Nickel-free copper alloy.

We found that many of the larger OSAT suppliers were reluctant to spend their limited engineering resources to engage with small clients like ours. This forced us to focus on 2nd tier OSAT suppliers. To our dismay, many of these smaller suppliers had fewer and less knowledgeable U.S. sales personnel. For example, we often had to wait for our technical requests to be relayed through the sales office to the factory and back again for responses. Although we explained our packaging goals to the sales personnel, it was disappointing that only a couple were able to recommend to us their lowest cost packaging solution. Essentially, we were forced to request massive amounts of information, select candidate packages, and request formal quotations. We can only imagine the formidable task it would have been for a customer less experienced and skilled in IC packaging.

The most surprising discovery was that nearly all of the OSAT suppliers were neither able nor willing to thin and dice GaAs wafers. Most preferred that we prep the wafers ourselves and provide them ready for die attach on dicing tape and metal film frames. Those that could backgrind and saw the GaAs wafers quoted prices exceeding the entire wafer fabrication price. Apparently, most IC companies offering GaAs devices have this die prep performed in their wafer fabs or at their foundries.

Finally, given the simple design and five-mask wafer fab process of this Hall sensor and considering the high yields (>98.5%) obtainable at wafer sort, die prep, and assembly, it would be cost-effective to omit the 100% wafer probe testing and simply assemble unsorted wafers – taking any electrical yield losses at final test just prior to tape-and-reel packing.

FUTURE WORK
Most of the future work is aimed at further cost reduction efforts. One initiative seeks to reduce the cost of the Hall-effect sensor die, not by designing it smaller, but by reducing the dicing street width to attain more gross dice per wafer. This will require a switch from wafer sawing to a diamond scribe-and-break process or perhaps laser dicing. Our current wafer foundry is capable of thinning 100 mm diameter wafers to 75 µm and singulating them by means of the traditional scribe-and-break process. The price of Gold has risen to nearly $1,200 per ounce, and if it continues to rise we must consider switching from Gold to Copper bonding wire.

A potential application for this new Hall sensor is to use it as a non-contact proximity sensor to detect keystrokes on a keyboard. To meet aggressive keyboard cost targets, additional functionality could be integrated into the GaAs sensor die itself thereby enlarging it slightly and reducing the keyboard component count. An alternative might be to assemble a Silicon amplifier die together with the GaAs Hall sensor die in a multi-chip package.

An automotive application for this new Hall sensor would require it to withstand extreme environmental conditions, especially high temperatures. Therefore, it is planned to assess the reliability of the packaged sensor at temperatures in the 200°C – 250°C range. Of course, this may require that we specify or develop a new package for the sensor.

ACKNOWLEDGEMENTS
The authors wish to acknowledge the efforts and assistance provided by Mr. B.K. Ng of Hana Microelectronics Inc. and Ms. Inna Zaliznyak of Silicon Turnkey Systems.

CAPILLARY UNDERFILL PHYSICAL LIMITATIONS FOR FUTURE PACKAGES

Horatio Quinones and Tom Ratledge
ASYMTEK
Carlsbad, CA, USA

ABSTRACT

The capillary underfill (CUF) although a well established manufacturing assembly process, is being challenged as die thickness diminishes, the interconnection (bumps) get smaller and their number increases. Denser populated packages demand very tight tolerances for keep out zones (KOZ); the total package thickness challenges the process throughput since die contamination from underfill fluid is not allowed and multiple fluid dispense passes may be needed. All this challenges translate in lower capillary surface energies, increase in fluid flow drag, smaller particle size fluid that often results in increase in viscosity and therefore slow flow-out-times. The present work addresses these issues. A series of mathematical models based on surface energy evolution for CUF accounting for these new geometries and processes is proposed. In particular the problem of component proximity is and the gap topology issues are studied. Experimental data for CUF in the presence of these future assembly demands is shown. Although there are practical physical limitations for the CUF as experienced today if one were to implement it for future packages, new hybrid CUF methods that overcome such shortcoming are recommended.

INTRODUCTION

The capillary action occurring in small ducts has been studied by several disciplines of science. Molecular forces of particles in a fluidic matrix are just an instance. In the electronic packaging the fluid dispensing for various applications including, potting, filling, component underfilling is of common knowledge. The c capillary kinetics plays an important role in several of these applications. Contrary to the traditional injection molding, where the fluid is mobilized by an induced relatively high pressure differential at the surface of the fluid wave front, the capillary action is a result of adhesion forces overcoming the cohesive forces of the moving fluid. A Variational approach to determine the fluid-air-solid surface shape of the moving front will guide us in determining various geometric boundary conditions including gaps sizes, and steps occurring in the corresponding capillary ducts.

THEORY AND BACKGROUND

The analytical approach to solve the problem of surfaces is that of Mapertuis principle. These solutions, coupled with mechanical adhesive and cohesive forces hat include Vander-Wall forces and London forces due to oscillation of electron clouds in molecules that are in close proximity, can give us a good description of the kinetics that takes place for slow fluid flow under capillary action. Given a definite integral with boundary conditions, its stationary value can be found by minimization of a functional using Variational calculus tool [1].

$$\delta F\left(y, y', x\right) = F\left(y + \varepsilon\phi, y' + \varepsilon\phi', x\right) - F\left(y, y', x\right) = \varepsilon\left(\frac{\partial F}{\partial y}\phi + \frac{\partial F}{\partial y'}\phi'\right) +$$

The Variational of the definite integral can be computed as follows:

$$\delta\int_a^b F\left(y, y', x\right)dx = \int_a^b \delta F\left(y, y', x\right)dx = \varepsilon\int_a^b\left(\frac{\partial F}{\partial y}\phi + \frac{\partial F}{\partial y'}\phi'\right)dx$$

Dividing by ε ad integrating by parts the second term of the r.h.s. of above equations we obtain

$$\int_a^b \frac{\partial F}{\partial y'}\phi' dx = \left[\frac{\partial F}{\partial y'}\phi'\right]_a^b - \int_a^b \frac{d}{dx}\left(\frac{\partial F}{\partial y'}\right)dx$$

We define I as the definite integral and since the $\phi(x)$ vanishes at the limits of integration (boundary conditions are satisfied exactly, $x=a$ and $x=b$)

$$\frac{\delta I}{\varepsilon} = \int_a^b\left(\frac{\partial F}{\partial y} - \frac{d}{dx}\left(\frac{\partial F}{\partial y'}\right)\right)dx$$

We now define the function $\xi(x)$ as

$$\xi(x) = \frac{\partial F}{\partial y} - \frac{d}{dx}\left(\frac{\partial F}{\partial y'}\right)$$

Combining above expression we can then write the stationary value of the corresponding definite integral as

$$\frac{\delta I}{\varepsilon} = \int_a^b \xi(x)\phi(x)dx = 0$$

It can be easily seen that the above expression would be satisfied for any arbitrary $\phi(x)$ if and only if $\xi(x)$ vanishes everywhere in the space [a,b]. We can, on the other hand make $\phi(x)$ vanish everywhere except in 'a small neighborhood around a point say, $x=\zeta$. Within this "small interval," $\xi(x)$ is "practically" constant and can therefore,

90

be taken out of the sum (integral operation) as a simple multiplier factor

$$\frac{u}{t} = i(t) \int_{i-\mu}^{i+\mu} \varnothing(x)dx$$

As the radius μ tends approaches zero, our "error" also tend to vanish. The first' variation or linear term of the expression must vanish, hence we can write the expression known as the Euler-Lagrange Equation

$$f(x) = 0 = \frac{\partial F}{\partial P} - \frac{d}{dx}\left(\frac{\partial F}{\partial y'}\right)$$

About two centuries ago Poisson wrote the equation for the free energy of a solid elastic membrane

$$F = \frac{k_c}{2} \int_M (2H)\sqrt[4]{2} \cdot dS$$

Where H and dS are the mean curvature and infinitesimal area element of the surface respectively, and k_c is the bending elastic modulus. The energy Euler-Lagrange equation corresponding to these functional can be written as

$$\nabla^2 H + 2H(H^2 - K) = 0$$

And the solution for such functional satisfying the minimization of surface energies is the critical curve known as the Willmore surface F, written as

$$W(f) = \int_M H^2 dS = C + \int_M (H^2 - K)dS$$

NUMERICAL AND ANALYTICAL RESULTS

We carry out some solution for the formulation presented above and using boundary conditions corresponding to the flow of a fluid in a capillary action, between two parallel plates, and in the presence of various surface topographies. The geometry to be treated consist of parallel plates

Figure1. Drawing of the geometry of two parallel plates separated by a gap in the presence of a fluid flowing by capillary action between them.

The capillary motion results, in the presence of a step, i.e., a sudden increase in the gap between the parallel surfaces, (see figure2) resulted from the numerical analysis is depicted in the sequence of pictures shown in figure 3.

Figure 2. A sudden step on the organic substrate.

There one can observe that the fluid front has a tendency to behave in a way that preserves state of symmetry about a virtual horizontal plane parallel to the surfaces, thereby avoiding the creation of voids in the fluid path. The assumption here of course is that all surfaces in contact with the fluid are wettable to it, and that the adhesion is about the same for all of them. This flow behavior is very different from that of the case where induced pressure differential (as it is in the case of injection molding) for instance, the propensity to create voids is rather high in similar geometries. For the case of holes, the flow around them is the primary cause of void formation and it is governed by the flow velocity field around the hole, tangent to the circumference of the hole, and the fluid velocity as the fluid goes to a larger gap.

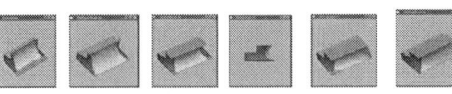

Figure 3. Capillary fluid flow in the presence of sudden step where the gap increases.

Other geometries analyzed can be depicted in figure 4, including grooves crossing and different grooves directions.

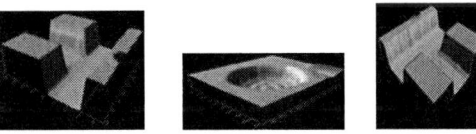

Figure 4. Various laser profile-meter geometries treated as spatial boundary conditions: grooves, hole and grooves intersections.

The case for underfilling components where the distance between them is comparable to the gap to be capillary underfill poses a whole new set of complications [3]. Aside from the fact that perhaps only jetting technologies can be of use (see figure 5) given that the needle dispensing requirements do not physically permits its use, there are other limitations imposed by the physics of the underfill and capillary equilibrium principles.

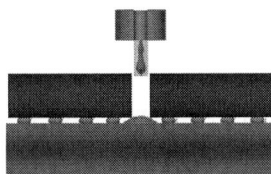

Figure 5. Jetting fluid from a distance above the components allowing for small distance between components.

Solutions to above problem indicate that when the distance between the components is the same as the gap or less, then capillary underfilling does not take place, this can be depicted in figure 6.

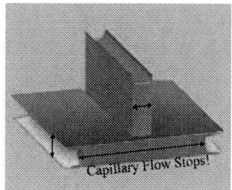

Figure 6. No capillary flow takes place for this geometry (gap same as distance of separation).

For cases where the distance separating the components is larger than the gap to be underfilled, (see figure 7), the solution indicates that capillary underfill occurs to completion, i.e., the fluid reaches the sides of the smaller of the two plates (in this case the glass plate).

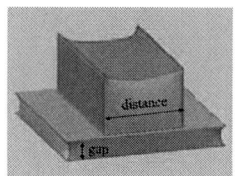

Figure 7. Drawing of two components side-by side separated by distance larger than the gap to be underfilled.

As depicted in figure 8, the fluid flows until it reaches equilibrium, a fillet around the plate including that area in between the components is formed. In figure 7 a sequence of event throughout time is shown where the fluid is moving under the influence of capillary effects.

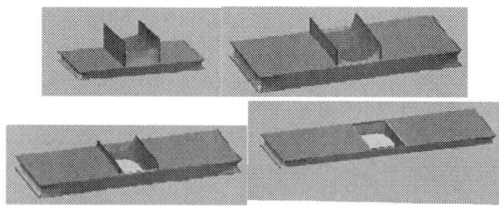

Figure 8. Sequence showing time slices of the fluid moving in a capillary manner in between the gap that is smaller than the distance separating the components.

EXPERIMENTAL RESULTS
Glass plates were mounted on organic boards with gaps varying from 75 to 300 μm. The surface of the board had different topographies including, grooves of different depth and widths as well as holes drilled to different diameters and depths from 250 μm to 1.5mm. Figure 9 depicts some of the samples used for the experiment.

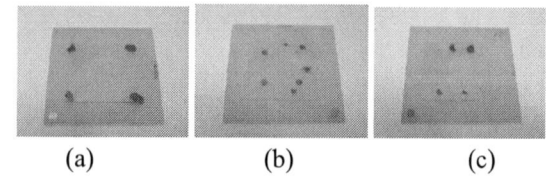

(a)   (b)   (c)

Figure 9. Samples used for capillary underfilling: (a) Drilled holes on organic substrate; (b) grove aligned parallel to the fluid flow direction; (c) Groove normal to the capillary fluid flow direction.

Underfill material was jetted along a side of the glass plate using Asymtek DJ9K jet[4]. Figure 10 depicts the capillary flow of the underfill in the presence of a hole with diameter about three times larger than the gap and about 1mm deep. It is noticed that a void is present, this resulted from thee fact that the fluid flows along the peripheral of the hole following a near perfect tangential direction to its circumference.

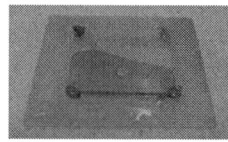

Figure 10. Capillary flow in the presence of a large hole on the organic substrate, void formation resulting from fluid flow tangential to the hole.

In the case of grooves present on the organic substrate the fluid flow in a capillary fashion without allowing void formation. When the groove is parallel to the direction of the flow void could be formed if such groove is very narrow (compared to the gap) and in particular, if its depth varies. However, for the case of grooves normal to the fluid flow, voids are not formed independent of the groove geometry. For a very deep groove, the flow will simply cease to continue, this can be understood from the point of view of pure equilibrium mechanisms, i.e., it can be simply seen as a boundary condition similar to the starting edge of the glass plate. In figure 11 depicts the case where both grooves, normal and parallel to the fluid flow direction were present.

Figure 11. Grooves are present in the organic substrate in both directions, normal and parallel to the fluid flow direction; no voids are formed during the capillary underfill.

CONCLUSIONS
A comprehensive analysis of capillary fluid flow has been presented and validated by actual data. The coupling of molecular forces adhesion and cohesive in nature and the minimization of surfaces, as in the case of elastic membranes that yield Willmore critical surfaces in the

differential geometry scheme gives an adequate way to model capillary fluid flow. These solutions were obtain in a way by perturbation theory where Variational of the functional derived from Poisson integral formulation mimic very closely the observation of capillary fluid flow under various boundary conditions that included several different geometries. For dispensing in presence of tight spacing and high density packages, the Jetting technology lends itself in a very transparent and practical manner. Above results can be used as guidelines in situations where uneven surfaces and arrays are present and capillary flow is used. Design rules for packaging design need to include the capillary physics present during underfilling and fluid dispensing in general.

ACKNOWLEDGMENT

The authors would like to thank Mr. A. Babiarz for his insights and comments on this work.

REFERENCES

[1] Martin Fuchs and Gregory Seregin, Variational Methods for problems from Plasticity Theory and for Generalized Newtonian Fluids.

[2] Christoph Bohle, G. Paul Peters, "Bryant Surfaces with smooth Ends", preprint, math.DG/0411480.

[3] Babiarz, Alec, "Proceedings of the Pan Pacific Conference, 2007."

[4] Horatio Quinones and Alec Babiarz, "Fluid Jetting for Next Generation Packages "Pac Tech, Berlin, April 2002

HIGH BRIGHTNESS LED ASSEMBLY USING
DPC SUBSTRATE AND *SUPER*MCPCB

Nick Wang[1], Allen Hsu[1], Andy Lim[1], Jerry Tan[1], Charles Lin[1],
Heinz Ru[2], Thompson Jiang[2], and David Liao[2]
[1] Bridge Semiconductor Corporation
[2] Tong Hsing Electronic Industries Ltd.
Taiwan

ABSTRACT

This paper is to evaluate the thermal performances of a Direct Plated Copper (DPC) substrates attached on a high performance MCPCB thermal board for high brightness LED (HBLED) applications.

The advantages of DPC substrate compared to traditional Direct Bond Copper (DBC) substrate include: more robust bonding between the substrate and copper metallization, ability to offer substrates with a wide range of copper thicknesses including thin copper for fine patterning, excellent electrical and thermal contact between copper plated vias and surface metallization and lower temperature processing which means much reduced residual stress in the copper metallization layers and copper/substrate interface.

The high performance MCPCB in this study has a unique structure, in which an integrated metal post is located directly underneath the LED device providing an effective thermal path for the assembled device. In addition, as this metal post is an integral part of the heat spreader, resin/metal delamination concern associated with most conventional MCPCB can be avoided and therefore provides the highest possible long-term reliability, especially when lead free solders are used as device attachment materials.

Using this DPC substrate and high performance MCPCB combined structure, it was established that the performances of LED devices–can be greatly improved. Investigations were conducted to establish the compatibility of the high performance Direct Plated Copper (DPC) ceramics substrate (Al_2O_3 and AlN) assembled on this improved thermal board structure and the findings will be presented in this paper.

INTRODUCTION

For double layered circuits (top and bottom layer metallization connected by vias), DPC has some very significant advantages over co-fired technologies such as HTCC and co-fired AlN for high volume assembly applications such as HBLEDs. The main advantage of DPC is large format size (6" x 6") and high location accuracy across the entire tile. Both of these attributes are critical for high volume automated assembly. DPC achieves this high location accuracy because the sample never shrinks during fabrication. Feature location accuracy is determined by laser drill hole location accuracy and metal pattern feature location accuracy. Since metallization layers are patterned using photo-

lithography, feature location is very tightly controlled. For co-fired technologies shrinkage on the order of 18% occurs during fabrication. Because of variations in this shrinkage across a substrate, typical location tolerances are +/- 0.5%, so two features separated by a distance L have a location tolerance of L(0.005). For larger format tiles, this variation in location leads to unacceptably high location errors for automated assembly operations.

In high power/high brightness applications, as the amount of heat that LED generated can be significant, metal-cored thermal boards have been developed to address this need. Conventionally, a thermal board consists of a layered structure; which consists of a routing layer, a thermal conductive dielectric layer and a thick metal sheet which acts as a heat spreader to dissipate the heat generated from the LED. Since the die pad or component pad is formed on the routing layer and separated from the heat spreader by the dielectric layer, the thermal dissipation path is obstructed since the thermal conductivity of the dielectrics is significantly lower as compared to that of the metal. Based on existing commercially available data, the thermal resistance of a board can range from 2 to 8 K/W depending on the selected dielectric material properties and thickness.

To address the above limitations, both Bridge Semiconductor Corporation and Tong Hsing Industries Ltd have developed a thermal management solution which combines DPC substrate and high performance MCPCB for new emerging applications in LED lightings, automotive headlights and even miniature projectors.

DIRECT PLATED COPPER CERAMICS SUBSTRATE

Tong Hsing has developed a patented [1] method for removing voids in a ceramic substrate, and more particularly a method including a chemical copper plating process after a sputtering titanium/copper step. The chemical copper plating step is able to successfully electrically connect both sides of the ceramic substrate so that when the copper pattern is formed on both sides of the ceramic substrate, communication is established. [2], [3]

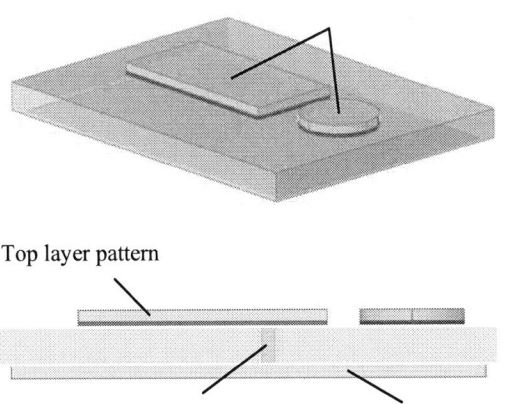

Top layer pattern

Metal plated via   Back layer pattern

Figure 1 Schematic of DPC ceramic substrate

· Unsurpassed thermal performance
· Low electrical resistance conductor lines
· Stable up to temperature > 340°C
· Accurate feature location, compatible with automated, large format assembly
· Fine line resolution allowing high density of devices and circuitry
· Proven reliability
· Lowest cost, highest performance ceramic solution

*Super*MCPCB

The high performance MCPCB, entitled *Super*MCPCB, aims to address the thermal and reliability issues associated with conventional MCPCB. Specifically, the unique features of *Super*MCPCB include an integral metal post located directly underneath the assembled devices to provide a direct and effective thermal dissipation pathway (figure 2).

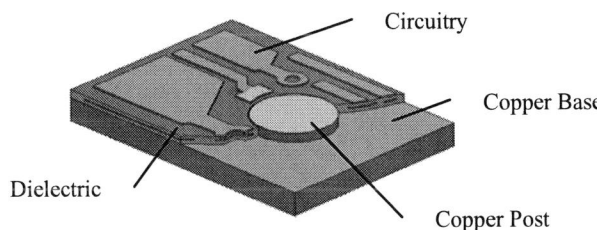

Circuitry

Copper Base

Dielectric

Copper Post

Figure 2 Schematic drawing of *Super*MCPCB

This *Super*MCPCB design offers many benefits for LED applications including:
• Allow higher forward current due to lower transient junction temperature
• Stable light output
• Lower optical loss
• Improved long-term lumen maintenance
• Alow higher design flexibility due to multi-layer routing
• Can be in either COB or SMT format
• Mechanically robust structure

In addition, it enjoys the highest possible long-term reliability at high temperature operations as there is no delamination concern underneath the LED devices since the copper post is an integral part of the heat spreader. A schematic structure comparison with conventional MCPCB is shown in figure 3.

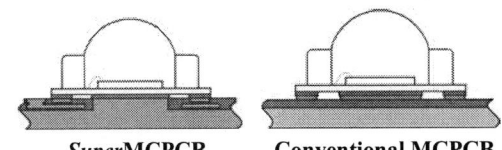

SuperMCPCB   **Conventional MCPCB**

Figure 3 Schematic Structure Comparisons

In general, for conventional MCPCB, the thermal dissipation is limited as the LED devices are not directly in contact with the heat spreader and the dielectric material and circuitry layers interferes with the thermal pathway.

*Super*MCPCB THERMAL PERFORMANCE

To validate that the improved thermal board performs better than conventional MCPCB, a measurement (figure 4) of the continuous temperature of a lighted LED was taken with results as shown in figure 5.

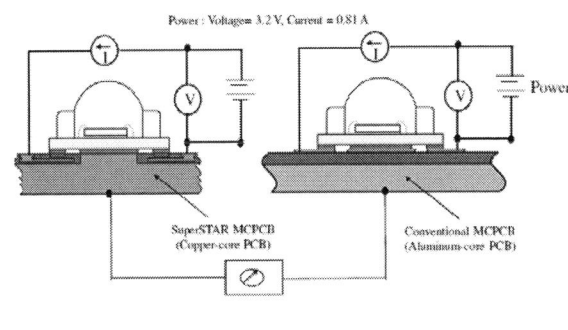

$\Theta jb=(Tj-Tb)/W$ Tj estimate by I and V data

Figure 4 Temperature measurement set up

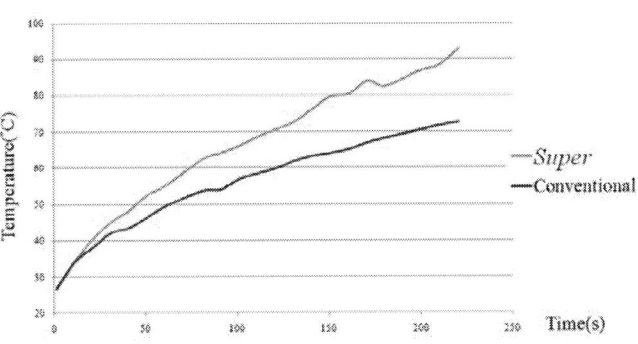

Figure 5 The result shows that the temperature of *Super*MCPCB rises quickly and exhibits much lower thermal resistance compared to that of conventional MCPCB.

COMPATIBILITY OF DPC CERAMICS SUBSTRATES WITH IMPROVED THERMAL BOARDS

To establish the compatibility of the *Super*MCPCB with high performance DPC ceramics substrates, total of 135 substrates with 15 and 20 mil thickness were manually soldered on the thermal board before being sent for reliability tests including thermal shock, thermal cycle and high temperature storage.

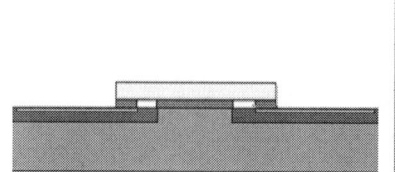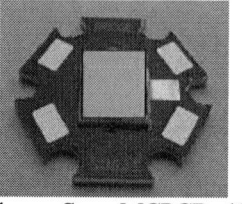

Figure 6 DPC substrates mounted on *Super*MCPCB (Cu STAR)

Reliability tests results by X-rays and SATS including selected cross sectioning of samples shows that there was no delamination for all 135 units (table 1).

Table 1 Tabulation of reliability test results

Reliability Testing	Test Samples	Test Results (15 Units)
Thermal Cycle -65 ℃ ~ 150 ℃ 1000Cycles	*Super*STAR+AlN 15mil	All passed
	*Super*STAR +AlN 20mil	All passed
	*Super*STAR+Al$_2$O$_3$ 20mil	All passed
Thermal Shock -65 ℃~ 150℃ 1000Cycles	*Super*STAR +AlN 15mil	All passed
	*Super*STAR +AlN 20mil	All passed
	*Super*STAR+Al$_2$O$_3$ 20mil	All passed
High Temp Storage +85 ℃/ 5%RH 1000 hrs	*Super*STAR +AlN 15mil	All passed
	*Super*STAR +AlN 20mil	All passed
	*Super*STAR+Al$_2$O$_3$ 20mil	All passed

EXPERIMENT OBJECTIVES / SAMPLES PREPARATION

Having established the compatibility of the *Super*MCPCB with DPC ceramics substrates, the next phase of investigation was:

1) To measure and compare the combined thermal resistance of the improved thermal board with conventional MCPCB by mounting a selected LED device (MPL 629 / Power: 1W with forward current 350mA) on the DPC ceramics substrate (see figure 7 for prepared samples)

2) To compare the effect of varying thickness and different types of DPC ceramics (AlN and Al$_2$O$_3$) substrates on the combined thermal resistances and different samples were prepared as shown in table 2.

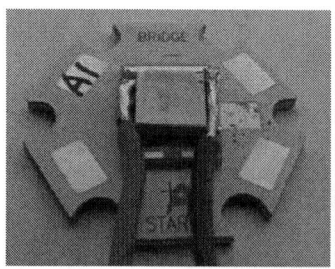

LED on *Super*MCPCB(Cu STAR) LED on conventional MCPCB

Figure 7 Pictures of LED on substrate assembled on MCPCB

Table 2 Groupings of Experimental Set-up

Group	Test Samples
A	*Super*STAR +AlN 15mil
B	*Super*STAR +AlN 20mil
C	*Super*STAR +Al$_2$O$_3$ 20mil
D	Al STAR+AlN 15mil
E	Al STAR+AlN 20mil
F	Al STAR+Al$_2$O$_3$ 20mil

THERMAL MEASUREMENT METHODLOGY

National Chung Hsing University Institute of Precision Engineering was engaged to measure the combined thermal resistances using the T3Ster thermal transient tester (set up as shown in figure 8).

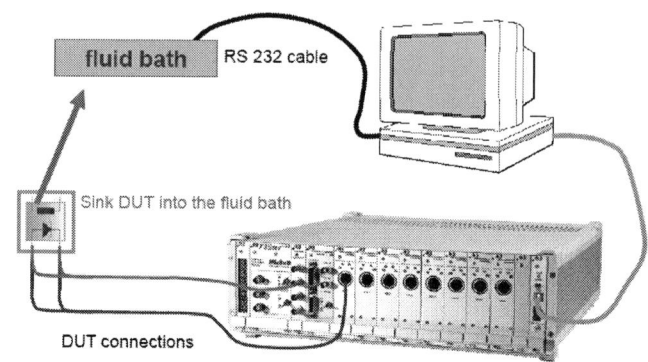

Figure 8 Sketch of experimental set-up (Courtesy of National Chung Hsing University Institute of Precision Engineering)

T3Ster Tester [4]

T3Ster (Mentor Graphics, USA) is a thermal transient tester used for measuring a precise junction temperature of a packaged IC or a LED chip by an electrical test method. It utilizes the thermal function to derive a thermal resistance of each layer of the chip package by a Network Identification by De-convolution (NID) method as shown in figure 9.

Once the thermal response obtained, T3Ster can derive the cumulative structure function (S.F.) and differential structure function of each device from the thermal response $T_{(t)}$.

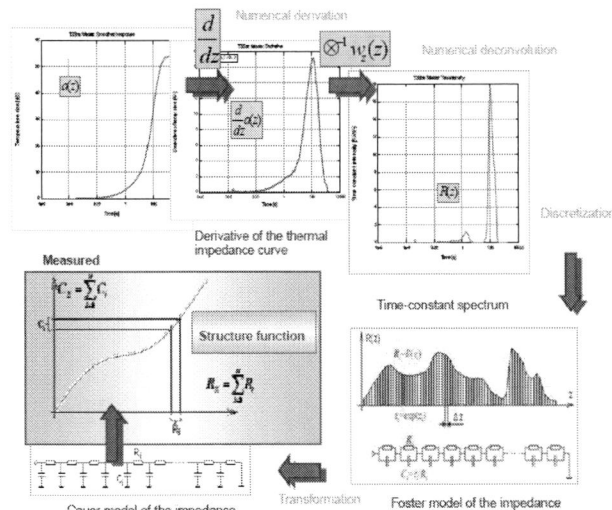

Figure 9 Network Identification De-convolution method (Courtesy of National Chung Hsing University Institute of Precision Engineering)

A total of six groups (A, B, C, D, E and F) were being studied and their schematics are shown in figure 10.

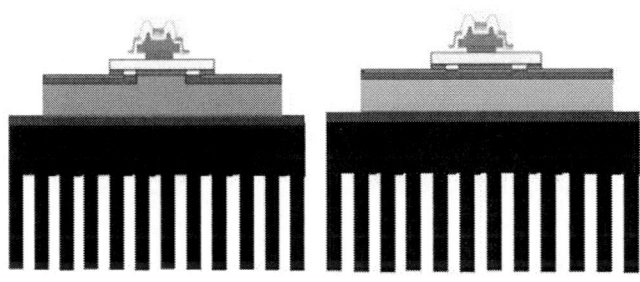

Groups A/B/C on *Super*STAR   Groups D/E/F on conventional MCPCB

Figure 10 Schematic of samples mounted on heat sink

R_{th} of LED Device + Substrate & MCPCB + Heat Sink
The cumulative structure function (S.F.) of the respective DUT is then obtained and figure 11 shows the S.F. and the differential S.F. of the DUT under group A.

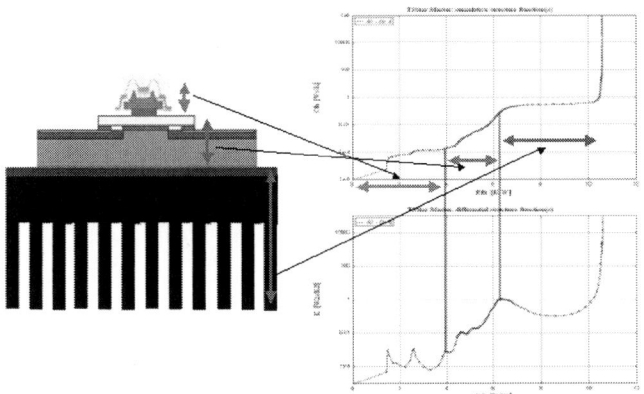

Figure 11 Cumulative structure function (S.F.) and differential S.F. of Group A (R_{th} of LED device + Substrate & MCPCB + Heat Sink)

RESULTS AND DISCUSSION

Table 3 shows the summary of R_{th} measurement.

Table 3 Combined R_{th} of DPC substrates and MCPCB

Group	Test Samples	Rth of DPC substrate and MCPCB (K/W)
A	*Super*STAR +AlN 15mil	2.30
B	*Super*STAR +AlN 20mil	2.47
C	*Super*STAR + Al$_2$O$_3$ 20mil	3.87
D	Al STAR+AlN 15mil	3.57
E	Al STAR+AlN 20mil	3.81
F	Al STAR+Al$_2$O$_3$ 20mil	5.16

Discussion on DPC Substrate Performance
- Using the same thermal board mounted with different thickness and material of metallized ceramic substrates, Group A (*Super*STAR + AlN 15 mils) has the lowest thermal resistance (R_{th}) compared with Group B (*Super*STAR + AlN 20 mils) and Group C (*Super*STAR + Al$_2$O$_3$ 20 mils). i.e. 2.3 versus 2.47 and 3.87 K/W. The result indicated that by replacing 20 mils metallized Al$_2$O$_3$ substrate with the same thickness of metallized AlN, the thermal resistance significantly improved by 1.4 K/W. This is mostly because the thermal conductivity of AlN (170 w/mK) is much higher than that of Al$_2$O$_3$ (~24 w/mK). Besides, the thermal resistance slightly further decreases 0.17K/W when reducing the thickness of AlN metallized substrate from 20 to 15 mils.

- Similar trend was obtained in the case of using conventional Al STAR. The thermal resistance dramatically decreased from 5.16 to 3.81 K/W with 1.35 K/W difference when comparing 20 mils of metallized AlN and Al$_2$O$_3$ ceramic substrates. The thermal resistance further decreased by 0.24K/W when reducing the thickness of AlN ceramic substrate from 20 to 15 mils.

Discussion on *Super*MCPCB Thermal Board Performance
- Using the same package substrate, Group A (*Super*STAR + AlN 15 mils) has much lower thermal resistance (R_{th}) compared with Group D (Al STAR + AlN 15 mils) i.e. 2.3 versus 3.57 K/W. The result shows that by replacing the conventional Al STAR with Bridge patented *Super*STAR, the thermal resistance improved by 1.27 K/W.

- To validate the above, using a different substrate thickness and material, Group B R_{th} is 1.34 K/W better than Group E, and Group C R_{th} is 1.29 K/W better than Group F respectively. This established the effectiveness of the *Super*STAR thermal board structure.

- Group C (*Super*STAR + Al$_2$O$_3$ 20 mils) has no significant thermal resistance (R_{th}) differences compared with Group D (Al STAR + AlN 15 mils) i.e. 3.87 versus 3.57 K/W - As the cost of AlN is significantly higher than Al$_2$O$_3$ versus the cost between copper and aluminum MCPCB, it will be more cost effective to adopt the *Super*STAR structure versus using a more expensive substrate to achieve similar R_{th}.

CONCLUSION

This paper shows how a superior HBLED thermal solution can be achieved using a smart combination of Direct Plated Copper (DPC) ceramics substrate and *Super*MCPCB structure compared to using conventional MCPCB.

Using this enhanced package and *Super*MCPCB structure, it was established that the reliability performances were also not compromised as no delamination was observed for both the mounted AlN and Al_2O_3 DPC substrates.

REFERENCES

[1] DPC Patent No.: US 6,800,211 B2

[2] http://metallized-ceramic.ready-online.com/company-story.html

[3] 先進微系統與構裝與技術聯盟 97 年度第三次會議

[4] V.Sze´kely, V.B. Tran, Fine structure of heat flow path in semiconductor devices: a measurement and identification method, Solid-State Electron. 31 (1988) 1363–1368).

HIGH THERMAL MASS, VERY HIGH LEAD COUNT SMT CONNECTOR REWORK PROCESS: PROCESS AND PROBLEM RESOLUTION

Jim Bielick, Brian Chapman, Mitchell Ferrill, Michael Fisher, Phil Isaacs,
Eddie Kobeda, and Theron Lewis
IBM
Rochester, MN, Endicott, NY, Poughkeepsie, NY, and Raleigh, NC, USA
pisaacs@us.ibm.com and kobeda@us.ibm.com

ABSTRACT

Very large high thermal mass surface mount connectors are difficult to rework. This paper will examine the difficulties of reworking a very high thermal mass, high density, 5040 lead count surface mount connector. As we approached this problem it was clear that the rework equipment available did not have the power required to perform this rework in an acceptable process window. A multi-functional team was assembled to develop, execute and qualify the connector rework process. This paper will describe the special equipment selected expressly for this process: Localized vapor phase reflow tool, site redress process, selective site solder paste printing process and the quality control processes.

To assure success multiple paths were developed in parallel. The local vapor phase process came out clearly superior to alternative processes, and was established and qualified within a few months. The process was verified through mechanical measurements, cross-sections, high-resolution X-Ray and stress testing and results met all requirements.

Key words: Very Large SMT Connector, Rework, high lead count

INTRODUCTION

Recent trends in the electronics industry continue to support the increase in performance and density of printed circuit board assembly designs, leading to further miniaturization. These trends include deep sub-micron CMOS I/O circuits and scalable processors meeting multi-gigahertz performance requirements. The industry group iNEMI[1] has similarly forecast increased frequencies and decreased footprint in future products, which will place substantial challenges on fine line technologies needed for assembly of electronic systems. As off-chip bandwidth requirements grow to meet the needs of high-end servers, and density of circuits becomes paramount to scalable architectures, the need to continue single-ended mode signals while achieving high frequency limits the interconnect solution to a custom design.[2]

We report in this work the development of a rework method that utilizes local vapor phase technology to attach a large 14 x 120 receptacle connector. This large connector supports 6 Gb/s single-ended mode signaling at a contact density of approximately 28 contacts per cm^2, and plugs a single processor node into one of eight mid-plane header connectors. Its approximate external dimensions are 28 cm length x 5 cm in width x 4 cm in. height. It supports 5,040 I/O and is surface mounted to a 4-5 mm thick printed circuit board. Figure 1 shows a picture of the connector at the edge of printed circuit board.

The connectors are made from individual wafers that are interference fit into stainless steel organizers. The individual wafers possess signal contacts on one face and ground contacts on the opposing face with the individual grounds are commoned within wafers. There are two ground connections for every signal connection. For the receptacle wafers, both signal and ground spring beam contacts are contained within insert molded apertures, while the header wafers possess signal and ground contact blades captured within molded plastic. This basic layout and design information will become important as we understand the defects which require rework.

The industry trends highlighted above present numerous challenges and limit the number of process choices we have in support of rework development. The original attach process for this connector utilizes vapor phase reflow, which is not widely used in the industry today. In fact, iNEMI[3] has highlighted that despite discussions around the return to vapor phase technology, at this time it is only being considered for niche applications.

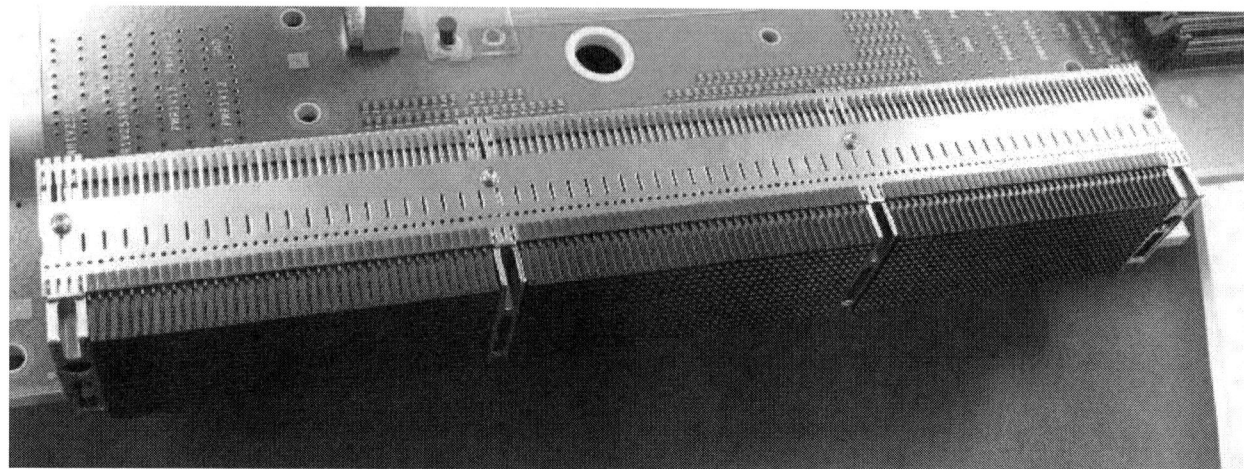

Figure 1: Receptacle connector showing wafer seating face and edge for plugging

In our case, vapor phase technology is well suited and its merits are well documented.[4][5] By boiling a liquid, energy is transferred through heat of condensation to the printed circuit board and its components. This method offers efficient heat transfer and is ideally suited for high mass component applications and reflow of solder that is independent of geometry and package density. With the advent of Pb-free solder paste and the need for higher temperatures to achieve melting points of new alloys, vapor phase reflow is being touted as an attractive alternative due to its control of maximum temperature limited by the liquid's boiling point.[6]

While vapor phase is ideally suited for initial attachment of this connector, a local vapor phase tool design offers an alternative for rework that "localizes" the process and limits the thermal exposure to the board. This simplifies the manufacturing process by minimizing the removal of additional components surrounding the connector region.

During qualification of connector rework we evaluated several methods to assess the quality and reliability of this interconnect solution. As noted, this application required increased bandwidth for leading edge performance and scalability. This limited us to new surface mount technology as opposed to through-hole soldered or press-fit connector attachment which affect both density and signal integrity characteristics. While this supports the objectives of our product electrically, the plugging of processor nodes into the mid-plane can create mechanical stresses on the solder joints even in the presence of additional guide and support features intended to protect the connector and solder joints during system actuation. During qualification of the rework process, we characterized solder joint integrity, intermetallic thickness and mechanical tolerance compliance to ensure that reworked connectors were capable of surviving these plugging stresses.

The number of leads, 5040, on the connector is extremely large compared to most surface mount connectors common in the industry. The large number of leads requires an excellent surface mount process yield. Table 1 has yield projections using a Poisson Distribution for 1, 2, 5, 10 and 25 ppm defects per lead assuming one connector per board.[7]

Table 1: Board yield projection based on 5040 leads and defects per lead

Average ppm defective/lead	% Yield with 0 Defects	% Boards with 1 Defect	% Boards with 2 or more defects
1	99.5	0.5	0.0
2	99.0	1.0	0.0
5	97.5	2.5	0.0
10	95.1	4.8	0.1
25	88.2	11.1	2.3

Twenty five ppm defects per lead is a good yield for surface mount leaded devices. In this case, the large connector yield is better. From this yield projection it can be seen even in relatively low defect rates per lead that there will be enough natural fallout to where a rework process will be required.

REWORK PROCESS

As was stated above this rework process uses vapor phase reflow technology.[8] It is important to note that this rework process applies a localized area of hot vapor to achieve selective reflow of components on the printed circuit board assembly, PCBA. The vapor phase reflow technology used to rework the large surface mount 14x120 daughter-card connector system is an off the shelf vapor phase rework machine which was adapted with special nozzles for this job.[9][10] Internally the machine is identified as a Local Vapor Phase, LVP, machine. The machine is shown in Figure 2.

Figure 2: Local vapor phase machine for large surface mount connector rework

There are several steps to the large surface mount connector rework process:

Step 1: Before the PCBA can go through any solder reflow rework process; the PCBA needs to go through a bake operation. The bake operation removes any moisture that might have been absorbed by the PCBA since going through the initial attach process. It is important to remove this moisture to avoid out-gassing induced damage to the printed circuit board laminate material or moisture sensitive components.

Step 2: After the bake process the next steps in PCBA preparation are taping of areas that need thermal or physical protection, removal of temperature sensitive components, and removal of components that interfere with the rework operation itself. See figure 3.

Figure 3: Yellow highlighted components will be removed prior to rework

Step 3: The large surface mount connector is removed using an internally designed nozzle connector removal system which engages the connector and attaches to the LVP machine through hoses and mechanical connections. During the removal process, the connector and PCBA are pre-heated with multiple sources of hot air and a heating strip, followed by full vapor phase reflow. At maximum reflow temperature the LVP lifts the large surface mount connector from the PCBA within the connector removal

system nozzle. Then the PCBA, apparatus and connector are cooled. The connector is disposed of and the PCBA is transferred to the site dress area. See figure 4.

Step 4: At site dress, the large surface mount connector site on the PCBA is inspected for de-wet pads, excessively high solder bumps, PCB delamination and lifted SMT Pads. Touch-up is done on all de-wet pads with solder, flux and a solder iron. SMT pads that have excessive solder height or icicles are flattened using a solder iron and flux. Then the SMT pads are cleaned with isopropyl alcohol, IPA, and a lint free cloth. The PCBA is transferred to a hot gas rework machine with a controlled z-height vacuum nozzle[11] used for site redress. See figure 5.

Figure 4: Mounting pads prior to site dress

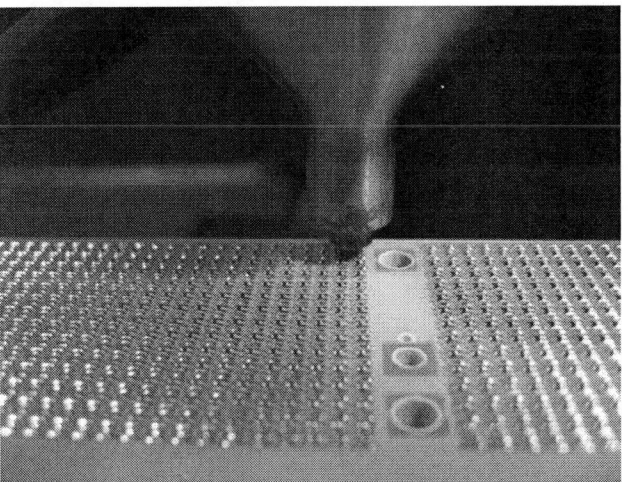

Figure 5: Site redress of the mounting pads after connector removal

Once the PCBA reaches the target pre-heat temperature the hot-gas rework tool will lower the site dress nozzle to the PCB, automatically sense the board surface, and set the vacuum for auto height adjustment. The site dress operation progresses slowly across the large connector SMT site. Once all the pads have been dressed, the SMT pads are cleaned using IPA and lint free cloth during the cool down cycle. Again, inspection is performed to examine the re-dressed SMT pads for any lifted pads, damaged pads, damaged solder mask, excess solder or solder bridges. If solder bridges are found, they are removed using a solder iron and braided copper wick.[12]

Step 5: The PCBA is transferred to a semi-automatic screen printing machine adapted to allow selective site solder paste printing. See figure 6. The stencil has cut-outs to accommodate the remaining components on the PCBA. See figure 7. Solder paste is applied to the stencil and printed onto the PCBA site in a single blade pass. Then the PCBA is removed from the solder paste screening machine and transferred to a solder paste measurement machine to verify the solder paste height and volume. After verification, the solder paste deposits are visually reconfirmed under a microscope by an operator. See figure 8.

Figure 6: Solder paste screen tool modified for connector site screening

Step 6: The PCBA is transferred back to the LVP Machine area and a new large surface mount connector is attached to the PCBA. An internal custom designed attach nozzle system is used to engage both the connector, PCBA, and the LVP system. Similar to the removal process, the connector and assembly are pre-heated with multiple sources of hot gas and a heating strip, followed by full vapor phase reflow and cool-down. Through the entire heating process the connector and the PCBA area directly under the connector are clamped under a spring load within the nozzle. This force is needed to balance between PCBA flatness and connector lead co-planarity.

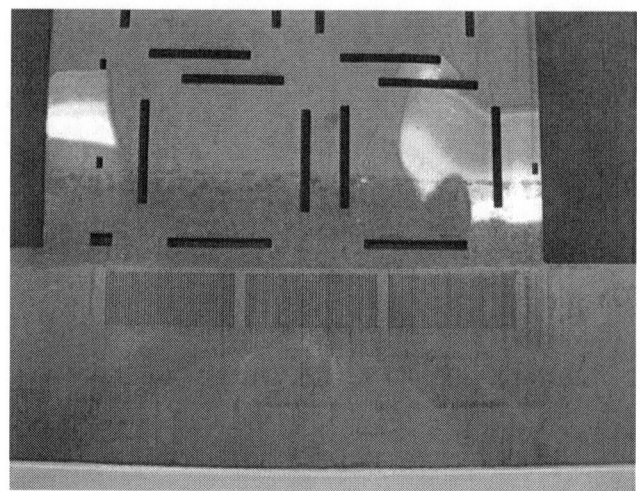

Figure 7: Cut-out stencil for site solder paste screening

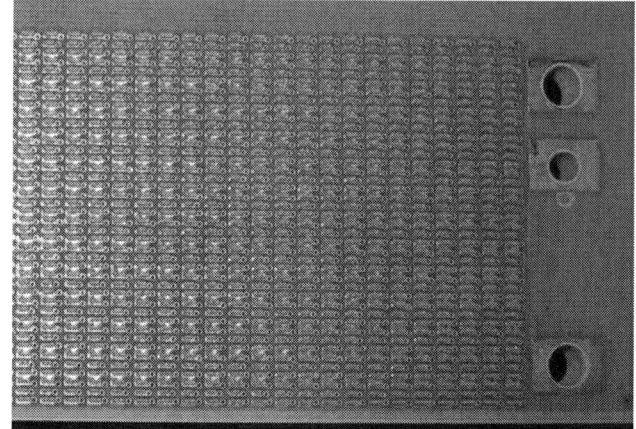

Figure 8: Connector site after solder paste screening

The PCBA and newly attached large surface mount connector system are now ready for verification by various optical inspection, mechanical measurements, X-Ray and electrical testing.

IN-LINE MECHANICAL VERIFICATION PROCESS
Due to the complexity and size of this connector the team decided on certain critical mechanical measurements and verifications as controls for the rework process. The chosen measurements for the process included connector site PCB flatness, wafer-face to guide-block true positional alignment and mechanical verification test.[13] These mechanical measurements and tests provide the ability to verify compliance with critical to function attributes of the connector for both qualification purposes as well as quality monitoring in production.

Localized PCB flatness at the SMT site was determined to be a key contributor to reliability during stress testing of assemblies. Samples measured during initial rework were deemed acceptable, however during the course of 2X rework qualification samples were found outside of the optimum range. Therefore an in-line measurement was implemented on all 2X reworks with an automated measurement device to control site flatness prior to reattach of the connector. During the plugging process it is extremely important that the mating interface of the connector meets certain dimensional criteria. Without meeting this criterion the plugging process can result in damage to the PCB as well as damage to the mating connector.

The most critical of these measurements is verification of the wafer-face to guide block dimensions. See figure 9. The wafer-face to guide block test measures the z-dimension of the top and bottom of each wafer with respect to a plane created by the connector guide blocks, e.g. that is in front of or behind the plane created by the guide blocks. The connector guide blocks act as a hard stop during connector plugging.

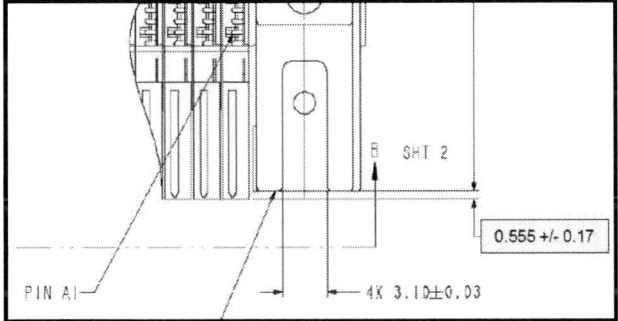

Figure 9: Wafer-to-Guide block tolerance

If the wafer is above the specification limit it will bottom out prior to guide block seating resulting in an overstressed condition of the solder joints. If the wafer is under the specified dimension it may not make good electrical contact with the mating wafer resulting in an electrical opens or intermittent connections.

Cross-Section
As part of the process development and qualification effort, reworked cards were cross-sectioned at various process stages to ensure product integrity. The IBM East Fishkill Materials Lab established a cross-section protocol to ensure consistent, comparable, valid results were achieved. Sample preparation and cross-sectioning were performed with an aluminum stiffener bolted to the backside of the connector site to avoid procedure-induced artifacts through unintentional mechanical damage caused by handling. A quantifiable figure of merit protocol was created to enable comparison and tracking of the solder joint quality from varying rework processes. Boards were sectioned at connector removal, site redress, connector re-attach, and after exposure to various stress conditions. Cross-sectioning included evaluation of solder joint integrity, lead alignment, solder balls, contaminates, pad-lift and solder intermetallic integrity. See figure 10. As part of the site redress qualification, solder thickness was also measured to ensure proper solder coating thickness. This was necessary to avoid de-wet issues associated with oxidation of thin solder or exposed copper/tin intermetallics.

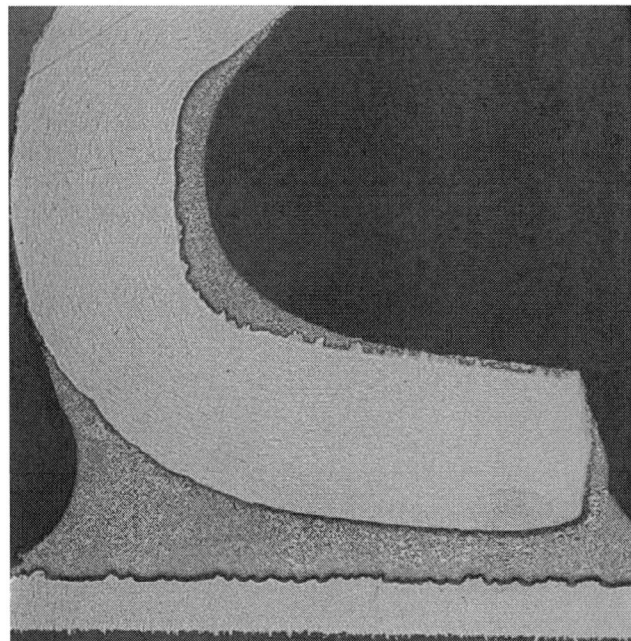

Figure 10: Cross-section of a solder joint

X-Ray Characterization
Reworked boards were inspected with 2D and 3D X-Ray protocol to ensure product quality. A 2D X-Ray auto-inspection algorithm was developed to detect solder defects and misaligned leads. The algorithm was tuned to ensure capture of significant defects, including low solder, non-wets and shorts. The program was tuned to minimize false calls which require operator interpretation. 3D X-Ray CT Scan was performed on any questionable solder joints flagged by the 2D inspection. For rework process qualification, full connectors were sampled with CT X-Ray, regardless of 2D X-Ray results.

Alternate Approach for Qualification
Due to the nature of the defects and the manner in which they fail an alternate test approach was developed. The test methodology sequence was as follows:

1. Mechanical verification at time zero
2. Multiple connector plug cycles in the system configuration
3. Mechanical verification
4. Accelerated thermal cycling

The Accelerated Thermal Cycling, ATC, followed by mechanical verification generally continued past the end of test until at least one solder joint had failed. By performing the thermal cycling and mechanical verification to detect the weak solder joints, it was possible to establish relative merit between the different rework methods.

Process selection activities began with samples from two different types of rework approaches. The first used the same basic tooling as the new build process and the second used a localized heat rework tool. Table 2 summarizes the results of this initial evaluation.

Based on this limited sample size, the two approaches appeared equivalent, so the localized vapor phase heating rework method was selected as the process to put through qualification testing. This process provided greater flexibility and less overall risks to the total assembly than the new build reflow process. The process was optimized and then samples were submitted to the reliability evaluation. The results of this testing are shown in table 3.

Table 2: A comparison of rework methods using mechanical and thermal stress methods

Process	Post Plug Cycle Result	1st ATC Interval
Localized Heating-Sample 1	Pass	Pass
Localized Heating – Sample 2	Pass	Not Performed
Localized Heating – Sample 3	Pass	Not Performed
New Build Process - Sample 1	Pass	Pass
New Build Process - Sample 2	Pass	Pass

All samples passed the initial system level multi-plug preconditioning as well as accelerated thermal cycling. Testing was continued past the end of test until at least one part failed. These failures were deemed acceptable because when the acceleration factor of the test is taken into account they failed late enough in testing as to be considered end of life fails and will occur at a much a higher number of on/off cycles than systems will see in a lifetime. Therefore it was concluded that the localized rework process yielded parts exceeded all quality and reliability objectives.

Table 3: Reliability assessment

Sample	Post Plug Cycle Result	ATC Test
1	Pass	Pass
2	Pass	Pass
3	Pass	Pass
4	Pass	Pass
5	Pass	Pass
6	Pass	Pass

SUMMARY
We report a very large surface mount connector rework process which was developed in a short period of time. Different process modules were evaluated and the local

vapor phase rework process proved itself to be the most stable and reliable process. As expected some defect modes were encountered during process development. Special tests were developed to find these defects through X-Ray analysis and mechanical verification testing, which led to further process optimization and elimination of these **defects.** A regime of plug testing and ATC testing was established to specifically stress the solder joints in such a way as to highlight the defects that were present thus enabling optimization of the process during development and selection of the best process to implement in manufacturing. Through this stress testing it was demonstrated that the process is repeatable and product produced is reliable.

ACKNOWLEDGEMENTS
We further wish to thank our co-workers who without their help this work would not have been possible. We wish to thank the IBM Fishkill Materials Analysis team for their analysis of the parts and our management who supported us in this effort.

REFERENCES

[1] iNemi is a registered trademark of International Electronics Manufacturing Initiative, Herndon, VA.
[2] H.Q. Le, et al., "IBM Power6 MicroArchitecture," IBM Journal of Research and Development, Vol. 51 No. 6, November 2007.
[3] iNemi Roadmap, 2009, M. Kelly et al.
[4] N. Heilmann, "A Comparison of Vaporphase, Infrared, and Hotgas Soldering," IEEE CH2629-4/88/0000-0070, pp. 70-72.
[5] L. Livosky, A. Pietrikova, and J. Durisin, "Monitoring of Temperature Profile of Vapour Phase Reflow Soldering," IEEE 978-1-4244-3974-4/08, pp. 667-669.
[6] A. Pietrikova, L. Livovsky, J. Urbancik, and R. Bucko, "Optimisation of Lead Free Solders Reflow Profile," IEEE 1-4244-0551-3/06, pp. 459-464.
[7] P. Isaacs and K. Puttlitz, *Chapter 20, Area Array Component Replacement,* Area Array Interconnection Handbook, 2001, Kluwer Academic Publishers, Norwell, Massachusetts, pp. 804-837,
[8] T. Lewis and J. George, "Vapor Phase Reflow for Reliable Assembly of a New High Performance Interconnect Technology," IMAPS Symposium on MicroElectronics, Nov 11-15, 2009, San Jose, CA, pp. 193-200.
[9] Vapor Works Machine is made by R&D Technical Services, Burnsville, Minnesota.
[10] The rework nozzles were developed within IBM for this application.
[11] Air-Vac DRS 25. Air-Vac's headquarters are in Seymour, CT.
[12] Soder-Wick is a registered trademark of the ITW Chemtronics Company, Kennesaw, GA.
[13] Mechanical Verification test is an IBM internally developed test.

HIGH SPEED TOUCH SCREEN ASSEMBLY USING ANISOTROPIC CONDUCTIVE ADHESIVES (ACAS) VERTICAL ULTRASONIC BONDING METHOD

Kyung-Wook Paik, Seung-Ho Kim, and Kiwon Lee
Nano Packaging and Interconnect Lab. (NPIL)
Department of Materials Science and Engineering
Korea Advanced Institute of Science and Technology (KAIST)
Daejeon, Korea
kwpaik@kaist.ac.kr

ABSTRACT

A novel anisotropic conductive film (ACF) bonding process using ultrasonic vibration was investigated in touch screen panel(TSP)-on-board (TOB) applications. The ACF temperature increased as the U/S power increased and the bonding pressure decreased. The ACF temperature was successfully controlled by adjusting both U/S power and bonding pressure. The significant meaning of this result is that the ACF bonding process can be remarkably improved by U/S bonding compared with conventional T/C bonding. Using the optimized U/S bonding parameters, the ACF interconnects showed significantly less thermal damages to PET substrates and similar bonding performances as T/C bonding in terms of the contact resistance and the adhesion strength. And the cure degree of adhesive resin was achieved 90 % at 3 sec.

In terms of thermal deformation, U/S bonding showed no severe thermal deformation of TSPs up to 120 °C which is much higher than T_g of PET substrates. This result indicates that ACA can be heated up to 120 °C without severe thermal deformation of the PET substrates by VUS. It is presumably due to the rapid heating rate and the short bonding time of VUS bonding. As a summary, the VUS method can be successfully used in touch screen assembly with high speed and good reliability.

Key words: touch screen panels, ACA, vertical unltrsonic bonding, assembly

ULTRASONIC ACA BONDING TECHNOLOGY

For ACF interconnection, thermo-compression (T/C) bonding is the most common method, however it is necessary to reduce the bonding temperature, time and pressure, because T/C bonding is often limited by high bonding temperature, slow thermal cure, uneven cure degree of adhesive, large thermal deformation of the assembly. Therefore, there are constant needs of lower bonding temperature and faster cure ACF bonding to replace the conventional T/C bonding.

Ultrasonic (U/S) bonding is one of alternative processes for ACF interconnection which have been suggested by Lee et al. [7][8]. In U/S bonding, ACFs can be rapidly heated by certain ultrasonic vibration, and it can be described by materials' complex young's modulus which consists of storage modulus and loss modulus under cyclic stress conditions. Storage modulus relates to elastically stored energy, and loss modulus relates to energy loss which converts to heat. In general, it is well known that visco-elastic materials such as polymers have large loss modulus. Therefore, it is expected that highly visco-elastic B-stage ACFs may generate a large amount of heat by ultrasonic vibration. As a result, the ACF layer can be rapidly heated and cured without additional chip/substrate heating.

Fig. 1 shows the schematic of an ultrasonic ACA bonder.

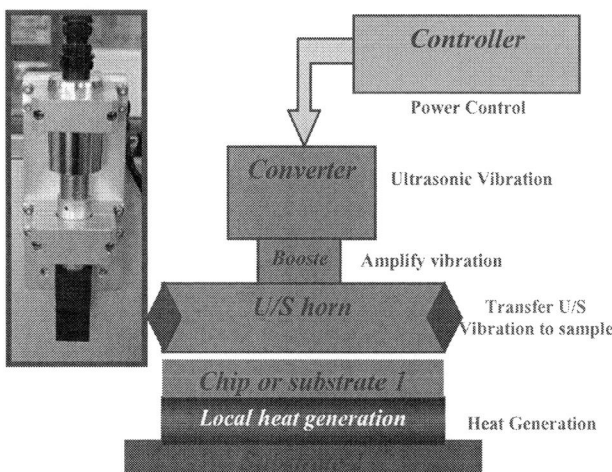

Fig. 1 Schematic and real photo of an ultrasonic ACA bonder with a converter, booster, and horn.

ACF temperature can be precisely controlled by the ultrasonic vibration amplitude as shown in the Fig. 2, FOB(Flex on Board) samples.

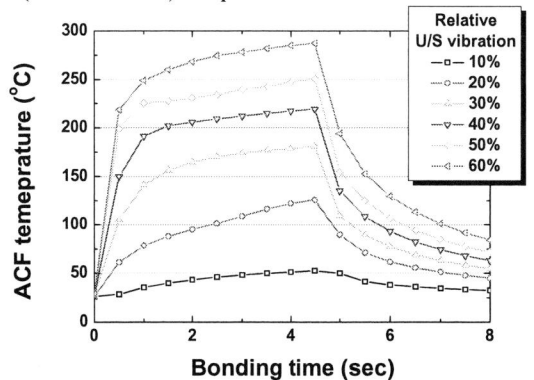

Fig. 2 ACF temperatures heated by the various ultrasonic vibration in FOB samples

The increase of ACF temperature can be explained with U/S vibration amplitudes. U/S vibration amplitudes increase with larger U/S powers at constant pressures, because the work by U/S vibration increases as the U/S power increases. According to the well-known equation which explains heat generation under cyclic deformation [9],

$$dQ = \frac{f(\Delta\varepsilon)^2 E''}{2}$$

heat generation (dQ) is proportional to cyclic strain ($\Delta\varepsilon$). And cyclic strain increases as U/S vibration amplitude increases. Therefore, the increase of U/S power causes the increase of U/S vibration amplitude, and cyclic strain of the ACFs resulting in more heat generation. Table 1 summarizes results of the optimized U/S ACF bonding in comparison with those of typical T/C bonding.

Table 1. The optimized U/S ACF bonding condition and their results in comparison with those of typical T/C bonding.

	T/C ACFs bonding at 190℃ for 15 sec	U/S ACFs bonding at room temperature for 3 sec
Peel strength (gf/cm)	622.47 (±28.59)	633.41 (±52.14)
Daisy-chain contact resistance (Ohm)	1.08 (±0.02)	1.13 (±0.04)

These results indicate that the process temperature of ACF bonding can be significantly reduced from a typical 190 ℃ to room temperature, and bonding time can be also significantly reduced from typical 15 seconds to less than 5 seconds by utilizing ultrasonic vibration.

TOUCH SCREEN PANEL ACA ASSEMBLY USING VUS

A touch screen is a display which can detect the presence and location of a touch within a display area. Due to its intuitive interface, the touch screen has been widely used in mobile device applications such as personal digital assistant (PDA), satellite navigation devices, and mobile phones. Recently, due to advantages such as low cost, light weigh, and shock resistance, polymer-based touch screen panels (TSPs) have been newly introduced in mobile device applications by replacing conventional glass-based TSPs.

However, in spite of its advantages, polymer-based TSPs still have several issues in interconnection processes using Anisotropic Conductive Adhesives (ACAs). Especially, the ACAs assembly process temperature should be limited below 150 ℃ to prevent thermal deformation in the conventional thermo-compression bonding because of the low T_gs of substrate materials such as polyethylene terephthalates (PETs). And the limited bonding temperature of below 150 ℃ results in significantly slow bonding times up to 15 seconds. Therefore, there have

been constant needs of alternative process methods which can eliminate thermal deformation and slow bonding time issues at the same time.

Vertical ultrasonic (VUS) bonding is an alternative ACA bonding process which utilizes spontaneous heat generation inside the ACA layer by vertical ultrasonic vibration. According to Lee et al, VUS bonding has various advantages such as less thermal deformation due to the room temperature US horn process and significantly reduced bonding times within 3 seconds in Flex-On-Board (FOB) applications. Thus, we can expect that VUS bonding method can eliminate both thermal deformation and reduce bonding times in polymer-based TSP applications.

In this study, the effects of VUS bonding parameters such as vibration amplitudes and bonding pressures were investigated and optimized in terms of thermal deformation of polymer substrates, electrical continuity, and pull adhesion strength of ACA joints. And the reliability of VUS bonded ACA joints were evaluated at various test conditions.

The test vehicles were resistive TSPs which consist of PET substrates, PC (Polycarbonate) substrates and FPCBs as shown in Fig. 3.

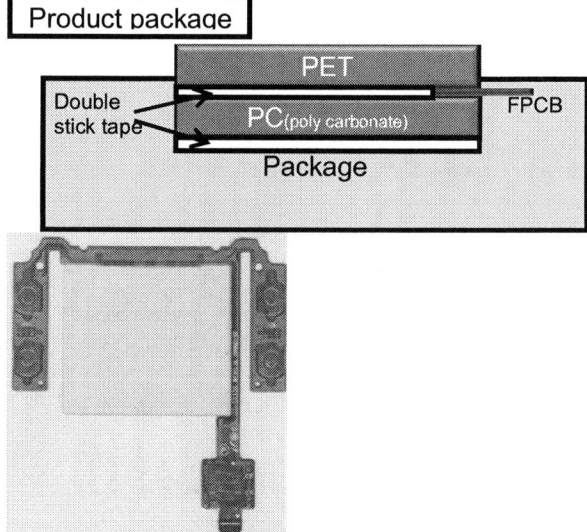

Fig. 3 Schematic diagram and real photo of assembled TSP in this experiment

There were screen-printed Ag paste pads on the Indium Tin Oxide (ITO) patterned PET and PC substrates, and Cu pads existed on both side of FPCB. The FPCBs were inserted between PET and PC substrates. And then, both PET-FPCB and PC-FPCB bonding were simultaneously conducted. Anisotropic Conductive Pastes (ACPs) were used as interconnection adhesives.

As the vertical ultrasonic vibration was applied as shown in Fig. 4, ACA temperatures rapidly increased from room temperature to above 200 ℃ within 2 seconds due to the spontaneous heat generation in the ACA itself and surrounding.

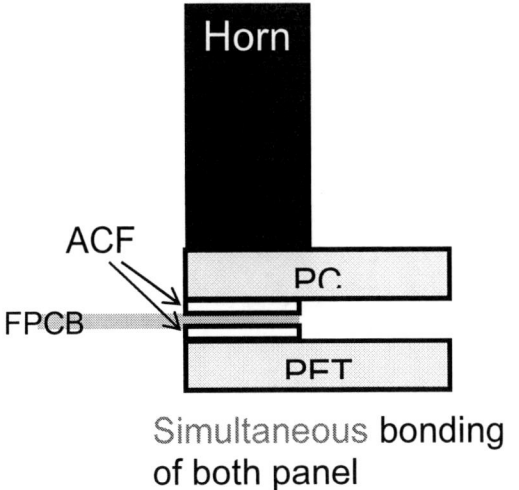

**Simultaneous bonding
of both panel**

Fig. 4 Schematic diagram of VUS assembly

The heating temperature profiles of conventional thermo-compression(TC) bonding and VUS bonding are shown in Fig. 5. TC bonding case, the hottest temperature occurs at the surface of the hot bar contacted to the PET Panel 1. Therefore, thermal damage can occur at the PET Panel 1. However, in case of VUS bonding, the hottest temperature occurs at the inside of ACA material resulting in no thermal damages to the TSP PET film.

T/C bonding

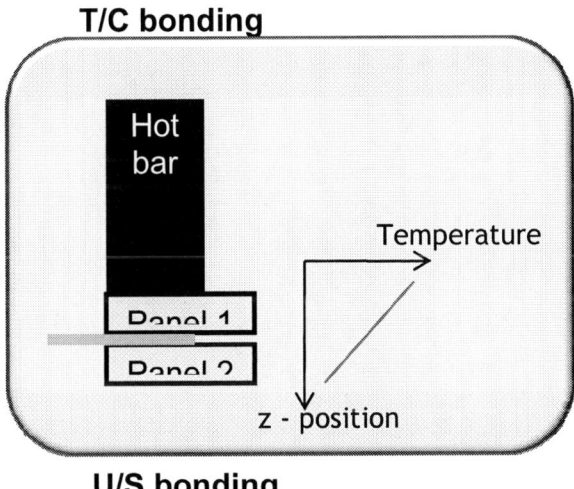

U/S bonding

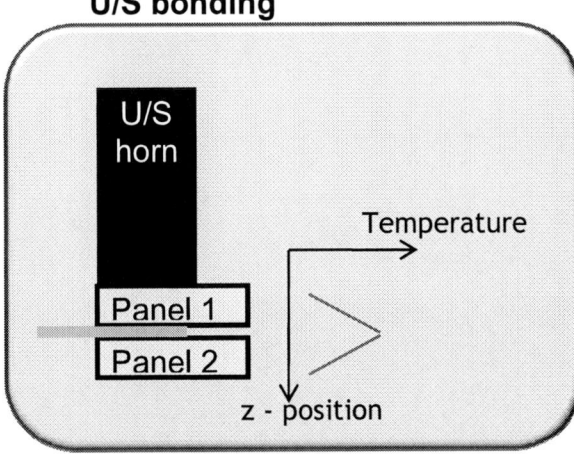

Fig. 5 Heating temperature profile difference between TC and VUS bonding methods.

In VUS bonding, the ultrasonic vibration showed significant effects on the peak temperature of the ACA layers. When the vibration amplitudes were 6 um and 9 um, the peak temperatures of the ACA layer were 93 °C and 172 °C at 2MPa bonding pressure, respectively. However, the bonding pressure only affected the heating rates during bonding. When the bonding pressures were 1 MPa and 3 MPa, the heating rates of ACA layer at 1 second bonding time were 82 °C/sec and 134 °C/sec with 7.5 um vibration amplitude, respectively. Considering both effects of ultrasonic vibration and bonding pressure, the ACA temperature could be successfully controlled during bonding.

In terms of thermal deformation, VUS bonding showed no severe thermal deformation of TSPs up to 120 °C which is much higher than T_g of PET substrates. This result indicates that ACA can be heated up to 120 °C without severe thermal deformation of the PET substrates by VUS. It is presumably due to the rapid heating rate and the short bonding time of VUS bonding.

In terms of electrical continuity of the ACA joints, VUS bonded TSPs showed stable electrical resistances at higher than 2 MPa bonding pressures and there were no significant effects of vibration amplitudes and bonding times on contact resistance. According to cross-section analysis, conductive balls in the ACA layer were well captured between electrodes at sufficient bonding pressures higher than 2 MPa.

At the same time, VUS bonded TSPs showed strong adhesion at the ACA joints with 1 second bonding time at 7.5 um vibration amplitude and 2 MPa bonding pressure. During the FPCB pull test, U/S bonded TSPs showed higher than 3 kgf strengths and the failure site was the FPCB itself. This result indicates that the adhesion at the ACA joints is stronger than the strength of the FPCB.

Therefore, VUS bonding parameters were optimized at 7.5 um vibration amplitude, 2 MPa bonding pressure, and 1 second bonding time in terms of thermal deformation, electrical continuity and pull adhesion strength of the ACA joints.

With the optimized parameters, various reliability tests were conducted, such as thermal shock test(-40°C/30min~80°C/30min), salt spray test(35°C, 5%NaCl, 48 hours), high temperature/high humidity test(60°C/90%RH, 240 hours), high temperature storage test(80°C, 240 hours) and writing test(250g, 200,000 times). After each test, the VUS bonded TSPs showed no significant changes in electrical resistance.
As a summary, the VUS method can be successfully used in touch screen assembly with high speed and good reliability.

ACKNOWLEDGMENT
This work was supported in part by MIC & IITA through IT Leading R&D Support Project.

REFERENCES

[1] Rao Tummala et al., <u>Fundamentals of Microsystems Packaging</u>, McGraw-Hill, pp400-409.

[2] Luu Nguyen, "Wafer-Level Chip-Scale Packaging", *Professional Development Course of 55th Electronic Components and Technology Conference*, Orlando, FL, no.16, May. 2005, pp. 4-19.

[3] Jean-Charles Souriau et al., "Development on Wafer Level Anisotropic Conductive Film for Flip-Chip Interconnection", *Proceeding of 54th Electronic Components and Technology Conference,* Las Vegas, 2004, pp. 155-158.

[4] R.W.Kay et al., "Stencil Printing Technology for Wafer Level Bumping at Sub-100 Micron Pitch using Pb-Free Alloys", *Proceeding of 55th Electronic Components and Technology Conference*, Orlando, FL, May. 2005, pp. 848-854.

[5] R.Wayne Johnson et al., "Wafer-Applied Underfill: Flip-Chip Assembly and Reliability", IEEE Trans. On Electronics Packaging Manufacturing, vol.27, no.2, (2004), pp. 101-108.

[6] K.W.Paik et al., "Flip Chip Assembly on Organic Boards Using Anisotropic Conductive Adhesives (ACAs) and Nickel/Gold Bumps", *Proceedings of 5th Electronic Packaging Technology Conference,* Singapore, 2000, pp. 378-384.

[7] Kiwon Lee et al., "Curing and Bonding Behaviors of Anisotropic Conductive Films (ACFs) by Ultrasonic Vibration for Flip Chip Interconnection", 56th Electronic Components and Technology Conference, San Diego, California, USA, May 30 – June 2, 2006

[8] Kiwon Lee et al., "Ultrasonic Anisotropic Conductive Films (ACFs) Bonding of Flexible Substrates on Organic Rigid Boards at Room Temperature", 57th Electronic Components and Technology Conference, Reno, Nevada, USA, May 29 – June 1, 2007

[9] J. Liu et al., "A Reliable and Environmentally Friendly Packaging Technology-Flip Chip Joining Using Anisotropically Conductive Adhesive", IEEE Trans. Comp. Packag., Manufact. Technol., Vol. 22, No. 2, pp.186~190, 1999

LEAD-FREE FLUX TECHNOLOGY AND INFLUENCE ON CLEANING

Ning-Cheng Lee, Ph.D.
Indium Corporation
Clinton, NY, USA
nclee@indium.com

ABSTRACT

Lead-free flux technology for electronic industry is mainly driven by high soldering temperature, high alloy surface tension, miniaturization, air soldering due to low cost consideration, and environmental concern. Accordingly, the flux features desired included high thermal stability, high resistance against burn-off, high oxidation resistance, high oxygen barrier capability, low surface tension, high fluxing capacity, slow wetting, low moisture pickup, high hot viscosity, and halogen-free. For each of the feature listed above, corresponding desired chemical structures can be deduced, and the impact of those structure on flux residue cleanability can be speculated. Overall, lead-free flux technology results in a greater difficulty in cleaning. Cleaner with a better matching solvency for the residue as well as a higher cleaning temperature or agitation are needed. Alkaline and polar cleaner are often needed to deal with the larger quantity of fluxing products. Reactive cleaner is also desired to address the side reaction products such as crosslinked residue.

Key words: lead-free, flux, flux residue, solder, soldering, cleaner, cleaning, SMT

INTRODUCTION

The electronic industry has been driven by miniaturization and low cost for many decades. Since mid 1990's, environmental consideration also joined the main theme. The resultant European RoHS and REACH essentially pushed the whole world toward lead-free soldering as well as halogen free. For industries which can not tolerate flux residue on the finished products, the combined constraints caused by the main drivers posed major challenges toward cleaning of post-soldering flux residue, as will be discussed below.

CHALLENGES

Temperature

Although some other alloys such as eutectic BiSn, eutectic SnZn or their modification are also in use, the main stream lead-free solder alloys adopted by electronic industry include Sn-Ag-Cu (SAC), Sn-Ag (SA), Sn-Cu (SC), and modification of those alloys, as shown in Fig. 1 [1].

The most commonly used alloys include Sn95.5Ag4.0Cu0.5 (SAC405), Sn96.5Ag3.0Cu0.5 (SAC305),

Sn98.5Ag1.0Cu0.5 (SAC105), Sn96.5Ag3.5, Sn99.3Cu0.7, with a melting temperature ranging from 217 to 227°C. This is about 40°C above the melting temperature of Sn63Pb37 (183°C) or Sn62Pb36Ag2 (179°C). Consequently, the soldering process employed for those alloys is also elevated to a higher temperature. For reflow process, the peak temperature ranges from 230 to 260°C. For wave soldering, the peak temperature ranges from 255 to 270°C. Thus, the soldering temperature typically is 20-40°C higher than that of SnPb.

The use of a higher soldering temperature inevitably results in (1) a greater amount of flux thermal decomposition and flux side-reaction, (2) a greater amount of flux burn-off, particularly at temperature above the melting temperature of solder, and (3) a higher extent of oxidation of both fluxes and metals. The resultant phenomena described above further induce poorer wetting and more voiding.

To avoid problems caused by the higher soldering temperature, fluxes with the following features are desired: (1) a higher thermal stability, (2) a higher resistance against burn-off, (3) a higher oxidation resistance, and (4) a higher oxygen barrier capability.

Wetting

The surface tension of lead-free alloys (0.55-0.57N/m for SAC) is about 20% higher than Sn63Pb37 (0.51N/m), as shown in Fig. 2 [2]. This higher surface tension results in poorer wetting, as reflected by the longer wetting time shown in Fig. 3 [2].

The impact of high surface tension, or poor wetting is tremendous, and may include symptoms such as (1) low solder joint strength, (2) high voiding, (3) poor solder joint reliability. The relation between high surface tension and high voiding rate is demonstrated in Fig. 4 [2].

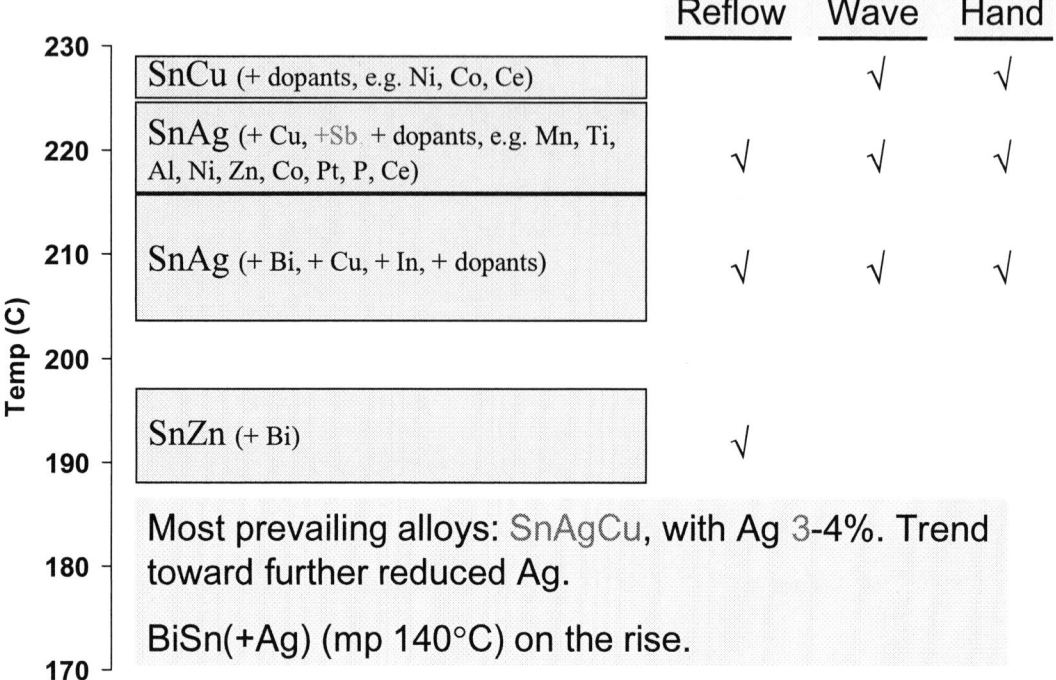

Fig. 1 Status of main stream lead-free solder alloys and their applications [1]

This deficiency in alloy wetting needs to be compensated with a flux with a better wetting. Thus, fluxes with the following features are needed: (1) a lower surface tension facilitating a better solder spread [3], and (2) a higher flux capacity and/or a higher flux strength.

Miniaturization
Oxide Thickness

The trend toward miniaturization has driven solder joints to shrink continuously. In general, the volume of soldering materials, including fluxes and solder, reduces in proportion with decreasing pitch. This proportional reduction trend is mainly driven by the (1) simplicity of design, (2) necessity of processing.

The first factor is easily understandable. The second factor is actually a result of physics. For instance, in SMT solder

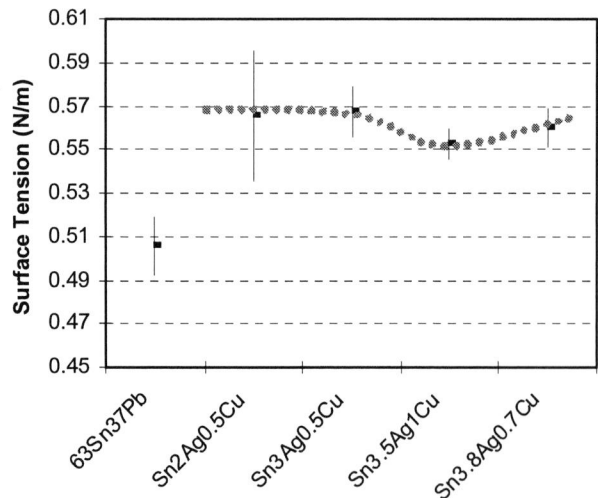

Fig. 2 Surface tension of Sn63Pb37 and SnAgCu alloys determined at 245°C and 260°C, respectively [2].

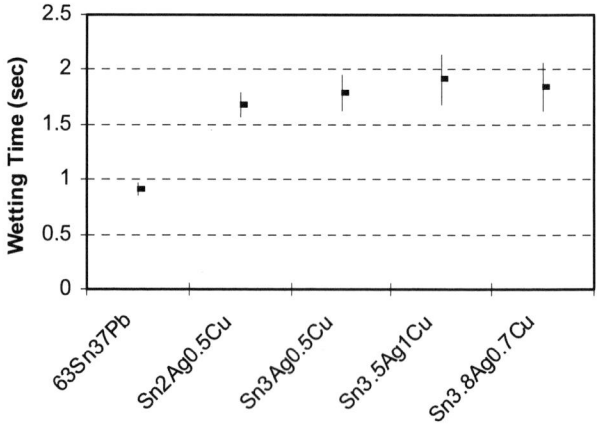

Fig. 3 Wetting time of Sn63Pb37 and SnAgCu solder alloys determined at 245°C and 260°C, respectively [2].

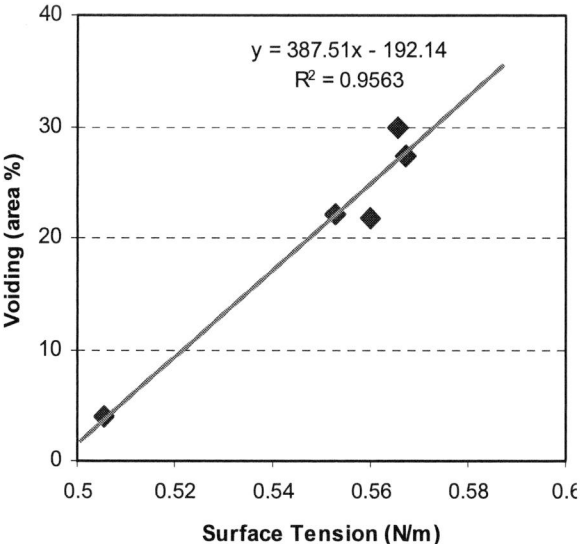

Fig. 4 Relation between surface tension and voiding rate at microvia for Sn63Pb37 and SnAgCu alloys [2].

paste printing, the amount of solder paste deposited is limited by the "area ratio" rule. If the area ratio of aperture opening to that of aperture side wall is smaller than 0.65, than the paste volume transfer efficiency decreases significantly. In other words, the adhesion of paste toward pads can not be pushed to be much smaller than the adhesion toward aperture side wall in order to have the paste properly released from stencil.

However, when the solder materials are shrunk in proportional to the pitch, the thickness of metal oxide does not shrink in proportion, as illustrated in Fig. 5.

The metal here refers to PCB pads, component leads, and solder powder. Consequently, the amount of oxide to be removed by unit volume of flux increases with decreasing pad dimension. To compensate for this increasing work load, the fluxing capacity per unit amount of flux needs to be increased.

Oxygen Penetration Path
Another challenge associated with decreasing pad

dimension is the decreasing oxygen penetration path through flux or solder paste, as also shown in Fig. 5. This inevitably results in a more rapid oxidation of both flux materials and metals covered by the flux if soldered under air. Hence, a flux with greater oxidation resistance as well as a greater oxygen barrier capability is desired for finer pitch applications.

Flux Burn-Off
To make things even more difficult, the flux burn-off increases with decreasing flux quantity deposited, as shown in Fig. 6 [4].

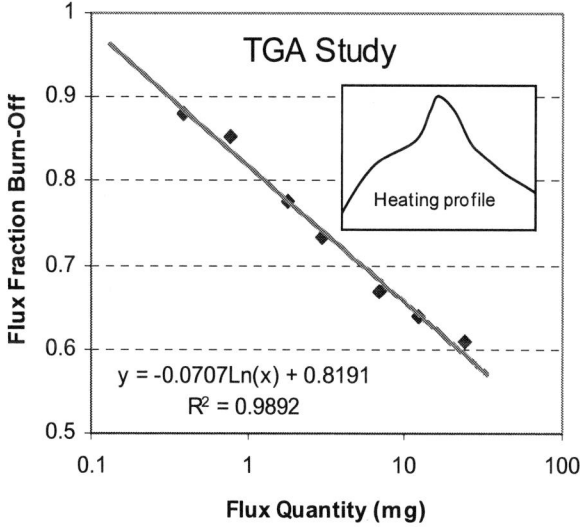

Fig. 6 Relation between flux fraction burn-off and flux quantity [4].

Apparently, the flux burnt-off during heating will not be able to participate oxide removal. To offset this unfavorable trend, the flux desired for finer pitch needs to be (1) more resistant to flux burn-off, or (2) higher in flux capacity per unit amount of flux.

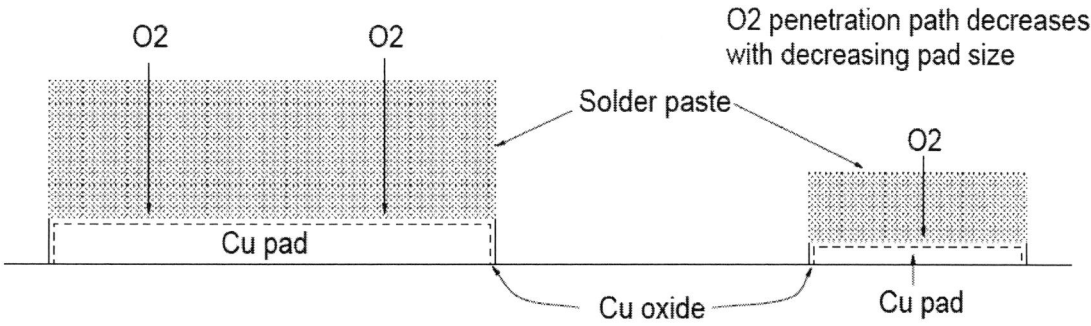

Fig. 5 Schematic drawing showing relation between oxide thickness, oxygen penetration path length, and pad dimension.

Wetting Speed

At SMT assembly, defects due to unbalanced wetting force, such as tombstoning or swimming, increases with decreasing component size. Under this situation, fluxes with a slower wetting speed would allow more time for the wetting force to be balanced. Fig. 7 shows tombstoning rate decreases with increasing wetting time [5]. Hence, fluxes with a slower wetting speed are desired with further miniaturization.

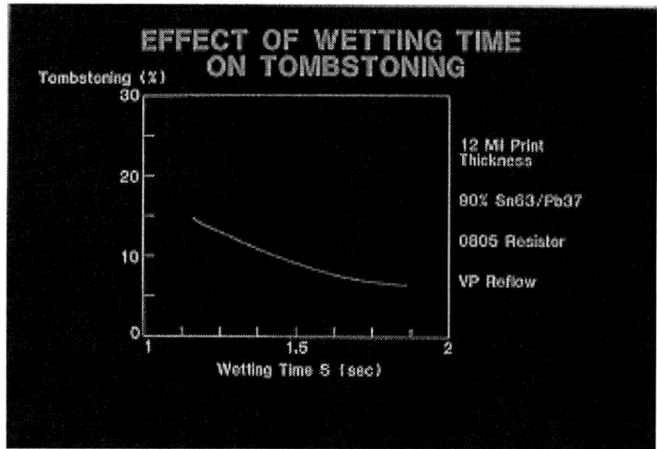

Fig. 7 Relation between tombstoning rate and flux wetting time [5].

Spattering

Miniaturization brings the solder joints closer to the gold fingers, hence is more vulnerable toward solder spattering.

Spattering can be caused by moisture pickup of the solder paste. It can also be caused by solder coalescence action. At reflow, the interior of solder powder melts. Once the solder powder surface oxide is eliminated by the fluxing reaction, the millions of tiny solder droplets will coalesce and form one integral solder piece. The faster the fluxing reaction rate, the stronger the coalescence driving force, and accordingly the more severe the spattering should be expected [3].

To minimize solder spattering, fluxes with low moisture pickup and slow wetting speed would be desired.

Slump

Bridging caused by slump of solder paste is another major concern at miniaturization. Fig. 8 shows the bridging rate increases with decreasing pitch dimension [5]. To prevent the slump from happening, fluxes with higher hot viscosity will be desired.

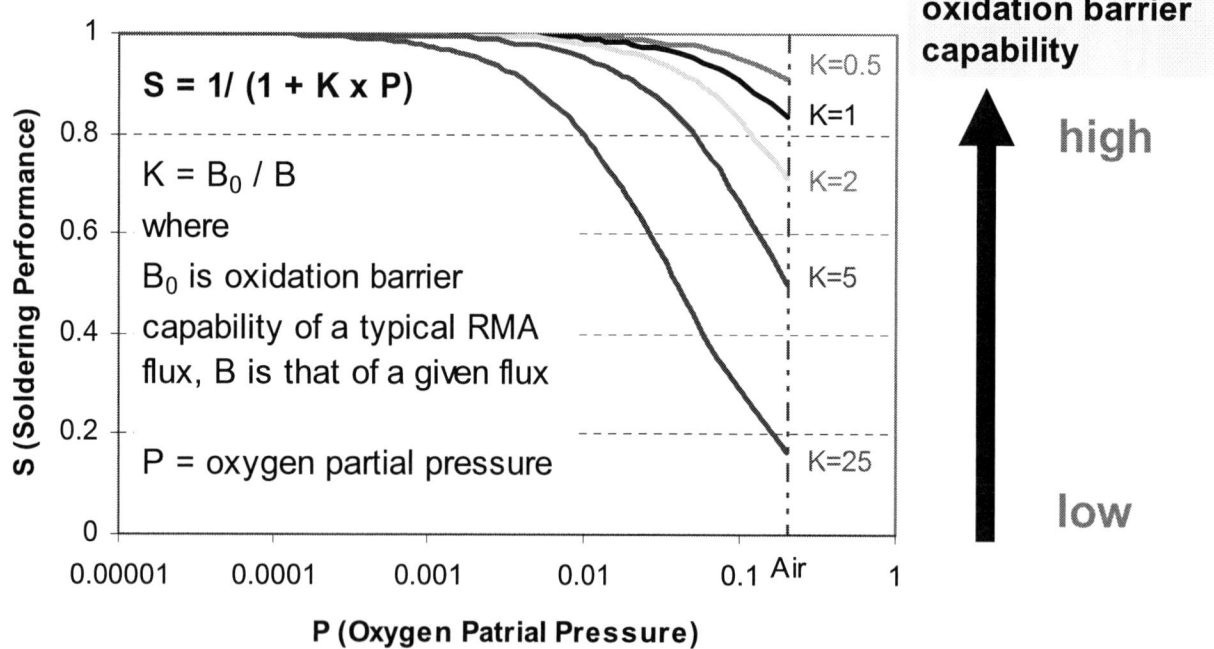

P. Jaeger and N.-C. Lee, "A Model Study of Low Residue No-Clean Solder Paste", in Proceeding of Nepcon West, Anaheim, CA, 1992.

Fig. 9 Soldering performance versus oxidation barrier capability versus oxygen partial pressure [6].

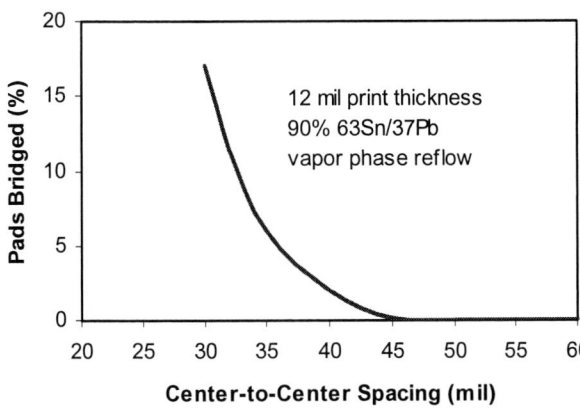

Fig. 8 Effect of center-to-center spacing on bridging [5].

Soldering Under Air

Low cost driver is pushing the industry hard to have soldering done under air. The immediate challenge is to achieve satisfactory soldering under air. Since oxidized surface is difficult to wet, the logical solution for this challenge is to employ fluxes with a good oxygen barrier capability so that oxidation can be minimized. The effect of oxygen barrier capability on soldering performance is illustrated in Fig. 9 [6]. It can be seen clearly that without the help of nitrogen atmosphere, a very high oxygen barrier capability of flux is needed in order to have high soldering performance.

biphenyls (PBB) and polybrominated diphenyl ethers (PBDE) as harmful to human health. On the other hand, REACH SIN (Substitute It Now) list contains 30 chemicals classified as EQUIVALENT LEVEL OF CONCERN, 17 chemicals as PBTs (persistent, bio-accumulative, and toxic substances), and 220 chemicals as Classified CMRs (carcinogenic, mutagenic, or toxic). Out of a total 267 chemicals, only very limited number of halogen-containing chemicals such as 1,2-dichlorobenzene and hexabromocyclododecane are listed. In the REACH SVHC (Substance of Very High Concern) list, 3 out of 15 chemicals are halogenated chemicals.

Although most of the halogen-containing flux chemicals used in the industry are not on the list of RoHS or REACH, the industry still moves rapidly toward halogen-free fluxes. Besides the questionable perception that halogen-free is less-corrosive than halogen-containing fluxes, another driver is the difficulty in differentiating the banned halogenated chemicals from those not on the banned list.

The trend toward halogen-free flux is here to stay.

Table 1 Flux feature desired in order to meet the challenges

Flux feature desired	Challenge				
	High temperature	High solder surface tension	Miniaturization	Air	Environment
High thermal stability	x				
High resistance against burn-off	x		x		
High oxidation resistance	x		x	x	
High oxygen barrier capability	x		x	x	
Low surface tension		x			
High fluxing capacity/strength		x	x		
Slow wetting speed			x		
Low moisture pickup			x		
High hot viscosity			x		
Halogen-free					x

In conjunction with the miniaturization trend, as discussed in Section 3, the flux desired should exhibit the following properties: (1) a great oxidation resistance to prevent the flux from being oxidized, and (2) a great oxygen barrier capability to protect parts and solder from being oxidized.

Halogen-Free

Halogen-free is a trend of the industry, mainly out of human's perception instead of actual impact of halogen on environmental consideration or reliability. RoHS specifies that brominated flame retardants, including polybrominated

REVIEW OF FLUXES DESIRED

Bu reviewing the challenges discussed above, the flux feature desired can be listed in Table 1.

The flux chemical structures implied which would support the flux features desired are shown in Table 2.

Table 2 Chemical structure of flux implied supporting the flux feature desired.

Flux feature desired	Chemical structure implied	Cleanability impact
High thermal stability	Strong chemical bonds such as C-F, Si-O, cyclic or aromatic structures	Nonpolar cleaner needed. For crystalline residue, may need cleaner with high solvency and high cleaning temperature or agitation.
High resistance against burn-off	High intermolecular force	
	- high molecular weight	Cleaner with high solvency needed. High cleaning temperature or agitation desired.
	- high polarity or hydrogen bonding concentration	Require polar cleaner.
High oxidation resistance	Flux may contain more oxidation resistant chemical bonds such as aromatic structures, hydrocarbons, or silicone materials.	Some nonpolar cleaner needed.
High oxygen barrier capability	Low molecular free volume desired	
	- high concentration of covalent bond, such as high MW, high cyclic or aromatic structure, high crosslink density	Generally the residue will be high in viscosity and difficult to dissolve. Cleaner with high solvency needed. High dissolution temperature or agitation desired. If crosslinked, reactive cleaner needed.
	- high conc. of hydrogen bonding, such as high concentration of –OH, -NH, -SH, -O-, -N-, -S- functional groups	The high polarity of residue requires some polar solvents to dissolve them.
Low surface tension	Low polarity, light element, such as hydrocarbon or silicone	Require nonpolar cleaner.
High fluxing capacity/strength	More polar functional groups, mainly more organic acids or halides, needed to react with oxides. Side reactions are very likely.	Side reaction products may be difficult to clean. Require more alkaline and polar cleaner.
Slow wetting speed	Chemicals with protective group on flux desired to slow down wetting.	Potential side reaction may cause greater difficulty in cleaning, reactive cleaner may be needed.
Low moisture pickup	Need nonpolar functional group	Need some nonpolar cleaner
High hot viscosity	Need some high MW ingredients	Cleaner with high solvency needed to dissolve the high MW residue. High cleaning temperature or agitation desired. If crosslinked, reactive cleaner required.
Halogen-free	High content of organic acids plus organic bases needed to react with oxides	Side reaction products may be difficult to clean. Require alkaline and polar cleaner

SUMMARY

Overall, lead-free flux technology results in a greater difficulty in cleaning. Cleaner with a better matching solvency for the residue as well as a higher cleaning temperature or agitation are needed.

Alkaline and polar cleaner are often needed to deal with the larger quantity of fluxing products. Reactive cleaner is also desired to address the side reaction products such as crosslinked residue.

CONCLUSIONS

Lead-free flux technology for electronic industry is mainly driven by high soldering temperature, high alloy surface tension, miniaturization, air soldering due to low cost consideration, and environmental concern. Accordingly, the flux features desired included high thermal stability, high resistance against burn-off, high oxidation resistance, high oxygen barrier capability, low surface tension, high fluxing capacity, slow wetting, low moisture pickup, high hot viscosity, and halogen-free. For each of the feature listed above, corresponding desired chemical structures can be deduced, and the impact of those structure on flux residue cleanability can be speculated. Overall, lead-free flux technology results in a greater difficulty in cleaning. Cleaner with a better matching solvency for the residue as well as a higher cleaning temperature or agitation are needed. Alkaline and polar cleaner are often needed to deal with the larger quantity of fluxing products. Reactive cleaner is also desired to address the side reaction products such as crosslinked residue.

REFERENCES

1. Ning-Cheng Lee, "Achieving high reliability lead-free soldering – materials consideration", ECTC, short course, San Diego, CA, May 26-29, 2009.
2. Benlih Huang, Arnab Dasgupta and Ning-Cheng Lee, "Effect of SAC Composition on Soldering Performance", Semicon West, STS:IEMT, San Jose, CA, July 13-16, 2004.
3. Ning-Cheng Lee, " REFLOW SOLDERING: Processing and Troubleshooting SMT, BGA, CSP and Flip Chip Technologies", Newnes, pp.288, 2001.
4. Ning-Cheng Lee, "Combining Superior Anti-Oxidation and Superior Print - Is it Really Impossible?", EPP EUROPE DECEMBRE 2007, pp.20-21
5. Gregory Evans and Ning-Cheng Lee, " Solder Paste: Meeting The SMT Challenge", SITE Magazine 1987
6. Paul Jaeger and Ning-Cheng Lee, " A Model Study of Low Residue No-Clean Solder Paste", Nepcon West 1992
7. Originally published in the Proceedings of the SMTA International Conference, San Diego, California, October 4 - 8, 2009.

DPBO – A NEW CONTROL CHART FOR ELECTRONICS ASSEMBLY

Daryl L. Santos, Ph.D.
SSIE Department
Small Scale Systems Integration and Packaging (S^3IP) – A NYS Center of Excellence
Binghamton University
Binghamton, NY, USA
santos@binghamton.edu

ABSTRACT

As six sigma (6σ) and better processes are demanded for higher yields and as organizations move from measuring defects in terms of parts-per-million (ppm) towards parts-per-billion (ppb), the resolution of extant control charts is becoming insufficient to monitor process quality. This work describes the development of a new statistical process control (SPC) chart that is used to monitor processes in terms of defects-per-billion-opportunities (dpbo). A logical extension of the defects-per-million-opportunities (dpmo) control chart, calculations used to derive the dpbo control limits will be presented and examples of in-control and out-of-control processes will be offered.

Key words: statistical process control, SPC, dpmo, dpbo, attributes data, ppm, ppb

INTRODUCTION

Long before Motorola made the term "six sigma" a buzzword in many industries, companies have strived to track their defects and improve their processes. Statistical process control (SPC) charts, developed by Shewhart dating back to the 1930s (see Shewhart (1931) for example) are useful for two purposes: 1) as a watchdog, an SPC chart can determine if a process is in-control; 2) data plotted on an SPC chart – particular for *variables types* of data (data that are measured on a continuous scale) – can be used as estimates for process parameters. For example, the centerline of a an *Xbar* chart can be used as an estimate of the process mean and the centerline of an *R* chart (or of an *s* chart) can be used to derive an estimate of the process standard deviation. For variables data, these estimates of mean and standard deviation can be used in process capability formulas such as C_p, C_{pk} and others to determine process yields (see Besterfield (2008) or Montgomery (2005) for more details).

When the types of data that are collected can be measured in discrete counts (such as the number of defects of a certain type (e.g., solder shorts on a PCB) or the number of bad printed circuit boards produced in a lot), then *attributes* control charts become appropriate for use.

The new SPC chart described in this work falls under the category of attributes control charts. The following is a list of the commonly used attributes control charts (others do

exist) and a description of problem types for where their use is appropriate:

- *p* chart – assesses the proportion defective and is used in the situation in which the entire product is considered to be good or bad. Per period in which the data are taken, subgroup sizes are greater than 1.
- *q* chart – assesses the proportion good parts, is also used in the situation in which the entire product is considered to be good or bad; it is the complement of the p chart. Subgroup sizes are greater than one.
- *np* chart – assesses the number of defective parts in a subgroup. This is a scaled version of the p chart where "n" is the subgroup size.
- *c* chart – assesses the number of defects per part (i.e., each part can have multiple defects). In each period of which data are collected, only one item is inspected (i.e., the subgroup size is 1).
- *u* chart – similar to the c chart, but is used for situations where the subgroup size is greater than one in each period.

The *c* and *u* charts are utilized when a product can have multiple defect opportunities. Consider a printed circuit board for example. The PCB can have solder defects, could have missing components, could have components misoriented, etc. If the interest is in determining what the total number of defects per PCB is, then the use of a c or u chart – depending upon sample size per period – would be appropriate. On the other hand, if we were only interested in determining if, as a whole, the PCB was "good" or "bad", then the use of the other charts (*p, q, np*, and other variations not discussed herein) would be appropriate.

Traditional control charts were developed in an era in which products were not as complex as they are in today's age. Taking the electronics industry as but one example, some products (e.g., printed circuit boards (PCBs)) have thousands and others have even tens of thousands of opportunities for defects. When it is of importance to accurately track how many defects that occur per product (as opposed to just making a determination if the entire product is "good" or "bad"), the increased complexity of today's products drives the need for new control charting techniques.

Relatively recently, a new control chart (not the even newer one which is the focus of this paper) was developed primarily for use in electronics manufacturing. This control chart is the defects per million opportunities, or *dpmo* chart. One of the first references for the *dpmo* chart is an electronics manufacturing application by Ngo (1995).

A Packard Bell researcher has suggested (Revelino, 1997) that "world-class" in electronics manufacturing would drop so much that defect levels will eventually be referred to in parts-per-billion (*ppb*). In fact, it was in part due to this statement that the defects per billion opportunities (*dbpo*) control chart was inspired.

DPMO AND DPBO CONTROL CHARTS – EXTENSIONS OF THE U CHART

The reader familiar with SPC calculations and the *dpmo* chart will realize that it is a chart that is used when it is of interest to keep track of the total number of defects per product. As such, it is an extension of (i.e., the control limits and plot point calculations are modifications of) the *u* chart as referred to above.

U Chart Calculations

In order to understand the calculations of the new, *dpbo* chart, it is of interest to show the calculations of first the *u* chart and then the *dpmo* chart.

For the *u* chart calculations, we need to define the following:

- k = number of subgroups
- n_i = size of subgroup i ($i = 1, 2, ..., k$); subgroups are typically constant, but may be allowed to vary
- c_i = count of defects in subgroup i ($i = 1, 2, ..., k$)

Given the above, the point to plot on this chart is the average number of defects per unit, typically denoted as u_i, and is calculated per each subgroup, i, as follows:

$$u_i = \frac{c_i}{n_i}$$

Once all subgroups are gathered, the centerline for the *u* chart is $\bar{u}$ and is calculated as follows:

$$\bar{u} = \frac{\text{total defects found in all subgroups}}{\text{total number of items inspected}} = \frac{\sum_i c_i}{\sum_i n_i}$$

Now that the centerline is established, the control limits can be calculated. As in other control limits, they are ±3 standard deviations of the process output. As process output for defect counts is assumed to follow a Poisson distribution (refer to Montgomery (2005)), the control limits are the following:

$$\bar{u} \pm 3\sqrt{\frac{\bar{u}}{n_i}}$$

DPMO Chart Calculations

In order to understand the calculations of the new, *dpbo* chart, it is of interest to show the calculations of first the *u* chart and then the *dpmo* chart.

One of the first references to the *dpmo* chart can be found in relation to an electronics manufacturing application by Ngo (1995). Although this reference is more than a decade old, most recently published quality control textbooks (see Montgomery (2005), DeVor et al. (2007), and Besterfield (2008) for examples) still do not include the *dpmo* chart in their review of attributes control charts.

The *dpmo* chart is particularly useful in the monitoring of electronics manufacturing operations or any other process that has products with large numbers of defect opportunities. For another practical application of the *dpmo* chart, see Santos et al. (1997) and see Yepez et al. (2008) for other SPC applications for electronics manufacturing. As electronics products become more complex, the number of defect opportunities per product has increased tremendously in recent years. Electronics products (e.g., backplanes, complex motherboards for server systems, etc.) can have as many as thousands of opportunities for defects per circuit board. The defects can be traced to improper solder joints (potentially thousands on a PCB), missing components, improperly placed components, and others.

Just as the *np* chart is a scaled version of the *p* chart, the *dpmo* chart is a scaled version of the *u* chart. The *u* chart assumes a few defect opportunities per product, but the *dpmo* chart assumes there are a substantial number of defect opportunities per product.

In order to calculate the control limits for the *dpmo* chart, we need to define the following:

- k = number of subgroups
- n_i = size of subgroup i ($i = 1, 2, ..., k$); subgroups are typically constant, but may be allowed to vary
- c_i = count of defects in subgroup i ($i = 1, 2, ..., k$)
- Number of defect opportunities per product (1)

Given the above, the average number of defects per unit, dpu_i, is calculated per each subgroup in a similar fashion as the plot points for the *u* chart. However, dpu_i, shown below, is not the plot point for the *dpmo* chart:

$$dpu_i = \frac{c_i}{n_i}$$

The plot point for this chart is $dpmo_i$, for each subgroup, i. The plot point is found by the following calculation which is a scaled version of the number of defects per unit provided in reference to one million defect opportunities:

$$dpmo_i = \frac{dpu_i}{\text{number of defect opportunities}} \times 10^6$$

Once all subgroups are gathered, the centerline for the *dpmo* chart is $\overline{dpmo}$ and is calculated as the average of all the *dpmo* values as follows:

$$\overline{dpmo} = \frac{\sum_i dpmo_i}{k}$$

Now that the centerline is established, the control limits can be calculated. The control limits for the *dpmo* chart will also be based upon the Poisson distribution as the purpose is to still measure defect counts which are assumed to be distributed in a Poisson fashion; but, as expected, the *dpmo* chart limits are a scaled version of the *u* chart's as follows:

$$\overline{dpmo} \pm 3\sqrt{\frac{\overline{dpmo} \; x \; 10^6}{n_i x \, (\text{number of defect opportunities})}}$$

And the above can be simplified to the following:

$$\overline{dpmo} \pm 3000\sqrt{\frac{\overline{dpmo}}{n_i \; x \, (\text{number of defect opportunities})}}$$

DPBO Chart Calculations

It may be obvious to the reader that the *dpbo* chart represents yet another scaling of the *u* chart. In this case, and keeping in line with the earlier statement by Revelino, the scale/resolution of a *u* chart is just too large to be of practical use for dealing with defect levels at six sigma or beyond (i.e., very low defect levels with applications that have high opportunities for defects). Like the *dpmo* chart, the *dpbo* chart is of benefit to the variety of manufacturing operations that have products with very large numbers of defect opportunities such as the multitude of organizations in the various levels of the electronics packaging industry – from Level 0 (semiconductor fabrication) through Level 2 (printed circuit board assembly).

Aside from a new variable, $dpbo_i$, the variables in this chart are the same as those appearing for use in the *dpmo* chart discussion, above. The calculations to follow have been provided in both in an unpublished presentation (Santos, 2008) and in another that was recently published (Santos, 2009a).

The plot point for this chart is $dpbo_i$ and is calculated (based upon the dpu_i calculation as above) as follows:

$$dpbo_i = \frac{dpu_i}{\text{number of defect opportunities}} \; x \; 10^9$$

The centerline is $\overline{dpbo}$ and is calculated as follows:

$$\overline{dpbo} = \frac{\sum_i dpbo_i}{k} \tag{6}$$

The control limits for the *dpbo* chart are the following:

$$\overline{dpbo} \pm 3\sqrt{\frac{\overline{dpbo} \; x \; 10^9}{n_i x \, (\text{number of defect opportunities})}}$$

The above limits simplify to the following:

$$\tag{7}$$

$$\overline{dpbo} \pm 94,868.33\sqrt{\frac{\overline{dpbo}}{n_i \; x \, (\text{number of defect opportunities})}}$$

DPBO CHART EXAMPLE

To illustrate the *dpbo* chart, a hypothetical example – but one indicative of a typical PCB assembly process – will be utilized. In addition to a *dpbo* chart, a *dpmo* chart for the same data will be presented. This is the same example as is demonstrated in Santos (2009a). This hypothetical example is intentionally designed to be of questionable (i.e., less than 6σ) quality levels so as to compare the *dpbo* chart with the *dpmo* chart and to demonstrate that it is, in fact, a scaled version of the *dpmo* (and, though not demonstrated, the *u*) chart.

For this example, assume the following are known:
- 100 PCB assemblies are inspected each day ($n_i = 100$ for all days)
- Each assembly has 3,000 opportunities for defects
- 24 days are used to establish the control limits ($k = 24$)

Table 1 lists, for each day, the total number of defects found (per 100 assemblies), the *dpu* values, the *dpmo* values, and the *dpbo* values.

Day	Defects	*dpu*	*dpmo*	*dpbo*
1	19	0.19	63.33	63333.33
2	19	0.19	63.33	63333.33
3	22	0.22	73.33	73333.33
4	19	0.19	63.33	63333.33
5	21	0.21	70.00	70000.00
6	17	0.17	56.67	56666.67
7	29	0.29	96.67	96666.67
8	13	0.13	43.33	43333.33
9	15	0.15	50.00	50000.00
10	17	0.17	56.67	56666.67
11	16	0.16	53.33	53333.33
12	17	0.17	56.67	56666.67
13	17	0.17	56.67	56666.67
14	15	0.15	50.00	50000.00
15	23	0.23	76.67	76666.67
16	22	0.22	73.33	73333.33
17	27	0.27	90.00	90000.00
18	17	0.17	56.67	56666.67
19	20	0.20	66.67	66666.67
20	22	0.22	73.33	73333.33
21	20	0.20	66.67	66666.67
22	23	0.23	76.67	76666.67
23	30	0.30	100.00	100000.00
24	24	0.24	80.00	80000.00
		Averages	67.22	67222.22

Table 1. Example Data for *dpmo* and *dpbo* Control Chart Calculations

To demonstrate the calculations, consider the first day (subgroup). In Day 1, 19 defects were found. From the earlier equation, $dpu_1 = 19/100 = 0.19$. The $dpmo_1$ value is 63.33 ($= 10^6 \cdot 0.19/3000$). Of course, $dpbo_1$ is 63,333.33 ($= 10^9 \cdot 0.19/3000$). Based upon the results in Table 1 and in the earlier equations, the control limits for the *dpmo* chart are approximately the following:

- UCL = 112
- Centerline = 67
- LCL = 22

The control limits for the *dpbo* chart are approximately the following, as based upon Table 1 and the earlier equations:

- UCL = 112,130
- Centerline = 67,222
- LCL = 22,315

Figure 1 displays the *dpmo* chart for this example and Figure 2 displays the *dpbo* chart for this example.

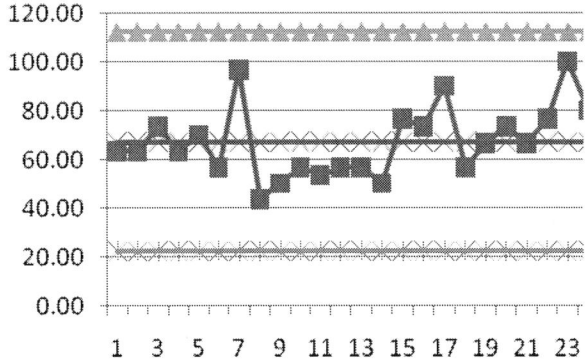

Figure 1. *dpmo* Chart for Example Problem

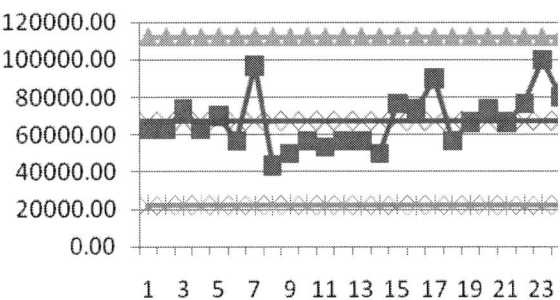

Figure 2. *dpbo* Chart for Example Problem

Analysis of DPBO Charts

Analysis of the *dpbo* chart is no different than the analysis of other attributes control charts. For this example, there are no plot points outside of the ±3σ bands. As such, it may be chosen to accept these control limits subject to periodic review. While there are no points outside the ±3σ control limits, there is an interesting situation in Days 8 through 14. Each of these plot points is below the centerline. Since these represent better-than-average defect levels as compared to the rest of the subgroups, it may be of interest (in a realistic situation) to investigate to determine if there are assignable causes for this.

An earlier point should be revisited that was discussed with this hypothetical example. This example intentionally does not reflect a 6σ process. So while the process appears to be in control, efforts should be made to reduce the defect levels. An additional comment can be made regarding high (whether out-of-control or not) *dpbo* values. Consider $dpbo_{23}$ which is the highest in this data set. Day 23 has the highest defect count of 30; even though this is in-control, one might be interested in determining why the value is high. What is not apparent from the data (and this is true whether we are using any of the other scaled charts – *dpmo* or *u*) is whether defects are indicative of all the parts in a subgroup (i.e., spread out among the 100 PCBs) or do these defects come from 1 or a few number of the PCBs on Day 23. If representative of the entire day, then investigation should be made to determine why that day was appreciably different from any other. On the other hand, the process, for that day, could be relatively similar to any other day, but the defects may have come from one bad (or very few) PCB(s) and perhaps the defects are traced more to supplied parts, than to process settings/parameters.

SUMMARY

From an examination of the example presented, it is apparent that the chart limits and plot points for the *dpbo* chart are 1000x their respective values for the *dpmo* chart – as they should be. The *dpbo* has been presented as a scaled version of not only the *dpmo* chart, but also of the *u* chart. The interested reader might then pose the question (which is similar to one posed in a recent graduate engineering class covering SPC concepts (Santos, 2009b): Why don't we just get the *dpmo* control limits and multiply those by 1000 instead of performing all of the calculations of the *dpbo* chart? The answer is simple – the author does not propose the use of both charts in realistic applications, only the use of one chart. Two charts were developed for demonstration and comparison purposes. Thus, if the *dpbo* chart is to be used, it is *not* suggested to be used *in addition to* the *dpmo* chart, but to *be used instead of* the *dpmo* chart.

The benefit from the use of the *dpbo* chart will ultimately become evident as processes with large opportunities for defect counts reach high quality (i.e., low defect) levels that are measured in terms of *ppb*, as opposed to *ppm*.

ACKNOWLEDGMENTS

This work is partially supported by the Small Scale Systems Integration and Packaging Center – a NYS Center of Excellence at Binghamton University.

REFERENCES

Besterfield, D. (2008). *Quality Control*, 8th ed., Upper Saddle River, NJ: Pearson Prentice Hall.

DeVor R.E, Chang, T.-H., and Sutherland, J.W. (2007). *Statistical Quality Design and Control: Contemporary Concepts and Methods*, 2nd ed., Upper Saddle River, NJ: Pearson Prentice Hall.

Montgomery, D.C. (2005). *Introduction to Statistical Quality Control*, 5th ed., Hoboken, NJ: John Wiley & Sons.

Ngo, P. (1995). Control Charts for Assembly Operations. *Circuits Assembly*, pp. 40–42 Sept.

Revelino, D. (1997). Achieving Single Digit DPMO in SMT Processes, *Surface Mount International Proceedings*, pp. 697-702.

Santos, D.L., et al. (1997). Defect Reduction in PCB Contract Manufacturing Operations. *Computers and Industrial Engineering*, 33(1-2), pp. 381-384.

Santos, D.L. (2008). Beyond Six Sigma – A Control Chart for Tracking Defects Per Billion Opportunities (dpbo). Invited but unpublished presentation at the 13th Annual International Journal of Industrial Engineering Conference, Las Vegas, NV, Sept.

Santos, D.L. (2009a). Beyond Six Sigma – A Control Chart for Tracking Defects per Billion Opportunities (dpbo). *International Journal of Industrial Engineering*, 16(3), 227-233.

Santos, D.L. (2009b). Classroom discussion in SSIE 561 – Quality Assurance for Engineers, Lecture 15, March 24.

Shewhart, W. A. (1931). *Economic Control of Quality of Manufactured Product*. New York: D. Van Nostrand Company.

Yepez, D. et al. (2008). A Comparison of Processing Techniques to Create Registration Holes on Continuous Roll-to-Roll Flexible Electronics Substrates Using SPC Techniques, *Proceedings: 13th Annual International Conference on Industrial Theory, Applications, and Practice*, Las Vegas, NV, Sept., pp. 763-772.

USING TEST OPTIMIZATION TO IMPROVE REWORK EFFECTIVENESS

Juan Coronado, Jorge Valle, Omar Garcia, Luis Manuel Zamora, Fernando Rodas, Mario Aguilar,
Federico Santos, Carlos Alberto Robles, Zhen (Jane) Feng*, Ph.D., Dason Cheung,* and Murad Kurwa*
FLEXTRONICS International Inc.
Zapopan, Mexico
Milpitas, CA, USA*

ABSTRACT

Currently with increasing PCBA density and complexity, it has become challenging to produce higher yields. Sometime we have to face a large number of rework boards in manufacturing. We use **A**utomated **X**-ray **I**nspection (AXI) and 2DX for finding the defective location and type for printed circuit boards, however most AXI machines were designed to test boards directly from the SMT line (not for rework board). We are faced with two questions: 1) How can we effectively use test machines to diagnose the defects? And 2) How can we reduce defects for the current SMT line with Real test data and results?

This paper will discuss the following:
1. The X-ray machine is a Non-Destructive method for detecting defects on PCBA boards. How can we effectively use different X-ray machines (AXI and 2DX)?
2. How can we reduce test time? Set right Algorithm and Threshold for programming; manual test mode only for critical parts; several programs are just for critical components.
3. Real time data feedback to line. Prevention is much more important than detection.

We have done the following:
1. Evaluated AXI, 2DX, AOI, and other test machines' capability in order to understand their advantages and disadvantages, and to use them efficiently.
2. Kept the machines at the optimized conditions. Maintained AXI programs with high detection coverage and low false call ratio.
3. Performed sampling test for production boards at AXI.
4. For some critical parts, used individual program to test them at AXI or examine at 2DX.
5. Studied and analyzed the AOI, AXI, 2DX data, provided REAL time feedback to SMT line.
6. Used SPC data, focused main existing defects daily, found root cause, and solved issue as soon as possible with process, test and SMT programming engineers.

Key words: AXI, 2DX, AOI, SPI, Efficiency, and Yields.

INTRODUCTION

As PCBA becomes more complex, the X-ray machine as a Non-Destructive test method is widely used in electronics manufacturing, especially for components under RFI (radio frequency interference) shield. This year we faced the challenge of having numerous rework boards, and in the meantime we still ran the similar products on the line. This presented the following questions: How can we effectively use different X-ray machines: AXI and 2DX? How can we reduce test time? How can we use real time data feedback toward our SMT process? We had to answer these kinds of questions more practically and seriously than we did before. In this paper we will review what we have been doing for the last year, and share it with our SMT field colleagues.

The board we worked with was a complex board which had CSP (pitch size = 0.4mm), QFN (0.5mm), BGA (0.8mm), 0402, 0201. There were about 60 solder joints per cm^2. The PCB thickness was about 1.23mm. More than 82% components were under RFI shield. With current test machines' capabilities, we focused on SPI, AOI, AXI testing data and analyzed data on a daily basis, and reduced defects as soon as possible. We used AXI to sample test the production board and not only used X-ray test for rework because prevention was most important. In the meantime, we also introduced new AXI machines, evaluated machines, and worked with the vendors to improve the machine's performance. Working closely with manufacturing engineers, the SMT yields were increasing obviously since we put the effort in these testing methods: SPI, AOI, and AXI that will be explained in the Methods & Improvement section.

In conclusion, we will list what we learned from our progress, and what we are planning to do with it. SMT Improvement is not a simple job, and there are many items we can consider for further actions.

METHODS AND IMPROVEMENT

All test machines do not increase value during the SMT manufacturing. However no one can ignore them before having zero defects products. How can we effectively use test machines? For this product, we had SPI and AOI machines for each line, and an AXI machine for almost 10 lines. Therefore we focused on SPI and AOI test data and used them to improve our process. We used AXI as sampling test and for rework boards.

SPI

Based on SPI DPMO data (Figure 1), we introduced Nano-Coating stencil to the product. The Nano-Coating was applied only to the walls of the apertures and the bottom sides (side facing the printed circuit board). Due to the chemical composition of the Nano-Coating, intrinsic and durable properties are guaranteed: Better solder paste release, enlarged area ratio, enlarged aspect ratio, reduced number of stencil cleaning cycles. Nano coated stencil provided in comparison to traditional stencils 6 to 18 % higher solder paste transfer relating to the nominal volume. The solder paste SPI images for Nano coated and laser cut stencil for CSP (pitch size = 0.4 mm) are shown in Figure 2. The stencil aperture reduction from 9.5 mils to 8.6 mils was possible with the Nano coated stencil and solder paste deposit was not smearing out of the pad with Nano coated stencil. Figure 3 indicates that the solder height Cpk increased to 1.55 from 1.25 after using Nano coated stencil replace laser cut stencil for CSP component. Figure 4 listed the paste height dispersion between stencil types: Laser Nickel, Nano, and Electro Forming Stencil which is a new type that is used hardened Nickel (NiEX). Figure 5 shows that the usage of the Nano coated stencil performed well and reduce the solder short for CSP.

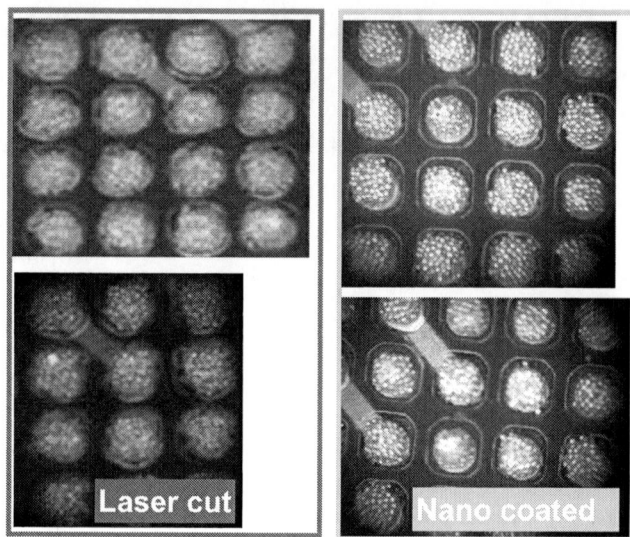

Figure 2. SPI images for CSP with Laser cut and Nano coated stencil.

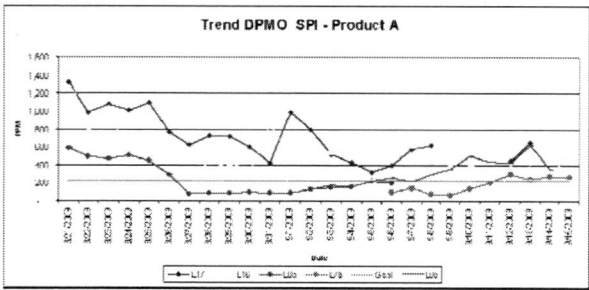

Figure 1. SPI DPMO data

Since SPI was the first test machine to detect defective solder joint with its height, area, volume, we preferred to detect as many as possible defects at the beginning of SMT process[2]. Therefore we worked with our SPI vendor, and installed the latest software version 7.7 for the Cyberoptics SE300 and inspected critical areas such as the CSP was restricted.

Our recommendations for SPI founding are as follows:
1. Any repeat defect continuously found three times should be reported to the line leader/supervisor and the line should be stopped.
2. The line leader/supervisor should justify whether the issue is due to the stencil, paste or any setup issue for paste printer or SPI, etc.
3. If stencil clogging is found then the stencil should be cleaned.
4. If the paste level is low then paste should be added.
5. Otherwise engineering should be notified for future solution.
6. SPI should scan 100% coverage and report to any alert operator for clarification.
7. For any insufficient solder less than 60% of the target volume, the board should be cleaned and re-printed especially on the BGA and QFN components.

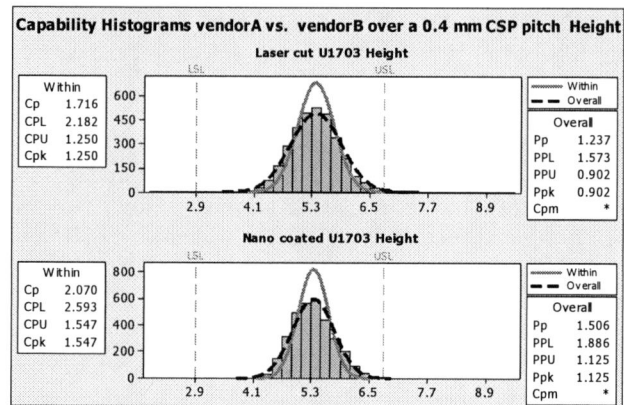

Figure 3. Cpk for CSP with Laser cut and Nano coated stencil.

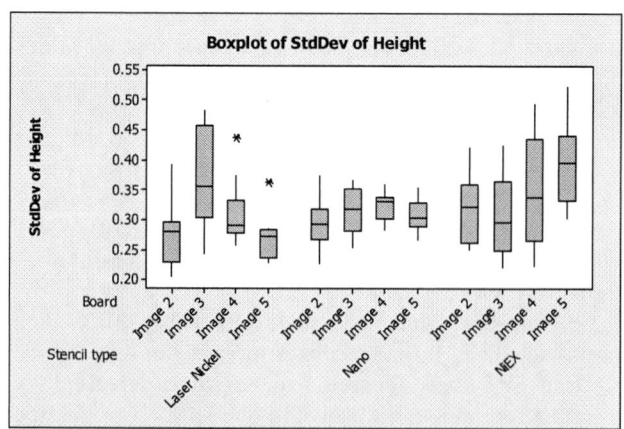

Figure 4. Paste height Variation between stencil types: Laser Ni, Nano and NiEX.

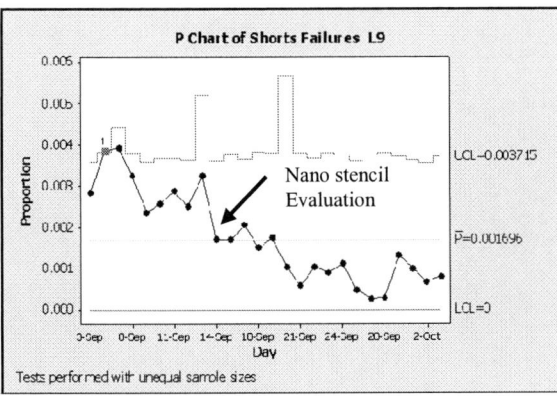

Figure 5. P chart of Short failure for CSP part

AOI

We used AOI data monitor SMT process daily and solved the defects issue. Figure 6 shows AOI weekly and daily DPMO sample data. The main issues identified were misalignment components which were U1703, L1001, and L140 as shown Figure 7a. This defect was mainly attributed to a PCB manufacturing issue, and stencil was re-designed to minimize the impact. However U1703 was impacting our yield on board level test even by keeping an eye on stencil cleanliness. U1703 was the top #1 defect based on Figure 7a (AQT of Agilent AOI data). Therefore we assigned this item to a team. After a deep analysis, we found that Head #2 on placement machine was placing the U1703 as misplaced. Acceptance tolerance on the AOI was reduced < 70 microns having this in place. We reduced to almost zero defects due to misplace U1703. Figure 7b indicates the defective location after our action, and U1703 disappear from the Pareto graph. So then we needed to focus defects on U801. Usually we focus top three defective locations every time. Figure 8 shows AOI DPMO was decreased after our actions for several weeks.

Currently we are working for these items in order to increase AOI capabilities.
1. Adding feeders' information on every single AOI program, so it will help and expedite SMT real defect.
2. Scanning system so defect can be uploaded on our data collection system: Further analysis can be made; having all the failed boards' information in Flex Flow system.
3. Our DPMO target < 100 or less for mobile boards

Our recommendations for AOI founding and work with IT engineers are as follows:
1. We linked the AOI system to the Alert Trigger system in order to stop testing if there were three consecutive failures with the same problem, plus an additional rule to react faster.
2. Make sure the AOI defective information is available for SMT technician and process engineer in real time.

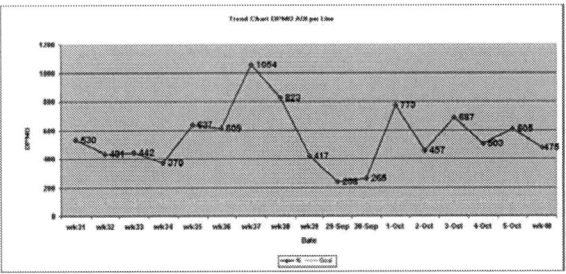

Figure 6. AOI weekly and daily DPMO

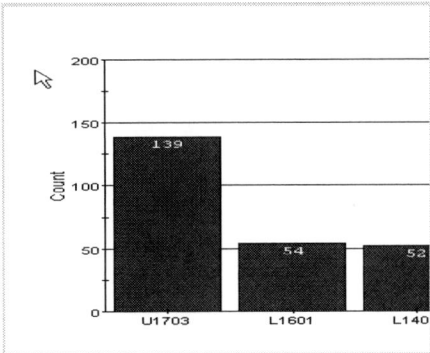

Figure 7a. Before our action.

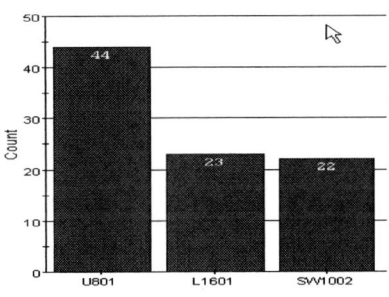

Figure 7b. After our action

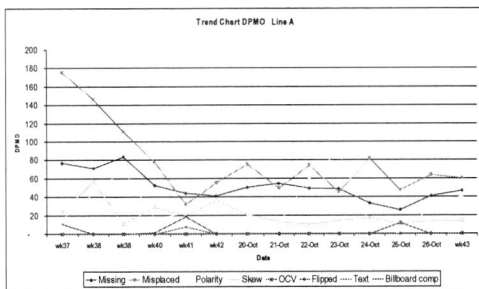

Figure 8. AOI DPMO Improvement

AXI

Because we did not have 5DX on each SMT line, we usually expected to develop 5DX program with full coverage for a new product if we had AXI machine time. This was the easy way to provide defects with real time to our process, especially for new process and package. Sometimes we also used measurement (variable) data to identify our process: We used BGA voids % to optimize oven profile[1]. After the process became stable with different and necessary actions, we preferred to reduce 5DX test time by using sampling or reducing test coverage.

We usually only tested BGA, PTH, and some critical components at 5DX. However we made sure that AOI and/or ICT covered the components which 5DX did not test. We used test tool software (Coverage Analysis System) for SPI, AOI, 5DX, and ICT test coverage study as shown in Figures 9a to 9d. Before optimization test, 10% of the components had repeat test and only 86% of the components had been tested as shown in Figure 9a. At the beginning, 5DX coverage was only 17% (Figure 9b).

Sometimes we adjusted 5DX program test coverage due to a process issue or other testing machine limitations. An example is shown in Table 1. QFN U5001 was tested at AOI at the beginning; however AOI was challenging to detect all QFN defects, especially for open and insufficient. So we put U5001 on 5DX test again. This part was also under RFI shield and it was also no access to test on ICT. This QFN tested on AOI too because AOI can detect short easily, so if it was short, the defective information would feedback to SMT line early. With this information, we saved the testing cost and also balanced the line better than before, and could make defective pin information feedback to line quickly. Figure 9c and 9d shows the test coverage as almost 100% after optimization, and repeat test coverage is 46%, 1% components (RFI shield) are not tested. The software can list all components' testing status.

In order to keep the 5DX machine at the optimized test conditions, we listed preventive maintenance as a regular schedule: Weekly, bi-monthly, and semi-annually. We also maintained 5DX program with good detection coverage and reasonable false call by using defective board from ICT with real defects escaped. Fine tuning 5DX program is on going.

As already stated, 5DX is designed to test boards from the SMT line directly. However we needed to check BGA after rework. Therefore we only generated BGA for the program whereby it may have more slices than the regular program. The false call rate may be higher than the regular program.

AXI is not a machine, it is a system and a test tool for solder joint inspection with attribute data, and it also can be a process improvement tool with measurement (variable) data[2]. Our recommendations are as follows:
1. Fine tuning is on going as process changed.
2. Suggest using golden defective board to test before and after bi-monthly confirmation and adjustment. Verify the same defective number and check the machine working conditions.
3. Feedback loop with repair station operator, ICT, Function test, and Final test technicians. AXI programmers need real defective board for further fine turning.
4. Monitor false call rate on going. It is easy to escape real defective call with too many false calls for rework station operators.

5. Use variable data of AXI to improve/optimize process.

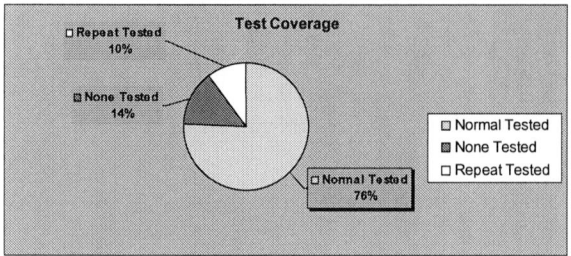

Figure 9a. Test coverage before optimization

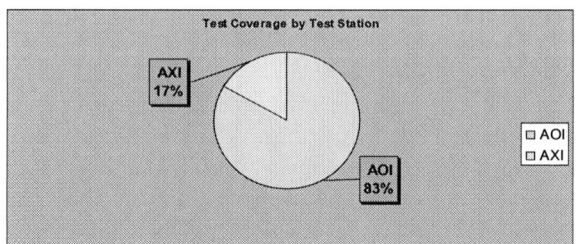

Figure 9b. AOI and AXI testing % before optimization

	BOM List	Package	AXI Test	AOI Test	ICT Test
Before	U5001	QFN65_9X9_5MM	NO TESTED	TESTED	NO TESTED
Now	U5001	QFN65_9X9_5MM	TESTED	TESTED	NO TESTED

Table 1. QFN U5001 testing information

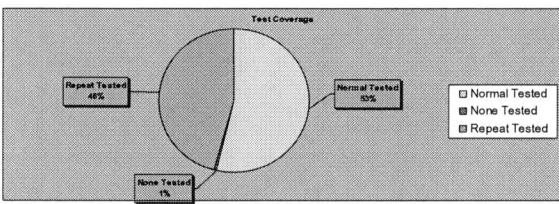

Figure 9c. Test coverage after optimization

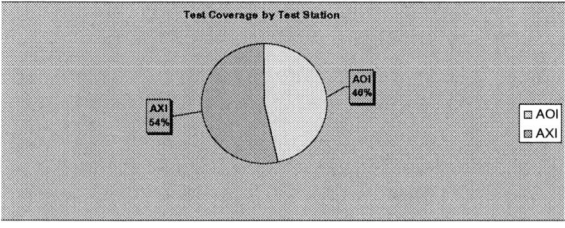

Figure 9d. AOI and AXI testing % after optimization

AXI Evaluation

Recently we evaluated some other AXI machines capabilities at our manufacturing line, and determined their advantage and disadvantage, and used the machines efficiently. We used the production board for programming speed, inspection speed, defect detection capabilities tests, and Gage R&R study. The boards have QFN, BGA, Fine pitch Gullwing, 0402, 0201, and the newest models contain POP components too.

Vendor A was 5DX, and we used it as the reference for the evaluation. Table 2 lists Gage R& R results. We used the same pins and components for the data collection. All results were obtained with the same analysis method and software (Minitab). At least 30 pins for each package, three operators, and each tested three times for the Gage R&R data collection. Our target was < 10% for Gage R&R, and 30% was acceptable. The vendor B and C data were better than 5DX, and only one item was > 30% for each vendor.

For AXI programming time, none of the vendors' machines could meet our expectation (< 10 hours). For inspection test time (tested number of pin/second), vendor B was slower than 5DX, and vendor C was much faster than 5DX; however vendor B had 100% components test coverage, and vender C did not.

For real defects escaped, we only used a small number boards for testing: None of the vendors met our target (defects escaped < 5%). Vendor B was similar to 5DX, and vendor C had more room to improve. For false call ppm, none of the vendors had good performance: its ppm was less than 2000. Figures 10-11 listed QFN void and Fine Pitch Gullwing insufficient defects detected images from each AXI. The images are not that clear for insufficient solder for Fine Pitch Gullwing as shown in Figure 11. That maybe the reason why all three AXI machines didn't meet our expectation: defects escaped < 5% with our evaluation boards.

Figure 12 shows the attribute Gage R & R results from software Minitab for vendor B. We tested nine times for one board with nine defective and 17 good components. The inspected and matched percentage was 96.15 for within Appraisers, and Appraiser versus Standard. Vendor A had better data than vendor B, but vendor C data was worse than vendor B.

While we are still working with these vendors at our manufacturing line and use their machines for testing our products, we are still looking for good AXI machines to meet our needs.

Gage R&R (variable data)	Vendor A	Vendor B	Vendor C
BGA ball diameter (U603)	4.26	1.98	5.67
BGA ball diameter (1703)	27.49	6.98	*0.89
FP Gullwing Fillet length	44.97	12.89	22.43
QFN Fillet length	11.38	32.68	102
Chip Fillet length	37.89	11.76	34.26

Table 2. Gage R&R for three AXI machines

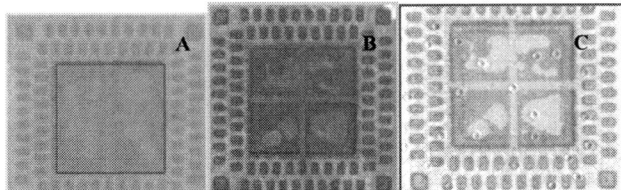

Figure 10. QFN Void (Vendor A, B, C)

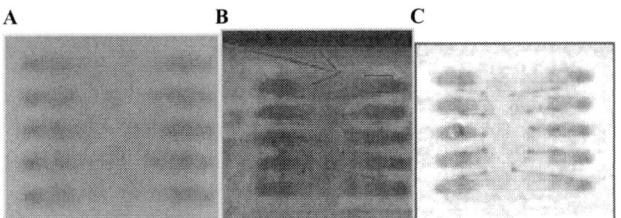

Figure 11. FP Gullwing Insufficient (Vendor A, B, C)

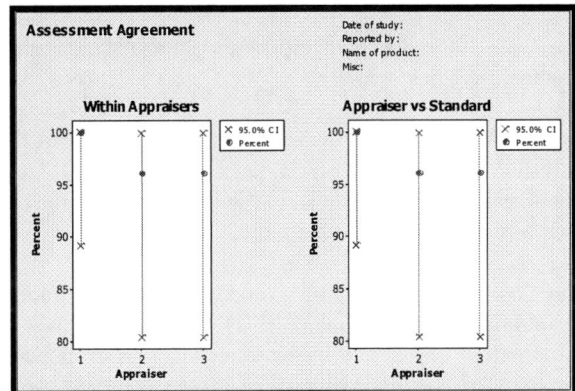

Figure 12. Attribute Gage R&R for Vendor B

2DX

2DX system can achieve the highest level of magnification up to 12,000 X and has the best resolution down to 100 nm. It is a very useful tool for failure analysis, but is also successfully used on the production line. The 2DX machine can tilt the image intensifier down 70 degrees in order to inspect BGA, QFN and other devices for voids, cracks, open joints, head-in-pillow and other defects. The highest level of magnification is available at oblique angle view and full rotation around the suspected joint is very easily accomplished. This is the key for finding and verifying open solder balls, joints and micro-cracks [3]

Compared with AXI, 2DX's disadvantage is that 100% auto inspection mode is not available. Also testing time is longer than AXI when we use 2DX for testing BGA with AUTO mode.

Therefore 2DX and AXI complement each other. For this assembly, we used 2DX for checking rework boards, and also for some critical components such as QFN and microphone. 2DX has a very detailed and clear image permitting to identify questionable solder joints which cannot be determined by AXI. Some packages (POP) and special defects like HIP require higher levels of experience for the 2DX operators.

Our recommendations are as follows:
1. Provide good training for 2DX operators.
2. Use 2DX to identify critical solder joints and packages and set the accurate threshold for the AXI program. Use 2DX for quick NPI delivery while we don't have an AXI program already in place as programming the AXI machine takes long time.

CONCLUSION

1. SPI, AOI, AXI and 2DX are Non-Destructive test machines that we need to use on RFI products due the poor testability, lack of test points and the heavy use of metals shields that complicate inspection and repair.

2. Use SPI and AOI as much as possible to inspect products on lines due to their short test time and early possible to detect defects.

3. Use AXI 100% for NPI and new package size in the new technologies.

4. Use 2DX image and data to set up accurate AXI Threshold for some critical packages. Use 2DX to examine components for which AXI testing cannot be used or the AXI data is unreliable.

5. The optimization testing method is still under observation with variable SMT products. Use a combination of test and inspection processes to inspect complicated boards with the most adequate and efficient manner.

6. We are still looking for an efficient and effective AXI machine which has an AXI advantage but also has a 2DX advantage which has a clear image for solder joints.

REFERENCES

[1] Zhen (Jane) Feng, Eduardo Toledo, Jonathan Jian and Murad Kurwa, "Reducing BGA Defects with AXI Inspection", Circuits Assembly, July 2005

[2] Zhen (Jane) Feng, Alex Garcia, Thomas Munnerlyn, Walid Meliane, Scott Kingery, and Murad Kurwa, "Automated X-ray Inspection: SMT Process Improvement Tool", SMTA Proceeding, September 2006.

[3] Zhen (Jane) Feng, Juan Carlos Gonzalez, Evstatin Krastev, Sea Tang, and Murad Kurwa, "Non-Destructive Techniques for Identifying Crack Defect in BGA Joints: TDR, 2DX, and Cross-section/SEM Comparison", SMTA Proceeding, August, 2008.

ACKNOWLEDGEMENTS

Flextronics Engineering and Production teams in Guadalajara, Mexico. X-ray vendor B and C support teams, Dage support team, and SPI Cyberoptics vendor support team.
Norma Viridiana Ojeda, Andy Zhang, JS Huang.

HOW IT CAN ENABLE ENERGY OPTIMIZATION:
A CASE STUDY ON TEXTILE FACTORIES

Eugenio Capra, Chiara Francalanci, Daniele Zagordi, and Alex Zazzera
Politecnico di Milano – Dipartimento di Elettronica e Informazione
Milano, Italy
capra@elet.polimi.it, francala@elet.polimi.it, daniele.zagordi@gmail.com, zazzeralex@gmail.com

ABSTRACT

This paper presents a case study that shows how state-of-the-art IT solutions can significantly reduce energy consumption and operative costs of an industrial process with a negligible investment. We analyzed the processes of textile factories and found out that the process responsible for the largest energy consumption is the drying of cloth rolls. We designed an IT system based on sensors and automatic controllers enabled by software optimization algorithms, and implemented it in an Italian SME. Our solution led to a reduction of 10% of the energy bill (electricity and gas).

INTRODUCTION

The attention paid to the environmental impact of industrial processes has significantly grown in the last few years, both because of corporate social responsibility issues and because of cost optimization opportunities.
Information Technology (IT), which on one hand consumes energy per se, on the other hand may play a significant role in monitoring and optimizing the energy consumption of industrial and business processes, as it has been recently suggested by many researches, including McKinsey Quarterly [6]-

This paper presents a case study that shows how state-of-the-art IT solutions can significantly reduce energy consumption and operative costs of an industrial process with a negligible investment. We analyzed the processes of textile factories (responsible of 7% of the Italian emissions amount, [7]) and found out that the process responsible for the largest energy consumption is the drying of cloth rolls. The energy bill of this process may reach up to 10% of total revenues. We designed an IT system based on sensors and automatic controllers enabled by software optimization algorithms, and implemented it in an Italian SME. Our solution led to a reduction of 10% of the energy bill (electricity and gas). If our solution were implemented in all Italian textile factories, we estimate a potential saving of 2,788 TWh per year, equivalent to 2 million tCO2eq, or 1 million trees.

The paper is organized as follows. Next section provides an overview of current researches on energy consumption, and in particular on the relevant role of IT. Then we describe our case study and the process analyzed, we present the implemented system and discuss the results. Finally, we propose some future work.

STATE OF THE ART
The Relevance of Energy Consumption

Former UK Government and World Bank Chief Economist Lord Stern, author of the Stern Review, makes it clear that ignoring now the rising carbon emissions, which will result in dangerous climate change, will damage economic growth in the future. According to the report, if no action is taken, the overall costs and risks of climate change will be equivalent to losing at least 5% of global gross domestic product (GDP) each year [1].

Thirty-four countries have signed up to the legally binding Kyoto Protocol, the agreement negotiated via the United Nations Framework Convention on Climate Change (UNFCCC), which sets a target for average global carbon emissions reductions of 5.4% relative to 1990 levels by 2012. Discussions for a post-2012 agreement are currently underway.

Individual regions and countries have also developed their own targets. In 2007, the European Union (EU) announced a 20% emissions reduction target compared to 1990 levels by 2020 and will increase this to 30% if there is an international agreement post-2012. The UK is aiming at a reduction of 60% below 1990 levels by 2050, with an interim target of about half that. Germany is aiming at a 40% cut below 1990 levels by 2020, while Norway will become carbon neutral by 2050. California's climate change legislation, known as AB 32, commits the state to 80% reductions below 1990 levels by 2050. China's latest five-year plan (2006-2010) contains 20% energy efficiency improvement targets to try to reduce the impact of recent fuel shortages on its economic growth.

As governments across the world wake up to the urgency of rising temperatures, they are increasingly focusing on how business is responding to both reduce their carbon footprints and to develop and supply the required innovations for a low carbon world [1].

ICT plays a role in carbon emissions, as according to Gartner research of 2007 ICT emissions are 2% of world's total [1]. However, focusing researches on IT energy consumption would exclude the remaining 98% of consumptions. According to recent researches [2], the optimization enabled by ICT in not-IT sectors could be as high as five times the emissions of the ICT sector itself.

The Role of IT in Monitoring Processes Energy Consumption

Because of its diffusion, ICT should be considered as a key factor for reducing global carbon emissions. ICT may enable new and smart ways to reduce energy consumption and, at the same time, reduce its own.

ICT sector is unique because, by means of its products and services, it is the only one that can gather and monitor important information about energy consumption of different processes. The needed changes are possible only through awareness of inefficiencies.

ICT can supply a wide variety of data, that could be used to understand and then change behaviors, abilities and processes.

ICT can enable reductions in different ways. We can summarize them in 5 different areas:

1. **Dematerialization:** the substitution of high carbon products and activities with low carbon alternatives e.g. replacing face-to-face meetings with video-conferencing, or paper with e-billing could play a substantial role in reducing emissions.

2. **Increasing the efficiency of productive processes:** motor systems (devices that convert electricity into mechanical power) lie at the heart of global industrial activity. Carbon emissions due to the use of these devices are continuously growing. The ICT could be applied to make a better use of these devices and to improve processes where these devices are used for. Moreover, all the processes that involve heating consume a lot of energy and may be not optimized.

3. **Smart logistics:** it comprises a range of software and hardware tools that monitor, optimize and manage operations, which help reduce the storage needed for inventory, fuel consumption, kilometers driven and frequency of vehicles travelling empty or partially loaded.

4. **Smart buildings:** a suite of technologies used to make the design, construction and operation of buildings more efficient, applicable to both existing and new buildings.

5. **Smart grids:** a "smart grid" is a set of software and hardware tools that enable generators to route power more efficiently, reducing the need for excess capacity and allowing two-way, real time information exchange with their customers for real time demand side management (DSM). It improves efficiency, energy monitoring and data capture across the power generation and T&D network.

All the initiatives in these areas aims at reducing energy consumption, because energy consumption and carbon emission are strictly bound. Our research in particular focused on the second area.

THE CASE STUDY CONTEXT
The Textile Industry
Textile industry comprises a lot of different processes, whose final goal is to process natural fibers to create textile fabric.
The most important phases are:
- Production of raw material
- Spinning
- Weaving
- Dyeing
- Printing
- Refining
- Packing and distribution

The Italian textile industry is composed by a large number of SME, whose purpose is to enforce only one or two of the phases describe before; no one of them is able to perform all the processes above together.
The whole textile sector is responsible for 7% of the amount of Italian energy consumption, i.e. it consumes about 9,75 TWh/year [3].

The Drying Process
A deeper focus on textile processes has showed that 77% of energy consumption is due to only three of the seven phases described above: dyeing, printing and refining. This is because each of these phases involves the drying process. After a tissue has been printed, it is completely wet, and it must be air-dried to fix the ink, or the picture, on it. This is because ink is primarily composed by water, in a percentage close to 90%.

To achieve this goal, factories use different methods. The most widespread one is to dry the fabric by means of boiler rooms. The tissue enters in a succession of boiler rooms (usually between 5 and 10) transported by a roller. Here it finds hot air, warmed up by a gas burner and diffused by a fan. The mass and thermal exchange is then achieved and the fabric comes out dry from the boiler rooms.

Gas burners warm the air up to 190°C and fans uses about 5,5 KW per each one when they're used at full power. It is clear that this process is very expensive, especially in Italy where the cost of electricity is among the highest in the world.The goal of our work has been to use ICT to improve this process by using the best combination of methane and electricity.

IT MODEL AND SYSTEM FOR INCREASING THE ENERGY EFFICIENCY FOR THE DRYING PROCESS
We analyzed the process of a specific company as a case study. The main goal of the project has been to optimize the drying process of cloth rolls, as in most textile factories it is responsible for the largest energy consumption and it usually has a high potential for optimization.
First of all, we have to split the drying process into two phases:
- set-up
- drying process at top speed.

The first one is the beginning of the entire process and it is not so relevant for the optimization proposed in this paper, because the speed of the cloth when it enters into the drying machine is very slow and there are a lot of stops due to many settings of the machine made by the technicians.

After this phase, the drying process reaches the top speed. We focused on this specific procedure. In fact we could do a lot of optimization in this phase, because it is well predictable and has an high consumption.

Problem Modeling
First of all, we have to define the *manufacture* concept, which is defined by:

- Cloth type (cotton, silk, wool, etc.)
- Cloth thickness
- Color type
- Coverage ratio

All these properties are important because they affect the drying time; for example a very thick cloth with a very complex drawing to stamp on it will introduce more water into the machine than the same cloth with a simpler drawing (for example some colored point on a white background) and this will cause an higher energy consumption as a consequence of a higher temperature to reach. When the cloth enters the drying machine we have to consider four variables:

- Gas jet temperature
- Fan speed
- Humidity level
- Cloth temperature at the second round

The first two variables affect the heat exchange on the cloth; they are not independent and, as we will discuss later on, there is a PID which adjust parameters to prevent damages to the cloth inside the machine. It is important to notice that variation of the fan speed can be obtained quicker than a variation of the gas jet temperature because of physical constraints.

The humidity level inside the drying machine is directly affected by the coverage ratio and by the cloth speed, as a faster cloth will introduce more water inside in the same time unit. Accordingly, it is very important to control humidity level to avoid condensation which could wet the cloth roll.

The last parameter to be considered is the temperature at the second round; as we explained in the previous section, at this point of the cloth path the wet surface will match with the roller surface. It is very important that there is no water inside the cloth to avoid dirtying the machine, but it is impossible to use humidity drill due to physical constraints For this reason we manually controlled the temperature on the cloth at this specific point.

During the problem modeling we decided to exclude some variables from the analysis. First of all, the humidity level

is not so important for cost reduction, as it never exceeds the minimum value for condensation. Moreover, the roller speed is not variable because it is fixed by technicians and setting a faster speed will cause a lot of stops, as we noticed during several experiments. Also the temperature at the second round cannot be controlled, as it is set by the technicians at the beginning and there is a PID controller that take care of it, by modifying the fan speed and the gas jet temperature accordingly to avoid some cloth damages. Now it is clear that a potential optimization should consider the exact combination of the fan speed and the gas jet temperature. However, these are not independent parameters, because when one of them is modified the PID controller adjusts the other one in order not to damage the cloth roll. As we discussed before, the fan speed is more responsive than the gas jet temperature and we decided to use the latter as control variable for the problem, letting the PID to adjust the other one.

The problem we just discussed can be considered as a minimum finding inside a cost function with the cost on the y-axis and the gas jet temperature on the x-axis. The cost can be calculated by this simple formula:

$$(KWh \text{ used} * €/KWh) + (m^3 \text{ gas used} * €/m)$$

and it should be calculated for a specific length of cloth roll in order to reach a steady state due to a temperature modification.

Algorithms for Finding Local Minima
In a lot of optimization problems it is not possible to find an exact solution for many reasons. For example there are a lot of situations where only an estimated value of a specific measure is available, but not its real value. In other cases there are real time constraints and the solution has to be computed in a very strict time. For these reasons it is very common to use *heuristic algorithms*. Generally, an heuristic algorithm is expected to find a *good* solution, i.e., a value comparable to the optimal one. However, there are some cases where an heuristic algorithm does not converge, which are called *NP-hard problems* [4].
These kind of algorithm can be divided into three main categories:

- *Constructive Algorithms*: starting from an empty solution they find some new elements to add to a complete solution in an iterative way.
- *Local Search Algorithms:* starting from a possible solution they try to reach the best one making some little adjustment in an iterative way until it does not change anymore.
- *Meta-Heuristic Algorithm:* they are similar to local search ones, but they try to avoid infinite loop.

The *hill-climbing algorithm* is an example of local search algorithm. Let us consider a set of discrete states called *x*; it tries to maximize (or minimize) a function *f(x)*. This kind of algorithm will compute the f value in each state x until a local maximum (or local minimum) x_m is reached. However, it can find a local maximum which is the highest value compared to its neighbors, but not the highest one compared to the entire set of states. There is an alternative version of this algorithm called *hill-climbing with random restart* that can solve this problem.

The Proposed Algorithm

In our study case we decided to use the *hill-climbing algorithm* to find not a local maximum, but a local minimum according to the function cost form, which we will discuss in the following section. This algorithm presents a very low complexity, so it can be used in a real-time context thanks to its rapid response time.

The proposed algorithm starts from the gas jet and the fan speed default values, derived from the technicians' experience and related to the cloth characteristics and the coverage percent. It starts decreasing the temperature inside the drying machine waiting for 400 meters of cloth roll, so to allow the conditions to stabilize. This waiting time is set to 400 meters for temperature variations of 5°C and 200 to meters for variations of 2°C , as in the latter case a shorter time is needed to reach a steady state. After this idle time, the algorithm starts to read the energy consumption for the next 400 meters of cloth roll and saves its value; then it compares this value with the previous one and if it is minor it continues decreasing the temperature. After two changes of directions it reaches the optimum state and saves this new combination into a database; the next time that a manufacture similar to this one will be produced, the algorithm will start with these settings.

The Developed IT System

The IT system to support these optimizations has been developed step by step, and finally tested, within the context of the textile factory analyzed in the case study.
We consider the case when the textile firm has no IT system to support its production process, and it has only the drying machine composed by at least one boiler with a gas jet and the fan. So the first intervention is to install some electrical inverter to control its motor systems. In particular, it is very important to control the fan speed and the humidity extractor fan.

Then it is necessary to install some temperature probes and some humidity probes to measure the parameters necessary to the algorithm. In particular the humidity probe has to be mounted inside the drying machine to control the humidity level and to avoid the damaging of the cloth roll. On the other hand, it is important to install a temperature probe for each boiler of the drying machine and three specific ones to control the cloth temperature at the end of the second round.

Another component to be implemented is a PID controller; this is a very common industrial automata

control. In our case it measures the temperature at the second step in input and compares it to a reference value. The difference between them, called error signal, is used to determine the output variables, which in our case are the fan speed and the gas jet.

Finally it is important to install a gas counter and electricity counter to monitor the instantaneous and the average cost of the manufacture. These values are very important to choose which action has to be performed.
All the manufacture data (the starting temperature, the starting fan speed, the temperature at the second step, etc.) have to be collected somewhere to permit some statistical analysis. Accordingly, it is necessary to install a simple database where all these kind of information can be stored and used to make some inferences.

The logical infrastructure of the controlling system can be seen in Figure 1. First of all, we can notice that the optimization algorithm has been placed ahead of the PID and that it controls the gas jet of each boiler. Accordingly, the PID adjusts the fan speed to maintain the temperature at the second round constant and to avoid any damage to the cloth roll.

The only intervention required to the technicians is to set the temperature desired at the second round; the other parameters, such as the initial temperature and the initial fan speed are taken from the database from the data of a similar manufacture. The electricity and the gas counters provide input variables for the algorithm; in fact, as we discussed before, the algorithm is based on the latest valus of energy consumption.

When the optimization algorithm has decided the action to take, it adjusts the boiler temperature reducing or increasing the gas jet; the PID controller adjusts the fan speed accordingly.

All the significant data are saved in the database at the end of the manufacture for two reasons: first, the algorithm at the beginning of the process will query for a similar manufacture in order to start from reasonable initial value; second, the data can be used for statistical analyses that could improve manual settings.

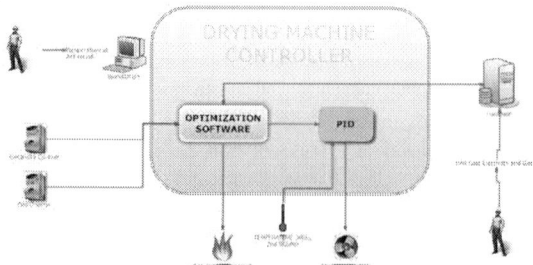

Figure 1 - The logical scheme of the developed IT system.

RESULTS

The Cost Function

First of all we conducted some tests to validate the initial hypothesis that the cost function, due to the variation of the temperature, has only one minimum. We conducted several tests on the drying machine of our case study firm. We report here four cases of these tests. We excluded from the analysis some manufactures with a low coverage ratio, as in these cases the machine work with minimum values of temperature and fan speed.

Table 1 reports the results of 4 tests specifying the temperature at the second round, the speed of the cloth roll, the coverage percent, the thickness of the cloth and the optimum consumption reached.

Figure 2 reports an example of test conducted to show the cost function. It is divided into two sections: the first part represents the total cost function, i.e., the sum of electricity and gas costs, and the second part shows some comparisons, in percent terms, between the minimum cost and the various costs at different temperatures.

Each graph is plotted with temperature intervals of 5°C. The test was conducted with the following parameters:

- Temperature at 2nd round: 80°C
- Cloth Speed: 55 m/min
- Coverage ratio: 80%
- Cloth thickness: 2.8 mm

In the worst non-optimized case, with a temperature of 160°C and fan speed at minimum level (40%) the cost is 9,19% higher than in the optimum case.

Test Case	Temp 2nd round [°C]	Cloth roll speed [m/min]	Coverage ratio [%]	Cloth thickness [mm]	T_{opt} [°C]	V_{opt} [%]
1	80	55	80	2.8	150	44
2	80	45	80	3.0	130	42
3	80	40	100	2.4	140	40
4	80	55	80	2.4	130	47

Table 1 - Test Cases.

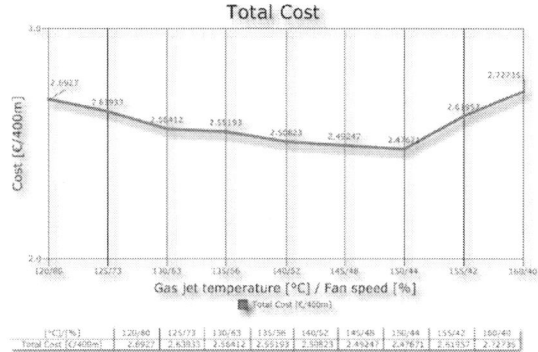

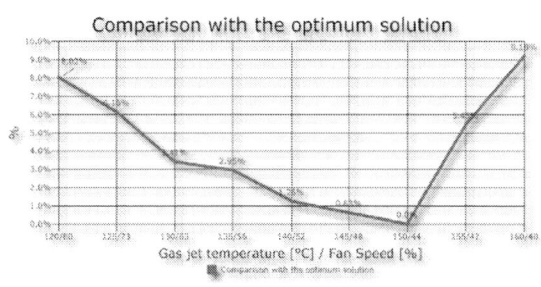

Figure 2 - First Test Case

Simulations

In this section we will discuss some simulations conducted to validate the algorithm. Unfortunately, at the time of writing this paper we have not yet managed to directly connect our system with the drying machine. However, we developed a Java simulator to validate our approach. We inserted the cost measured by the counters manually, and on the other hand we manually set the temperature and the fan speed according to output of the optimization software.

Figure 3 and Figure 4 presents the results of two simulations. Is can be noticed that the algorithm starts from the 1200th cloth roll due to the setup phase (about 400 meters). Temperature and fan speed do not rise proportionally: when one of them rises, the other one decreases. This is due to the PID action that tries to maintain the temperature at the second round constant. In both cases the algorithm needs six steps to reach the optimum combination.

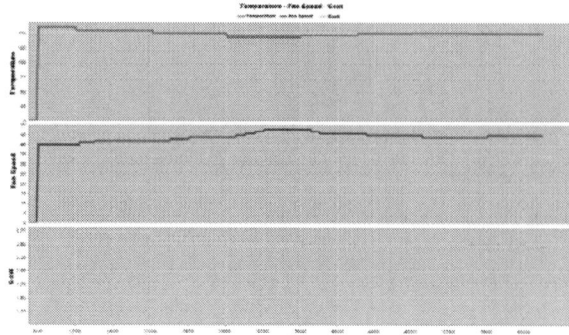

Figure 3 – Results of the first simulation (temperature, fan speed, and cost).

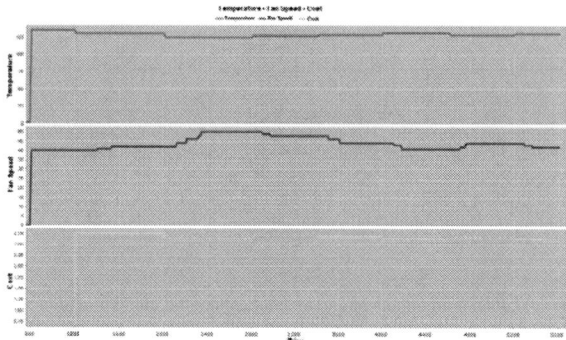

Figure 4 – Results of the second simulation (temperature, fan speed, and cost).

Discussion

In both simulations the algorithm reached the optimum solution in a really short time compared to the entire production time. The optimal combination are 150°C/44% in the first case, and 130°C/42% in the second case. These are different from the combinations used in most of the firms that we visited, as they tend to maintain as low as possible the fan speed, because they are misled by the fact the in Italy electricity is cheaper than gas. Our systems lead to average savings of 9-10%. The firm that we used as a case study has total revenues of 10 million €/year, and the energy bill impacts for the 8%. Saving up to 9-10% on energy could enable about 80k€ of saving per year, approximately 1% of the entire turnover.

CONCLUSIONS AND FUTURE WORK

IT technologies can help reduce energy consumption and the CO_2 emissions of other processes. The results of our case study seem to confirm the previsions of the *Economic Forum* [8], which foresees a reduction of CO_2 emissions between 5% and 20% thanks to the ICT. However, it could be hazardous to generalize this thesis because the optimization of processes through IT need a lot of research and analyzes in order to be "verticalized" to a specific process [5].

As future work, it is possible to collect more data about manufactures and populate the database to allow more in-

depth statistical analyses. Moreover, the temperature at the second round now comes from the technicians experience and it is considered as a constraint; we suppose that it could be optimized by testing it continually. Finally, the startup phase could be optimized as well by using fuzzy logic statements to model it.

ACKNOWLEDGMENTS

We wish to express our thanks to Mauro Miele, CEO of the textile firm where we conducted our tests, for his comments and help during this research work.

REFERENCES

[1] The Climate Group. *Smart 2020: Enabling the low carbon economy in the information age.*
Paper from http://www.theclimategroup.org/, 2009
[2] Gartner. *Green it: The new industry shockwave.* Presentation at Symposium/ITXPO conference, 2007.
[3] ISTAT. *I consumi energetici delle imprese industriali.* informazioni n.5, 2004.
[4] Dip. Ingegneria dell'Informazione di Padova Michele Monaci; *Algoritmi Euristici*; 2006
[5] World Economic Forum; *Green technology: Driving economic and enviromental benefits from ict*; http://www.weforum.org, 2009
[6] McKinsey Global Institute. *The carbon productivity challenge: curbing climate change and sustaining economic growth*, June 2008.
[7] Italian Ministry for Economic Development. *Energy statistics.* http://dgerm.sviluppoeconomico.gov.it/dgerm/, 2008
[8] World Economic Forum. *Green technology: Driving economic and environmental benefits from ICT.* Document from http://www.weforum.org, 2009.

A SURVEY ON IT MANAGERS GREEN AWARENESS

Eugenio Capra, Ph.D., and Alessandro Caleffi
Politecnico di Milano – Dipartimento di Elettronica e Informazione
Milan, Italy
eugenio.capra@polimi.it and ale.k6@hotmail.it

ABSTRACT

The energy consumed by IT systems has become significant, both from the point of view of CO2 footprint and of economic impact. Even though Green IT, i.e. the discipline that studies the environmental impact of IT, is gaining momentum, most of the IT managers have not yet built a mature and fact-based position with reference to these issues. This paper presents the results of a survey conducted on 138 Italian companies to evaluate the Green awareness of their IT Managers. Our data shows that 89% of IT managers do not know the electric consumption of the systems they manage and do not take into account energy efficiency parameters when evaluating new purchases. This may be due to the fact that in 81% of the analyzed companies IT managers are not responsible of the electricity budget. The paper also presents statistics on Green IT strategies and budget, and a success case that we analyzed in details.

Key words: Green IT; Green KPIs; Green awareness.

INTRODUCTION

The study of the environmental impact of IT is gaining momentum (Forrester 2007, Preimesberger 2007, Standage 2008). On one hand, IT producers are forced by recent legislation to monitor and control the environmental impact of IT components lifecycle, from production to dismissing, re-conditioning, and recycling (Krikke 2008). On the other hand, IT users are becoming concerned about the environmental implication that is consequential to the use of ICT. This new investigation area is commonly indicated as "Green IT" (Murugesan 2008).

According to Forrester Research (2007), 33% of North American and 48% of European IT procurement and operations professionals think that environmental and energy-related issues are very important in planning their company's IT operations. Green IT is regarded as a generally important topic by 94% of European and 85% of North American IT practitioners.

However, the behavior of IT managers appears to be still in a preliminary phase, when decisions are not led by an overall vision, but by momentarily trends and fears. Most of the IT-user companies that are investigating Green IT are led by image and marketing motivations, as the public is getting generally aware of and sensitive to environmental issues. Statistics and estimates on the CO

footprint of IT, which is responsible from 2 to 3% of global CO_2 emissions (Murugesan 2008, Brown and Lee 2007, Kumar 2007), have been published and are now quite widespread, leading companies to consider IT energy consumption within their social corporate responsibilities strategies. IT-user companies are also influenced by IT vendors, which are often leveraging Green IT to sell new products and services. According to Gartner (Kumar 2007), 44% of the world top 500 Computer, Telecom, and IT Service Provider companies think that climate changes represent a commercial opportunity for both new and existing products.

On the contrary, IT-user companies do not seem yet fully aware of the cost impact of the energy consumed by IT. Data on the operative costs of data centers collected by IDC (2007) show that energy on average accounts for 13% of total costs. Moreover, while the global spending for new servers has been more or less constant in the last ten years, the spending for power and cooling has grown almost 4 times (Josselyin et al. 2006). These statistics are elaborated at world level, but the economic impact of IT energy consumption is significantly higher in countries where energy is more expensive (e.g., in Italy, where the cost of energy for industrial use is four time as expensive as in the US, International Energy Agency 2007).

Notwithstanding these worrying reports, a lot of IT executives do not know the energy consumption and relevant cost of the infrastructure they manage. A recent survey conducted by EILT (Restorik 2007) has shown that 86% of IT departments in the UK do not know their CO_2 footprint.

How can IT managers reduce the energy consumption of the IT infrastructure that they manage if they do not know the current level of consumption and how this consumption is distributed in the different layers and modules of the IT system? Starting from the assumption that it is impossible to optimize what is not known, we conducted a survey on a sample of 138 Italian companies to evaluate the level of "green-awareness" of IT managers. Although green-awareness may be a very wide topic, in this particular context we associate green-awareness to a detailed knowledge of the energy consumption of IT, its costs, the importance of energy efficiency parameters in the purchase process, and the presence of an overall vision on Green IT.

The paper is organized as follows. Next section describes the sample of companies that we have selected and our empirical methodology. We will then present the results of the survey, and discuss one of the success cases we came across. Finally, we will draw some preliminary conclusions and propose future work.

SAMPLE AND METHODOLOGY

Since our goal was to evaluate the *green awareness* of IT Managers, we created a large database of firms homogeneous for size and sector. Our analysis was limited to Italy, because it was easier for us to get contacts and operate there. We think that Italy can be representative of Europe for a conservative analysis. In fact, the cost of electric energy for industrial use in Italy is the highest in Europe (just to cite some examples ,the unit cost in 2007 was 24 $cent/KWh in Italy, 13 $cent/KWh in UK, 9 $cent/KWh in Spain, 8 $cent/KWh in Germany, 5 $cent/KWh in France, IEA 2007) and the economic impact of Green IT should thus be more evident and significant for IT managers than in other countries.

We randomly selected 1,635 firms from the list of all Italian firms of the Chamber of Commerce, thus assuring a consistent sample. The Italian Chamber of Commerce provides the largest and most updated public repository of information about firms in Italy. We chose to use only one repository to have homogenous data, as different repositories may be structured according to different rules. We selected the firms according to two criteria: i) business sector, and ii) number of employees. We chose half of the 1,635 firms from the ICT sector, whereas the other half from other fields. Fig. 1 shows the distribution of the sample according to business sectors. As regards the size, we tried to select a sample equally representatives of Small Enterprises (less than 10 employees), Medium Enterprises (between 10 and 250 employess) and Large Enterprises (more than 250 employees) according to the EU definition (European Commission 2006). Fig. 2 shows the distribution of the sample according to the number of employees.

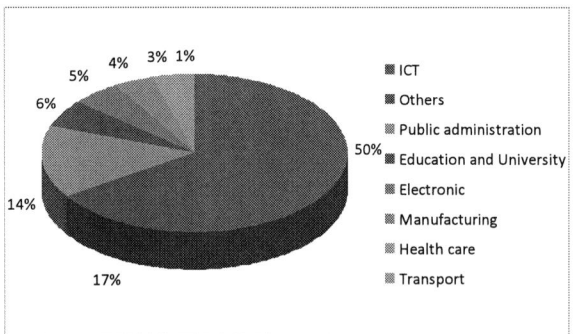

Fig. 1 – Distribution of the sample according to business sectors.

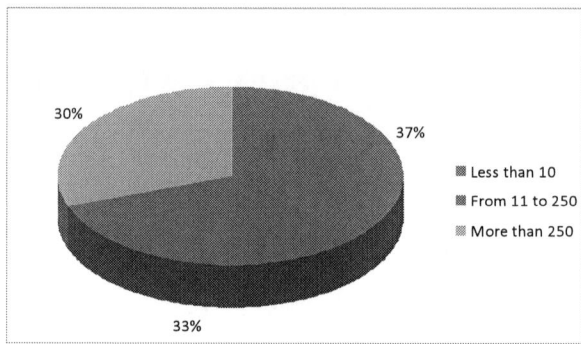

Fig. 2 – Distribution of the sample according to number of employees.

The IT Managers of the firms in our sample were invited to answer to an on-line survey, that we managed by means of the platform myK (http://myk.concept.it). The campaign started with a pilot phase on March 21st, 2008. We sent the questionnaire to 100 firms additional to our sample, selected with the same criteria described above. We obtained 11 replies and contacted by e-mail and phone some of the respondents. We leveraged their comments to clarify the questionnaire and make sure that all the questions were not ambiguous. We also decided to eliminate any references in the cover e-mail and on the cover page of the on-line survey to guarantee casual responses and eliminate a possible non-response bias (Green 2008).

The real campaign started on May 21st , 2008 and ended, after two follow-ups, on August 1st, 2008. The final statistics about the campaign are presented in Table 1.

	First Mailing	1st Follow-up	2nd Follow-up	Total
Sent e-mails	1,635	1,392	960	-
Answered	87	39	12	138
Declined	156	393	541	1,090
No answer	1,392	960	407	-

Table 1 - Summary statistics of the online survey campaign.

Whenever inconsistencies were found, we contacted the respondents by e-mail and asked for clarification, even though we did not always receive answers. We excluded from the sample the answers for which we were not able to solve inconsistencies, thus resulting in a final sample of 138 answers. We had a final response rate of 8.4%.

The distribution of the respondents according to business sectors and number of employees is essentially the same of the original sample of 1,635 firms.

The questionnaire was in three parts. The first part contained general questions about the firm and about the interviewee. This part had the purposes to confirm the information on size and business sector, and to classify the firm according to revenues and ICT spending. Fig. 3 shows the results of these questions.

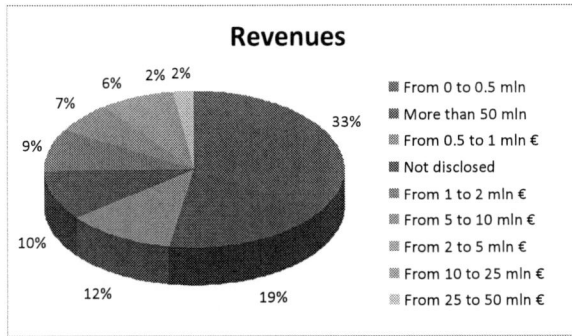

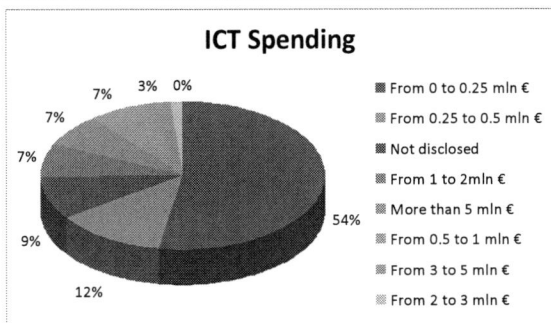

Fig. 3 – Distribution of the respondents according to revenues and ICT spending.

The second part of the questionnaire focused on the awareness of IT managers of Green IT issues and of the electrical consumption of the infrastructure they manage. Originally, this part also aimed at evaluating the impact of electric costs on total costs, but as will be discussed in the following section, we did not gather enough data on this topic to be statistically significant.

The third part of the questionnaire focused on the eventual strategy for Green IT, including questions on the budget. Finally, we decided to make an in-depth analysis on some firms that emerged as success cases. This analysis was conducted by personally interviewing the IT and Energy Managers.

SURVEY RESULTS
First of all, we asked IT managers what was in their opinion the meaning of "Green IT". We encouraged open and free replies, and then classified all the collected replies as shown in Fig. 4. According to our results, 43% percent of Italian IT managers never heard about Green IT. These IT managers are not yet aware of Green IT issues, and their firms are not yet worried about the environmental impact of IT. 16% of the IT manager have a confused or partial perception of Green IT, limited to e-waste or to the management of energy utilities, whereas only 44% of the respondents clearly see Green IT as an eco-compatible and eco-responsible attitude towards the use and the design of IT. This shows a substantial lack of maturity of Italian of IT managers when compared to the European average (see the Forrester Research statistics cited in the Introduction Section, 2007).

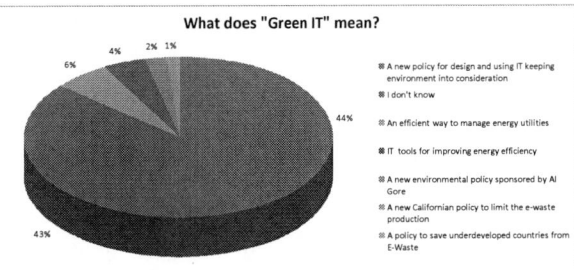

Fig. 4 – Replies to the question "What does Green IT mean?".

Starting from the assumption that is impossible to analyze and optimize what is not known, we asked IT managers if they knew the electrical consumption of their companies and of their IT infrastructure, and obtained appalling results. 84% of IT managers do not know the energetic bills of their organization, and 89% of them do not know the electric consumption of IT either. If we consider the consumption of the cooling systems for the IT infrastructure (which accounts for approximately 40% of the total energy consumption of IT and for 60% of the annual operating spending for servers, Josselyn et al. 2006), it is unknown to 90% of IT managers. The last two percentages have been computed considering only the firms that manage IT in-house (81% of our sample).

Table 2 presents data on the awareness of IT managers on total consumption, IT consumption, and IT cooling consumption, classified according to firm size.

Size	Firm total consumption	IT consumption	IT cooling consumption
Small	12%	8%	4%
Medium	15%	11%	13%
Large	21%	16%	13%
Average	16%	11%	10%

Table 2 – IT managers who are aware of electrical consumptions classified by firm size.

Even though awareness slightly rises with the size of the firm, as it is quite natural to think, the situation in large firms is not much better than in SMEs.

The respondents who knew the electrical consumptions were invited to indicate the values of actual consumptions of their firms. However, these results are not statistically significant, since they are based on the answers of the very few IT managers who know the consumption of their firm.

IT systems are quite widespread in the organization of a firm, as they include PCs, laptops, network routers and switches, printers, UPSs, and other devices. Total IT consumption can be estimated by specific analyses or measured by a monitoring infrastructure built ad-hoc. However, data centers (i.e., mainframes or server farms) are more concentrated and their consumption should be easily tracked. Consequently, we performed an in-depth analysis on data centers, which are present in 46% of the firms in our sample.

Fig. 5 shows the average consumption of data centers declared by the respondents. More than half (51%) of IT managers do not even know the electric consumption of the data center they manage, although the relevant cost is significant. According to our data, a data center in Italy consumes on average 2 MW, which if the data center works 24h per day accounts for approximately 2.5 million Euro per year. Even if we consider the most frequent values of consumptions, i.e. approximately 300KW, their impact is significant. Just to have a term of comparison, 300KW is equivalent to the consumption of a large complex with 100 apartments.

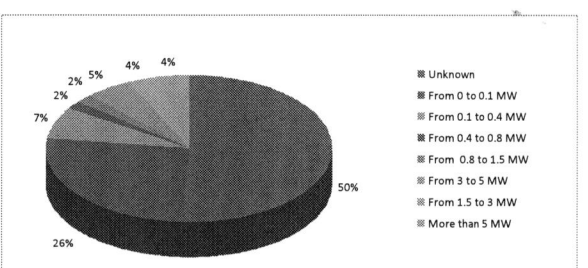

Fig. 5 – Electrical consumption of data centers declared by IT managers.

Why IT managers do not know (and do not care about) the electric consumption of the infrastructure they manage? Our research shows that only 19% of IT managers are responsible for the cost of the energy consumed by IT. In the other 81% of the cases electricity costs are charged on the budget of other functions independently from the appliances that caused the consumption. This evidence provides a rationale for another result of our survey. We asked IT managers to what extent they take energy efficiency into consideration when they buy new IT components. Fig. 6 presents the results, from which it is evident than in 70% of the cases environmental parameters are scarcely or not at all taken into consideration.

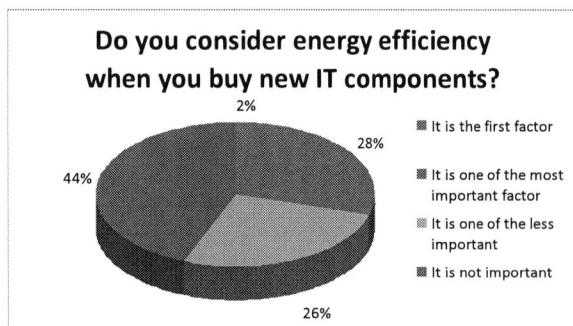

Fig. 6 – Relevance of energy efficiency in the purchase process of new IT components.

This may indicate shortsightedness of IT buyers, as there is plenty of evidence that shows that small premium prices invested for more energy efficient infrastructure are paid back by savings in less than 2 years (Samson 2008, Sturgeon 2008), especially in countries where the energy cost is high. Of course, the fact that IT managers do not pay the electricity bill provides a justification for this behavior.

Our survey shows another worrying result, in fact most of the firms in our sample do not have an energy efficiency plan and actually, the firms that are developing a plan, are not investing money in it. Firms do not have an overall strategy to improve the energy efficiency of IT and do not yet consider it a priority. Table 3 presents the replies of IT managers to the questions relevant to this topic classified according to firm size.

Size	Do you have an energy efficiency plan?			Do you have a budget for energy efficiency?		
	Yes	No	No, but I will have in the future	Yes	No	No, but I will have in the future
Small	15%	46%	39%	1%	88%	11%
Medium	11%	47%	42%	1%	87%	12%
Large	39%	48%	13%	4%	71%	25%
Average	19%	47%	34%	2%	82%	16%

Table 3 – Data on energy efficiency plans and budgets for energy efficiency classified according to firm size.

Large firms are leaders in this area, as 39% of them already have an energy efficiency plan, whereas SMEs are followers.

A SUCCESS CASE

In our analysis we also came across some success cases, i.e. companies that are aware of Green IT issues and that monitor energy consumptions by means of ad-hoc information systems. In this section we present the case of Bank Alpha, as an example that winning green strategies can be easily and cheaply implemented.

Bank Alpha is a large Italian bank, which thanks to a on-line monitoring system and an optimization project has reduced its electricity bill of 5% (4.2 mln Euro saving are expected in 2008).

The project started with the investigation of critical branches and buildings, which were identified by analyzing monthly consumptions. Critical structures are monitored by ad-hoc systems, which have required a little investment between 600 and 2,000 €, depending on available options. Each system is able to monitor the consumption of every IT component, and of every single appliance for heating and lighting. Data are measured in real time, made available online and integrated by an executive information system.

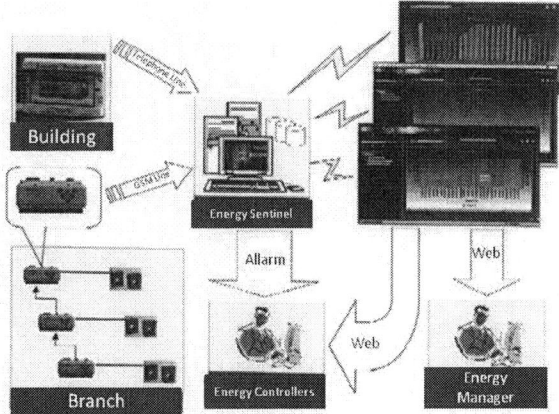

Fig. 7 – Bank Alpha energy consumption monitoring system

The executive information system is able to send sms and e-mails to the Energy Manager or to other deputed persons (Energy Controllers) in case that the consumptions are different from average values or from what expected (e.g., somebody forgets to switch off a machine). Furthermore, it allows to store and analyze data, to define KPIs and compare them with benchmarks. Fig. 7 presents an overview of the architecture of the system. The Energy Manager is a new role created within the company with the specific goal of implementing policies for improving energy efficiency.

The project has led not only to economic returns, but also to a better environmental performance. In fact, thanks to the project Bank Alpha saved 5% of the electricity cost in 2007, but it also reduced the consumption of electricity of 1.3 MWh and the emission of CO_2 of 650 ton.

This case study shows that a person or a function within a company aware of IT consumptions, and in general in charge of Green issues, may easily lead to implement effective systems to monitor consumptions and to optimize them, resulting in significant savings.

CONCLUSIONS AND FUTURE WORK

Our survey has shown that 84% of the IT managers do not know the consumption of the infrastructure they manage, and that only 30% of them consider energy efficiency as an important factor when new IT components are purchased. This evidence can be easily interpreted by noting that energy costs are charged to the IT budget in less than 20% of the companies analyzed. This introduces an important misalignment between corporate and personal objectives: it is unlikely that executives invest part of their budget to improve the energy efficiency of a system or to pay a premium price for "greener" IT components if they are not responsible for energy costs. On the other hand, IT energy costs can be charged to IT budgets only if they can be clearly identified and if the consumption of each appliance is monitored and registered.

Statistics shows that the energy consumption of IT has a deep impact on TCO (IDC 2007, Josselyin et al. 2006). In addition to that, Green IT has a moral relevance, as it affects the health of our planet. As these issues needs to be addressed, we would recommend companies to design and implement executive information systems to monitor the environmental performances of their IT systems. These technologies are quite mature: new servers and PCs automatically provides energy consumption through built-in functions (EMA 2008), several typologies of sensors capable of measuring the energy consumed by an appliance are available (Emerson 2008), including ammeter clams that do not require to cut wires, and data may be stored and analyzed by means of datawarehouse and KPIs dashboard tools. The case of Bank Alpha shows as this can be implemented with state-of-the-art technologies. Other studies (The Green Grid 2007) shows that a energy-aware re-engineering of data centers, based on the monitoring of consumptions and of green KPIs, may reduce consumptions from 5% to up to 20% with a payback, in most of the cases, shorter than one year (Caleffi 2008). Moreover, these projects may – and should – be extended to all the business processes and electric appliances, not only IT.

What is really needed is a corporate commitment to define a set of Green IT KPIs, to monitor them, and above all to take them into consideration. This last goal can be achieved only if IT managers are made responsible of the green performances of IT. This in turn may be obtained by specific incentives and by changing accountability rules, so that IT managers are responsible of IT energy costs, even though this involves significant organizational changes.

The natural continuation of our work is the definition of a general methodology and guidelines for identifying the most significant "Green IT KPIs" that a company should monitor. We propose to conduct another survey and to analyze some success cases in the near future to obtain this goal.

REFERENCES

E.G. Brown, and C. Lee (2007). Topic Overview: Green IT. Forrester Research report,

A. Caleffi (2008). Elaborazione e validazione empirica di un modello funzionale per un sistema informativo direzionale per monitorare i consumi e l'impatto ambientale delle aziende italiane. Master thesis. Politecnico di Milano.

EMA (2008). Green computing: using IT autonomation to achieve energy efficiency. March 2008. www.informatioweek.com.

EMEA Power and Cooling Study (2007). IDC.

Emerson (2008). Soluzioni perl'efficienza energetica per le applicazioni IT e Data Center in Europa. pp 1-31.

European Commission (2006). The new SME definition. Enterprise and Industry Publications.

W.H. Green (2008). Econometric Analysis. Prentice Hall.

S. L. Josselyin, B. Dillon, M. Nakamura, R. Arora, S. Lorenz, T. Meyer, R. Maceska, and L. Fernandez (2006).Worldwide and regional server 2006-2010 forecast. IDC report.

Key world energy statistics (2007). International Energy Agency.

J. Krikke (2008). Recycling e-Waste: the sky is the limit. IT Pro: pp. 50-55.

R. Kumar (2007). Important Power, Cooling and Green IT Concerns. Gartner report.

S. Murugesan (2008). Harnessing Green IT: Principles and Practices. IT Professional: vol. 10, no. 1, pp. 24-33.

C. Preimesberger (2007). How is green is IT's future?. www.eweek.com.

T. Restorick (2007). An Inefficient Truth. Global Action Plan Report. www.globalactionplan. org.uk/research.aspx
T. Samson (2008). Are green IT premiums worth the cost?. InfoWorld. http://weblog.infoworld.com/ sustainableit/archives/2008/06/prices_green_pc.html

T. Standage (2008). More silicon, less carbon. The Economist, Nov 19th, 2008.
W. Sturgeon (2008). Green IT: do it for the money, if nothing else. CNET. http://news.cnet.com/2100-1022_3-6137822.html

Tapping Buyers' Growing Interest in Green IT (2007).

The Green Grid (2007). Guidelines for energy effient DataCenters. www.thegreengrid.org.

The Green Grid (2007). The green grid opportunity. www.thegreengrid.org.

The Green Grid (2007). Green grid metrics. www.thegreengrid.org.

ON IMPROVING OPERATIONS SCHEDULING IN ELECTRONICS MANUFACTURING

Daryl L. Santos, Ph.D.
Systems Science and Industrial Engineering Department
Small Scale Systems Integration and Packaging (S³IP) – A NYS Center of Excellence
Binghamton University
Binghamton, NY, USA
santos@binghamton.edu

ABSTRACT

There is no doubt that manufacturers schedule operations – without doing so, product would never be shipped. However, what seems to receive comparatively little attention in many organizations is discussion on how to assign or release jobs throughout the facility in order to optimize the scheduling activity. This paper describes a general scheduling environment, known as the hybrid flow shop (HFS), which exists in many facilities. For example, surface mount assembly is one type of production flow that can be modeled as a hybrid flow shop. Fundamental concepts of HFS scheduling, performance metrics tied to costs and facility utilization, and developments towards improved scheduling techniques for HFS are reviewed in this work.

Key words: Production scheduling, scheduling complexity, hybrid flow shop, flow shops with multiple processors.

INTRODUCTION

Herein, "production scheduling" is defined as the allocation of limited resources (machines) to tasks (jobs or orders) over time with the goal of optimizing one or more performance measures. Little known in most industries, including electronics manufacturing, is just how complicated the production scheduling function is, mathematically speaking. This paper is organized by the following discussion areas:

- a brief overview of baseline production scheduling terminology,
- common assumptions used to generate a baseline schedule,
- the fundamental and common scheduling models or environments and a hierarchy of their complexity,
- common performance measures used to evaluate schedules and a hierarchy of their complexity, and
- mathematical approaches used to obtain feasible schedules.

The paper then devotes a section to the complexity of "static problems" in each of the following scheduling environments as they are found in a variety of electronics manufacturing settings: single machine scheduling, parallel processor (a.k.a., multiprocessor) scheduling, flow shop scheduling, and hybrid flow shop (a.k.a., flow shop with multiple processor) scheduling. Hybrid flow shop (HFS) scheduling will receive additional attention in this paper as, compared to the others, it is a "newer" problem as found in the literature.

SCHEDULING TERMINOLOGY

The following is a brief glossary of common terminology used in scheduling problems and in this paper:

- Job – a "job" is considered an entity and, for practical purposes, represents an order. A job may have one or more operations that need to be performed on it, depending upon the environment. In this paper, we assume that a job can only be processed on one machine at a time.
- Environment – also known as a model, the environment describes the number of machines in the facility and the routing of the jobs through the facility.
- Static problems – refers to situations in which the number of jobs "n" is fixed, and the processing times are known for each job. These types of problems are also known as deterministic scheduling problems. *Static problems are the focus of this paper.*
- Dynamic problems - in these problems, there is more uncertainty in the environment – the number of jobs is not known in advance, when they arrive is random, etc.
- Performance measure – described in more detail to follow, the performance measure (a.k.a., objective function) is used to assess the quality of the schedule.
- Technological constraints – generally refers to the processing order of the jobs, the ready times of the jobs, and the processing times of the jobs.
- Feasible schedule – denotes a schedule that does not violate any technological constraints. E.g., a job is not scheduled to take longer than its processing time, is not scheduled to start before it is ready, etc.
- Optimal schedule – denotes the best feasible schedule, in terms of the performance measure. Optimal schedules are difficult, if not impossible to obtain in a practical time frame due to the complexity of most scheduling problems.
- Preemption – describes a situation in which a job, once started on a machine, can be interrupted for another job, and then continued on the machine at a later time.
- Instance – refers to a specific scheduling problem in which the environment is known, the performance

measure is known, the number of jobs is known, the processing times are known, the ready times are known, and the job routings are known.

COMMON SCHEDULING ASSUMPTIONS
To follow is a list of common scheduling assumptions used to obtain a baseline schedule. These assumptions can be found in most texts related to production scheduling such as those by French (1982) and Pinedo (2002). It is important to note that in actual practice, some of these may be violated; nonetheless they are usually invoked to generate a baseline feasible schedule:

- Each Job is an Entity – for jobs with multiple operations (multiple-machine problems), no two operations of the same job may be processed simultaneously
- No Preemptions – an operation, once started, must be completed before another job is started on that machine
- One Visit – each job has one or more distinct operations, one on each machine stage, and cannot be processed more than once on a specific machine stage
- No Cancellation – each job must have all its operations performed
- Times Independent of Schedule - Two assumptions are usually followed: a) Set-up time is sequence independent (and is included in processing times) , b) Transfer times are either negligible or are included in processing times. Problems where the sequence significantly affects the set-up times are not considered in this work.
- In-Process Inventory Allowed – jobs may wait, in a queue, for next machine
- Idle Machines are Okay
- A machine may only process one operation at a time
- No Break-Downs will Occur During Scheduling Period
- Technical Constraints Known - Technical constraints are known in advance and they are fixed (e.g., processing orders)
- Static problems are studied/assumed in this paper. What are known and fixed are the following:
  - # Jobs (n)
  - # Machines (m)
  - Processing Times
  - Ready Times
  - Other quantities needed to define a particular *instance*

COMMON SCHEDULING MODELS
The common production scheduling models are the following:

- Single Machine Scheduling – in single machine scheduling, there is only one machine to be scheduled.
- Multiprocessor Scheduling – in multiprocessor scheduling (a.k.a., parallel machine scheduling), there is only one stage of processing but there are multiple machines available to perform that task. Thus, a job can go to any one of the machines. Unless otherwise stated, the machines are assumed to be identical, so that

a job's processing time is the same regardless of which machine it visits. In situations where the machines are not identical (either unrelated (no relationship between the speeds of the machines) or uniform (the speeds of the machines are fixed in relationship to each other)), the complexity of the problem increases.

- Job Shop Scheduling – in job shop scheduling, n jobs are to be processed on m machine stages (each stage having just one processor). However, each job has a unique routing through the shop that may or may not be the same as one or more other jobs. Job shop problems where there are multiple machines at one or more of the m stages are possible. While job shop scheduling does occur in electronics manufacturing, most electronics manufacturing examples are not job shops and will not be further discussed in this paper.
- Open Shop Scheduling – in open shop scheduling, there are multiple machines, but there are no routing restrictions on the jobs. Open shop scheduling is not common in electronics manufacturing and will not be further discussed in this paper.
- Flow Shop Scheduling – in flow shop scheduling, there are multiple machine stages and each job follows the same routing through the stages as all other jobs.
- Hybrid Flow Shop Scheduling – HFS scheduling is an extension of the flow shop problem wherein, at one or more stages, there are multiple (parallel) machines available to process the jobs.

Not surprisingly, the hierarchy of the complexity of scheduling problems in Figure 1, below, begins with the single machine problem with the HFS being one of the more complex environments.

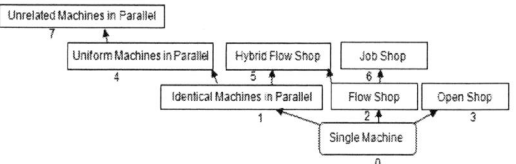

Figure 1. Complexity of Scheduling Environments (modified from Pinedo, 2002)

Later in the discussion of the mathematical complexity of many of these scheduling problems (single machine, multiprocessor, flow shop, and HFS), it will be easier to understand the structure of Figure 1.

PERFORMANCE MEASURES
This section gives a brief review of the common performance measures broken down into three different categories: Completion Time based criteria; Due-date based criteria; and Inventory & Utilization criteria. More detailed descriptions and other performance measures can be found in Pinedo (2002) or other texts.

Completion Time Based Performance Measures
These criteria are used to evaluate a schedule when the quality is gauged by when jobs complete. Some of the

common performance measures in this category are the following:

- Maximum flow time (F_{max}) – flow time is defined as the difference between when a job completes its last operation and when it was ready to begin its first operation. Like most scheduling criteria, it is desired to *minimize* the value of the largest flow time. This is done in situations where the cost of the schedule is directly related to the job that takes the longest time in the system.
- Maximum completion time (C_{max}) – also known as the makespan. Used to define when the last job exits the system. This is used when the cost of a schedule depends on how long the processing system is devoted to the entire set of jobs.
- Average flow time and average completion time – when the schedule's cost is related to average performance (instead of an individual job's performance), then minimizing the average flow time or average completion time of all of the jobs can be assessed. Minimizing average flow time is equivalent to minimizing the sum of all of the flow times; similarly, for average completion time and sum of all completion times.
- Weighted based versions – when different jobs have different weights (or priorities) – such as when a cost penalty is assessed by a customer, then average weighted flow time, average weighted completion time, or other variations may be used.

Due-Date Based Performance Measures

These criteria are used when a schedule is tied directly to how far due-date targets are missed. Prior to describing some of the common due-date based measures, it is important to define some values. The lateness of a job, i, is calculated as follows:

$$L(i) = C(i) - d(i)$$

Where $C(i)$ represents when the job completes and $d(i)$ represents the due date. Lateness, by the above calculation, can therefore be positive (it is tardy) or negative (it is early).

Lateness-based performance measures (e.g., minimizing the maximum lateness (L_{max}), minimizing the average lateness (which is the same as minimizing the sum of total lateness), etc.) are appropriate when there is a positive reward for completing a job early and that reward increases the earlier the job is.

When we are interested in whether a job finishes before its due-date, then tardiness is of importance. The tardiness of a job is calculated as follows:

$$T(i) = \max \{ 0, L(i) \}$$

Sometimes a binary variable, $U(i)$, is created for a job based upon its tardiness. If it is tardy, $T(i) > 0$, then $U(i) = 1$. Otherwise, $U(i) = 0$.

Tardiness-based measures (minimizing maximum tardiness, minimizing the number of tardy jobs, n_T or $\Sigma_i U(i)$,

minimizing the average tardiness, etc.) are appropriate when early jobs bring no reward; there are only penalties for jobs with positive lateness.

In some situations, the earliness is important to assess and we may not want jobs to be too early. The earliness of a job is calculated as follows:

$$E(i) = \max \{ 0, -L(i) \}$$

In the descriptions, above, many due-date based criteria were mentioned. In addition to the above, weighted based versions of many of the above have also been studied.

A hierarchy of the complexity of some of the scheduling criteria from the above two categories appears in Figure 2, below.

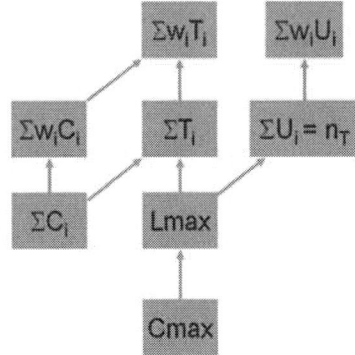

Figure 2. Objective Function Hierarchy (modified from Pinedo, 2002)

Inventory and Utilization Based Criteria

Most of the production scheduling research is aimed at criteria in the two aforementioned categories: completion based criteria and due-date based criteria. When schedule costs are tied to inventory and machine-utilization, then other criteria can be important to gauge a schedule including, but not limited to the following:

- Mean Number of Unfinished Jobs
- Mean Number of Jobs Waiting for M/C's
- Mean Number of Jobs Being Processed at Any Time

SCHEDULE GENERATION TECHNIQUES

A hierarchy of the complexity of scheduling environments has been provided in Figure 1. Discussion to follow will provide the reader an understanding of the "combinatorial explosive" nature of scheduling problems. Throughout the history of the study of scheduling problems (with most of the important findings beginning in the 1950s), techniques for finding a schedule for a particular problem instance can be broadly categorized in these two areas: optimal techniques and heuristic techniques.

In this paper, an optimal technique is one that is guaranteed to find an optimal solution by considering (explicitly or implicitly) all possible schedules and choosing the best from among them. Most optimal techniques are algorithms in the sense that they find, in a certain amount of iterations, the best feasible schedule. Examples of optimal techniques are mixed integer linear programming (MILP) models, branch-and-bound (B&B) models, and dynamic programming (DP) models. Other optimal techniques take advantage of the structure of the problem and construct an optimal schedule. A classical example of this is Johnson's Algorithm (1954) which finds optimal makespans in two-stage flow shop problems. In rare cases, some optimal techniques are quick (such as Johnson's Algorithm) wherein an increase in the problem size – whether in number of jobs, number of machines, etc. – does not cause a large increase in solution times. In most cases, due to the total possible number of schedules to evaluate – examples of these calculations to follow – optimal search techniques like MILP, B&B, and others may fail to find an optimal solution due to constraints. As such, the search for an optimal solution may need to be abandoned in lieu of finding a good, hopefully near-optimal solution but in a quick time frame.

For the purposes of this paper, a heuristic is not guaranteed to find an optimal solution. Heuristics are designed to give good, hopefully near-optimal (if not optimal) solutions in a relatively quick time frame. Some heuristics are designed to take advantage of specific scheduling environments; for example, there are many heuristics developed for the minimum makespan flow shop scheduling problem (e.g., the NEH heuristic (Nawaz et al. (1983)), the CDS heuristic (Campbell et al. (1970)), and others).

In recent years, there are more generalized heuristics that have been developed that can be used regardless of the scheduling environment or even the performance measure. The Shifting Bottleneck Method (SBM) – for reference, see Pinedo (2002) – and the Exchange Heuristic – see Yang and Ignizio, 1987 and subsequent advancements – are methods that can be applied in many scheduling environments and with many different performance measures. Also, the newer "metaheuristics" such as Simulated Annealing, Tabu Search, Genetic Algorithms, Ant Colony Optimization, and Particle Swarm Optimization are not restricted to specific models. Generally speaking, the techniques mentioned in this paragraph are improvement methods; wherein, the heuristic begins with one or more feasible schedules and, based upon modifications of these schedules (i.e., searching in the neighborhood of known solutions), tries to find other schedules with improved performance measures.

In the descriptions of scheduling environments to follow (single machine through hybrid flow shop), their comparative complexities will be demonstrated by using a situation in which the number of jobs to be scheduled is not a large number: 10. Further, for simplification purposes, it is assumed that the ready times for all jobs in each of the examples are the same and no other special constraints (e.g., strict predecessor relationships) are considered.

SINGLE MACHINE SCHEDULING

Single machine scheduling is manifested in a variety of ways in electronics manufacturing. First of all, in some organizations, perhaps there is just one machine that is the bottleneck or "Herbie" (see Goldratt (1984)). Other machines in the same facility may have very little utilization. Thus, it may be more important to expend resources to effectively schedule that single bottleneck. In another case, from a macro-perspective, there may be some facilities that have multiple machines, but when the entire facility is devoted to one job at-a-time, then the entire facility can be treated as a single machine and scheduling the facility is thus a single machine scheduling problem. In yet another situation, surface mount assembly lines (for simplicity's sake, let's consider 3 steps: stencil printing, pick-and-place, and reflow), can be scheduled as a single machine model. Why? Because in most instances, these three steps are connected via conveyors and once the decision is made to schedule a particular board type, then the entire SMT line is devoted to that board type, before the next board type is sequenced to be assembled.

The most fundamental of all scheduling problems, the single machine model, can be a non-trivial problem, depending upon the performance measure. For example, average tardiness, total weighted tardiness, and others, are said to be NP-hard problems, a mathematical and computer-science term meaning that as the problem size increases, the time needed to search all possible schedules is non-polynomial in nature. For simplicity's sake, if the number of jobs increases by a factor of 2, not only would the time needed to search all possible schedules not only double, but more likely the increase in time would grow exponentially or at another quick rate. For single machine scheduling, if we want to schedule n jobs (with all jobs having the same ready times), then the total number of sequences to consider is $n!$.

For example, if we have 10 jobs in a single machine problem instance, then the total number of possible sequences is 10! or over 3.6 million schedules. Finding the best schedule from among the over 3.6 million possibilities would be like finding a needle in a haystack.

MULTIPROCESSOR SCHEDULING

There are multiple examples in which multiprocessor scheduling can appear in electronics manufacturing. Consider one example in which we have 3 SMT lines. As described above, an SMT line can be modeled as a single machine problem. If we have n jobs to be scheduled among m parallel machines, then the total number of schedules to consider is

$$_{n-1}C_{m-1}\frac{n!}{m!}$$

In the above, "C" represents the combinatorial operator from mathematics. So, in an example where we have 10 jobs to be scheduled among 3 SMT lines, the total number of schedules is not small:

$$_9C_2 \frac{10!}{3!} = 21,772,800$$

FLOW SHOP SCHEDULING

In some electronics facilities where the material flow is not highly conveyorized (e.g., material moves from station to station on carts or pallets and other situations), the environment may be modeled like a flow shop. While not common, some facilities do not conveyorize SMT operations, thus, since the job orders can change in going from one machine (e.g., stencil printing to pick-and-place or pick-and-place to reflow), then the SMT line can be scheduled as a flow shop. Or, in another example, and generally speaking, in boards that require THT operations, then SMT operations (even if the SMT operation can be modeled as a single machine), and then an inspection stage, then this can be considered as a flow shop with three stages: THT, SMT, and inspection. The new roll-to-roll (R2R) flexible electronics facility at the Center for Advanced Microelectronics Manufacturing (CAMM) at Binghamton University can also be modeled as a flow shop. In the fundamental process steps (see Figure 3, below), a web can go from vacuum deposition, to photolithography, to wet etch.

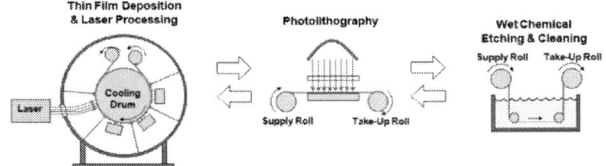

Figure 3. Flex Electronics Assembly: A 3-Stage Flow Shop

The number of possible schedules in a flow shop with n jobs and m stages of processing (with only one processor at each stage) is given by the following: $n!^m$.

So, if we consider 10 jobs that need to be scheduled in a 3-machine flow shop, then the total number of possible schedules is $10!^3$, which is an astronomical $\sim 4.8 \times 10^{19}$.

HYBRID FLOW SHOP (HFS) SCHEDULING

Compared with other scheduling environments that have been studied in the literature for over 50 years, HFS literature is rather in its infancy. HFS environments are prevalent in many industries including, but not limited to the following (Santos et al., 2001): PCB assembly, flexible manufacturing systems, and petrochemical production. The process flow for a HFS is shown in Figure 3.

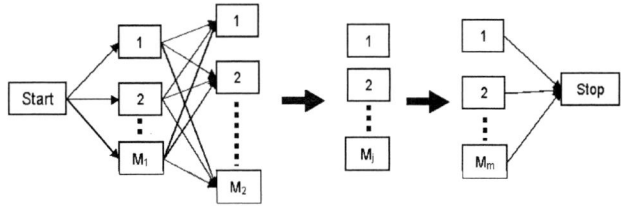

Figure 3. HFS Process Flow

Although one of the first articles on the environment did appear in the early 1970s (Salvador, 1973), most of the HFS literature did not begin to appear until the late 1980s (e.g., see Wittrock (1985), Brah (1988), and Gupta (1988), among others) and it began to grow in the 1990s (e.g., see Hunsucker and Shah (1994), Santos et al. (1995a, 1995b), Gupta and Tunc (1991, 1994), and others). If it is assumed that the machines are identical in processing speed at stages where multiple processors exist, then the number of possible sequences that can be generated for an HFS is provided by the following formula (Brah and Hunsucker, 1991):

$$\prod_{j=1}^{m} {}_{n-1}C_{M_j-1} \frac{n!}{M_j!}$$

In the above, n is the number of jobs, m is the number of processing stages, and M_j is the number of identical machines at stage j. Given a problem instance of 10 jobs with 3 stages with 2 identical machines at each of the stages, the total number of schedules that can be generated based upon this formula is over 4.3×10^{21}.

Ongoing Work in HFS Scheduling

In addition to some of the HFS work referenced above, our current efforts are devoted to improving the scheduling of HFS problems wherein the multiple processors at a stage are not identical. When the processors are not identical, then the complexity, in terms of the number of possible schedules is determined by the following formula:

$$\prod_{j=1}^{m} {}_{n-1}C_{M_j-1} \cdot n!$$

With 10 jobs, 3 stages, and two non-identical machines at each stage, the total number of schedules is approximately 3.5×10^{22}, which is noticeably larger than the prior example with a similar arrangement of stages and numbers of machines, but wherein the multiple machines were not identical processors.

All of the aforementioned references and discussion on the HFS and other environments have assumed that each machine can only process one job at a time (e.g., discrete parts processing) – an assumption that is safe to assume in many scheduling problems. When the environment includes processors that can process multiple jobs at a time (e.g., batch processing), then the problem complexity further increases. The HFS with a combination of discrete parts processing and batch processing machines is a problem that we are currently studying. Consider an electronics assembly operation as depicted in Figure 4.

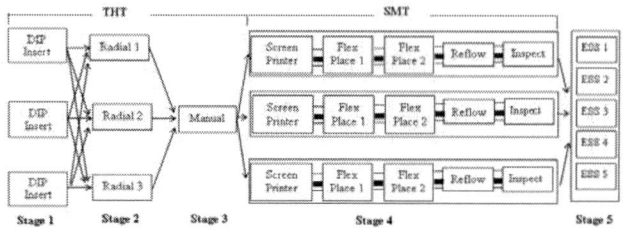

Figure 4. HFS with Discrete and Batch Processing

In the above, there are three stages of THT processing, one stage (of 3 lines) of SMT processing, and the fifth stage is a single stage of environmental stress screening (ESS). For the first four stages all boards are processed one at-a-time in each of the machines (or SMT lines). At Stage 5, each ESS chamber can process multiple boards at-a-time.

We have begun to apply common flow-shop based dispatching rules in empirical simulations of this environment (see Vishwanathan et al., 2007). We are also building theoretical foundations for the HFS with discrete and batch processors and wherein multiple processors at a stage are non-identical. As findings are discovered, they will be published. Only a few of our HFS related publications have been referenced in this work; the interested reader is welcome to contact the author for a more thorough listing of our prior efforts in the HFS environment.

ACKNOWLEDGMENTS
This work is partially supported by the Small Scale Systems Integration and Packaging Center – a NYS Center of Excellence at Binghamton University. Thanks also go out to the Integrated Electronics Engineering Center (IEEC) and Center for Advanced Microelectronics Manufacturing (CAMM) at Binghamton University. The author would also like to thank Dr. Purush Damodaran of Northern Illinois University for our ongoing collaborations in the HFS scheduling arena. For including him into his research group ('87-'93) and introducing him to the research area of production scheduling, the author would also like to personally thank Dr. J.L. Hunsucker – Industrial Engineering Professor Emeritus of the University of Houston, and Founder and President of the National Aquatic Safety Company (NASCO), Dickinson, TX.

REFERENCES
Brah, S.A., *Scheduling in a flow shop with multiple processors*. Ph.D. Thesis, University of Houston, Houston, TX, 1988.

Brah, S.A., and Hunsucker, J.L., 1991, "Branch and bound algorithm for the flow shop with multiple processors", *European Journal of Operational Research*, 51(1), pp. 88-99.

Campbell, H.G., R.A. Dudek, M.L. Smith, "A Heuristic Algorithm for the *n* Job, *m* Machine Sequencing Problem *Management Science*, Vol. 16, No. 10, pp. B-630-B-637 (1970).

French, S., *Sequencing and Scheduling*, Prentice-Hall, 1992

Goldratt, E.M., *The Goal*, North River Press, 1984.

Gupta, J.N.D., "Two-stage hybrid flowshop scheduling problem", *Journal of Operation Research Society*, Vol. 39, 1988, pp. 359-364.

Gupta, J.N.D., and Tunc, E.A., 1991, "Schedules for a two-stage hybrid flowshop with parallel machines at the second stage", *International Journal of Production Research*, 29(7), 1489-1502.

Gupta, J.N.D., and Tunc, E.A., 1994, "Scheduling a Two-Stage Hybrid Flowshop with Separable Setup and Removal Times", *European Journal of Operational Research*, 77, 415-428.

Hunsucker, J.L. & Shah, J.R., "Comparative performance analysis of priority rules in a constrained flow shop with multiple processors environment", *European Journal of Operational Research*, 72(1), 1994, 102-114.

Johnson, S.M., "Optimal two- and three-stage production schedules with setup times included," *Naval Res. Log. Quart.*, 61-68, 1954.

Nawaz, M., Enscore, E.E., and Ham, I., "A heuristic algorithm for the m-machine, n-job flow shop sequencing problem", *OMEGA, International Journal of Management Science* 11(1), 1983.

Pinedo, M.L., *Scheduling: Theory, Algorithms, and Systems, 2nd Ed.*, Prentice-Hall, 2002.

Salvador, M.S., "A Solution to a Special Case of Flow Shop Scheduling Problems", in: S.E. Elmaghraby (ed.), *Symposium of the Theory of Scheduling and Applications*, Springer-Verlag, 1973, New York.

Santos, D.L., Hunsucker, J.L., & Deal, D.E., "Global lower bounds for flow shops with multiple processors", *European Journal of Operational Research*, Vol. 80, No. 1, 1995a, pp. 112-120.

Santos, D.L., Hunsucker, J.L., & Deal, D.E., "FLOWMULT: permutation sequences for flow shops with multiple processors", *Journal of Information and Optimization Sciences*, Vol. 16, No. 2, 1995b, pp. 351-366.

Santos, D.L., Hunsucker, J.L., & Deal, D.E., "On Makespan Improvement in Flow Shops with Multiple Processors", *Production Planning & Control*, Vol. 12, No. 3, 2001, pp. 283-295.

Vishwanathan, K., Kulkarni, N., Pachamuthu, A., Santos, D.L., and Damodaran, P., "A Comparative Study of Dispatching Rules in Hybrid Flow Shops with Discrete and Batch Processors," *Industrial Engineering Research Conference Proceedings*, Nashville, TN, May 2007.

Wittrock, R.J., "Scheduling algorithms for flexible flow lines," *IBM Journal of Research and Development*, 29(4), 401-412, 1985.

Yang, T. & Ignizio, J.P., "An algorithm for the scheduling of army battalion training exercises", *Computers and Operations Research*, Vol. 14, 1987, 479-491.

EMERGING COMPETITORS IN EMERGING MARKETS:
WHAT PATENT APPLICATIONS REVEAL

H. J. Neuhaus, Ph.D., R. A. Fillion, and C. E. Bauer, Ph.D.
TechLead Corporation
Portland, OR, USA
herb.neuhaus@techleadcorp.com

ABSTRACT

Strategic business decisions begin with assessments of market need, value proposition and differentiation, profit potential and sustainability. Predicting the future implies uncertainty and emerging markets often push forecasting into the realm of guesses and hunches. However, structured analysis of published patents and patent applications provides valuable insight into strategies, aspirations and expected competitive positions long before the first sale in an emerging market.

Using MEMS packaging as an example, the authors show how in emerging markets pending patent applications often out number issued patents and then demonstrate how careful study of public databases for published patent applications yield a detailed picture of anticipated competitive environments as well as quantification of market trends and growth expectations. Finally, the authors apply the IP landscaping method to developing a strategic framework useful for investment, market development and strategic alliance planning.

Key words: intellectual property, patents, patent applications, MEMS packaging, emerging markets, competitive environment.

INTRODUCTION

Business planning requires judgments regarding market need, value proposition and differentiation, and profit potential and sustainability. Forecasting (predicting the future) either extends existing trends or seeks out similar situations and assumes history will repeat itself. However, markets for emerging technologies generally offer no reliable track record to extend. Emerging markets may match the patterns of other developments but scarce data for choosing one historical situation over another forces decision maker to rely on guesses and hunches – a risky and uncomfortable approach.

International patent law creates a window into strategies, aspirations and expected competitive positions long before the first sale in an emerging market. Because sale or other disclosures limit patent rights, hopeful competitors file patent applications as early as possible. After a statutory waiting period, the patent offices publish once-secret patent applications. In addition, issued patents immediately become part of the public record.

A systematic analysis of pending patent applications and issued patents offers a detailed intellectual property (IP) landscape useful for investment, market development and strategic alliance planning.

IP ANALYSIS METHODOLOGY

Key questions addressed in analyzing an (IP) landscape include: Who filed the patents? What areas of technology coverage past the scrutiny of the examiner? When and where did filing of the patents take place? Do coverage or technology gaps exist? What opportunities remain unexploited?

The methodology developed by the authors consists of the following steps:

1. Develop a comprehensive database of patents and patent applications by searching available on-line databases by patent classification, keyword and key inventors and assignees.

2. Assess the individual relevance of each patent and patent application using recursive and quantitative algorithm based on US and international classifications, keyword matches and proximity, relevance of other patents and applications with the same inventors and assignees, and the relevance of references and citations.

3. Follow reference and citation trails for each patent and patent application to identify additional significant patents, applications, inventors and assignees. The authors define references as mentions of older patents within the text of a particular patent. Required by law, references highlight related prior art. Both the inventors and examiners may add references to a patent. In contrast, the authors define citations as mentions of a particular patent in other, subsequent patents.

4. Classify or sort each patent and patent application by key subject matter. Review claims and group by key idea or type.

5. Assign a relative value to each patent and patent application by counting and characterizing all citations to each patent. Several authors[1,2,3] demonstrate the correlation between citations and patent value.

6. Identify competitive participants and analyze their strategies as reflected in their patent filings. Aspects of revealed strategies include long term investment, short

term exploitation, technical strengths, and capability gaps.

7. Identify opportunities by assessing subject matter, geographic, and/or assignee trends. Also, weaknesses and gaps in the portfolios of weaker participants may create the potential for strategic relationships.

8. Define strategic options based on the existing IP landscape. All participants seek cost and/or performance advantages, sustainable barriers to competitors, and freedom to operate (avoid infringement). Strategic options include patent filings, licensing, acquisitions, and various forms of strategic relationships.

PATENTS AND PATENT APPLICATIONS

Patents represent a set of exclusive rights granted by a government to an inventor or their assignee for a limited period of time in exchange for a public disclosure of an invention. Like a property title, patents contain an explicit definition of the invention (claims). Patents conform to a rigid format. Patents enjoy the assumption of validity. Although intangible, patents are unambiguous.

The process of obtaining a patent begins with filing a patent application. Next, the application undergoes examination during which the government verifies compliance with the statutory requirements of a patent. During the examination process, the government and inventors often negotiate the breadth and specific language of the claims which define the invention. Ultimately, the revised patent is denied or allowed (issued).

On average, 40 months elapses between filing of the application and patent issue or denial. Moreover, substantial revisions frequently occur between filing of the application and patent issue.

As a result, a pending patent application contains important uncertainties compared to an issued patent. Specifically:

1. The application may never become a patent. If rejected, a patent application becomes prior art without any exclusive rights for the inventors or their assignees. (Common.)

2. A patent may issue with claims identical to those contained in the application. (Very rare.)

3. A patent may issue with similar but narrower claims than contained in the application. (Very common.)

4. A patent may issue with claims revised so heavily as to be unrecognizable from the application. (Rare.)

5. The length of time between filing and patent issue or denial varies greatly from a few months (very rare) to many years (not unusual). Markets often move much more quickly than the patent examination process.

6. Patent applications do not conform to any standard format. Some applications contain complete and detailed information while some simply outline key concepts. For example, patents normally indicate assignees and references but applications often have neither.

7. If filed in multiple countries, an application may result in widely differing (or no) patents in each country.

8. Patent applications generally have few, if any, citations. As a result, the valuation methodologies cited previously do not apply to applications.

Finally, patent applications entail little expense compared to issued patents. Smaller entities may elect to abandon patents rather than pay the associated fees after considering market conditions and cash flow.

In the case of new and emerging technologies, the possibility of patent application rejection (denial), the length of the examination process, and the low relative cost of patent applications often result in pending patent applications significantly out numbering issued patents.

Thus, in spite of the associated uncertainties, pending patent applications often dominate the IP landscape for emerging markets. Far from an impediment, patent applications reveal the world as the applicant sees it, rather than as the patent office sees the world. Patent applications reveal the aspirations and expectations of competitors in ways patents cannot.

CASE STUDY: MEMS PACKAGING

While MEMS market reports and technology surveys are available, no systematic, industry-wide reviews of the MEMS packaging intellectual property landscape exist. This paper reports a preliminary analysis of 2,877 US patents and patent applications related to MEMS packaging. Ongoing analysis of the World IP Organization (WIPO) and European Patent Office filings targets a global understanding of the MEMS packaging IP landscape with expected completion in mid 2010.

MEMS packaging currently enjoys considerable attention as represented by industry publications[4,5,6], conferences[7,8,9], industry groups[10] and patent filings. Typical of emerging markets, 1,673 pending US patent applications related to MEMS packaging outnumber granted 1,204 US MEMS packaging patents.

While sharing many requirements with conventional semiconductor packaging, MEMS packaging differs in several key attributes. The most fundamental distinction is the need for non-electrical feed-throughs permitting interaction with the environment. The fragile, often movable microstructures cannot be encapsulated with molding compound and often require cavities within the package. Finally, MEMS structures typically require extra protection during assembly operations.

A number of enabling technologies, including low-damage wafer dicing, wafer bonding techniques, sealing

146

technologies, wafer-level packaging and through-silicon vias, form the basis for much of the competitive differentiation among industry participants.

The combination of unique requirements of MEMS packaging and availability of multiple technical approaches create a rich intellectual property landscape.

FILING DATE ANALYSIS

Figure.1 shows the number of US patents and patent application in the MEMS packaging arena by filing year from 2000 through 2009. As expected, the number of granted patents falls off sharply after 2005 reflecting the average patent examination period of 40 months.

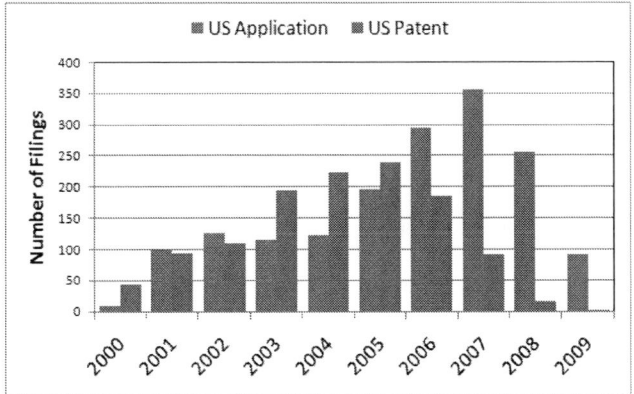

Figure 1. # of MEMS packaging patents & applications vs. filing year

The 18 month delay in publishing US patent applications may account for the fall off in patent applications in 2008 and 2009. However, the recession of 2008-2009 likely also contributes to the decrease in filings shown. CNN cites US Patent and Trademark statistics showing the number of US patent filings fell 2.3% in 2009 marking the first decline in filings since 1996 and attributes the decline to the recession[11] (Figure 2).

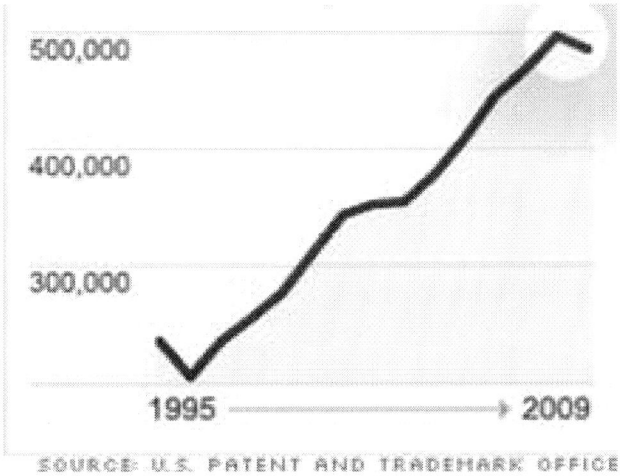

Figure 2. US Patent filings decline in 2009 due to recession[11]

ASSIGNEE ANALYSIS

Figure 3 shows the ten largest MEMS packaging portfolios (patents plus applications). Figure 3 shows MEMS packaging IP ownership dominated by large vertically integrated and research driven corporations with the notable exceptions of Silicon Genesis and Tessera, both IP companies. High-profile start-ups in the MEMS area, while present, generally hold relatively few patents and patent application assignments.

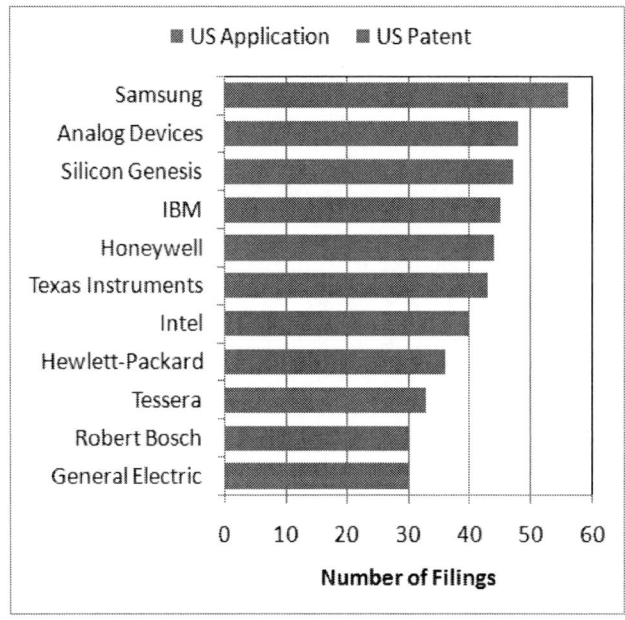

Figure 3. Ten largest MEMS packaging portfolios

Figure 4 shows the geographic distribution of MEMS patents and patent applications. The second bar label "na" refers to filings without assignment indicated in the filing. In the case of patents, "na" usually refers to cases in which the inventor is an individual that does not assign the patent to an employer. On the other hand, patent applications frequently bear no assignee information as noted above.

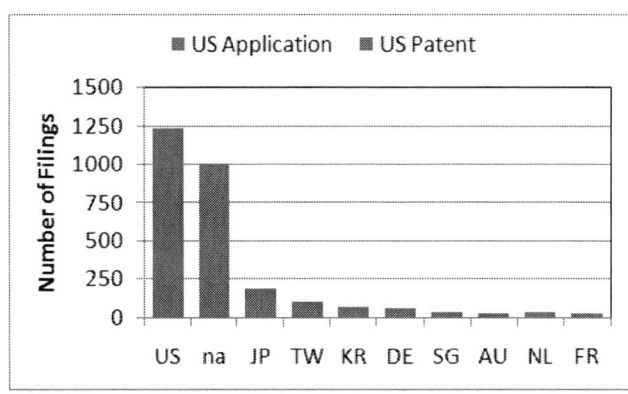

Figure 4. Geographic distribution of MEMS packaging patents and applications

Figure 4 clearly indicates a US lead, even vis-à-vis Japan, in MEMS packaging assignments. Arguably, the reliance on the US patents and patent applications biases these findings toward US assignees. However, previous IP landscape studies by the authors[12,13,14] demonstrate the minimal impact of such bias as the vast majority of patent applicants worldwide quickly file corresponding US patents to ensure protection in the largest markets in the western world. Overall, an analysis of US filings represents a valid and reasonable indicator of the world-wide IP landscape.

COMPETITOR STRATEGY ANALYSIS

Given the uncertainties associated with patent applications, treating patents and patent applications separately provides additional insight to the competitive positions (and strategic aspirations) of the leading participants. Figure 5 replots the data of Figure 3.

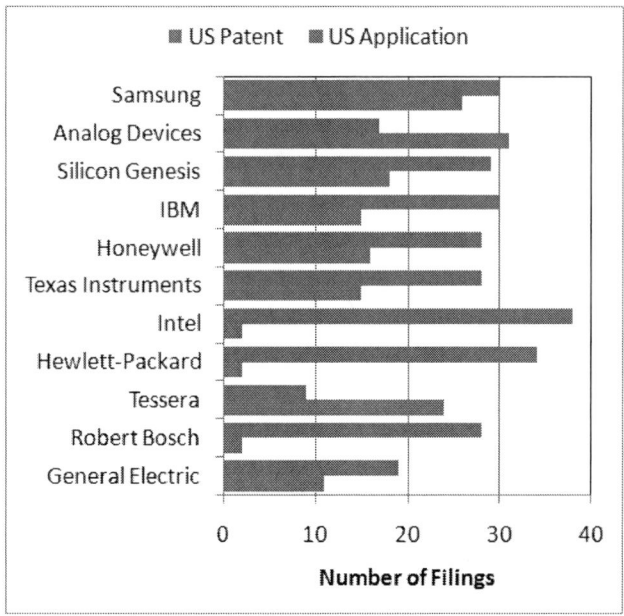

Figure 5. Ten largest MEMS packaging portfolios with patents & patent applications separated

Immediately, we note that some portfolios consist of mostly granted patents with relatively few pending applications (Hewlett-Packard, Bosch); some consist of mostly pending applications (Tessera, Analog Devices); and others have a nearly even mix of patents and pending applications (Samsung).

Analysis of the targeted market application of each patent and patent application in these portfolios reveals even more about strategies. For example, Figure 6 shows the distribution of targeted technologies/applications in the portfolios of the several leading participants. Identified targeted market applications include:

- Packages for MEMS devices
- Microphones and MEMS Pressure Sensors
- MEMS Switches
- Capped and Lidded Devices
- Optical MEMS and Sensors
- MEMS Mirrors

- MEMS Devices with particular electrical functionality (oscillators, resonators, etc)
- Other Methods & Structures
- Other Sensors and
- Other Actuators

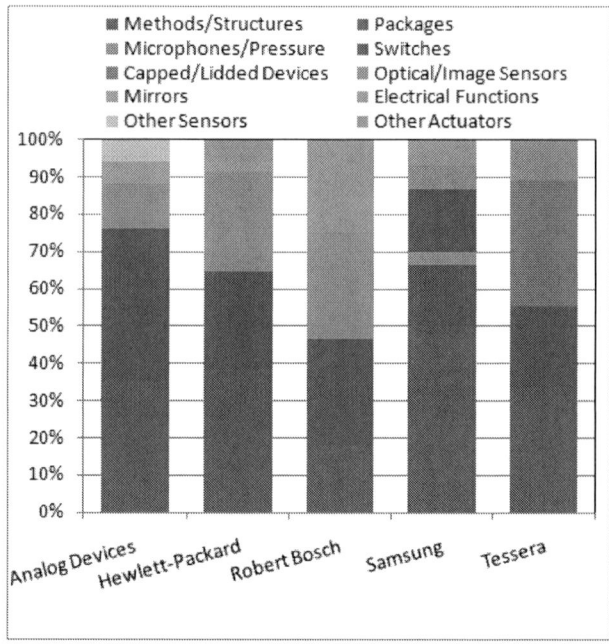

Figure 6. Distribution of market applications targeted in leading MEMs packaging portfolios (patents plus pending applications)

In Figure 6 one observes a strong emphasis in resonators and oscillators by Bosch while Tessera focuses on IP for capped or lidded devices. HP and Samsung show widely distributed efforts for differing target market applications.

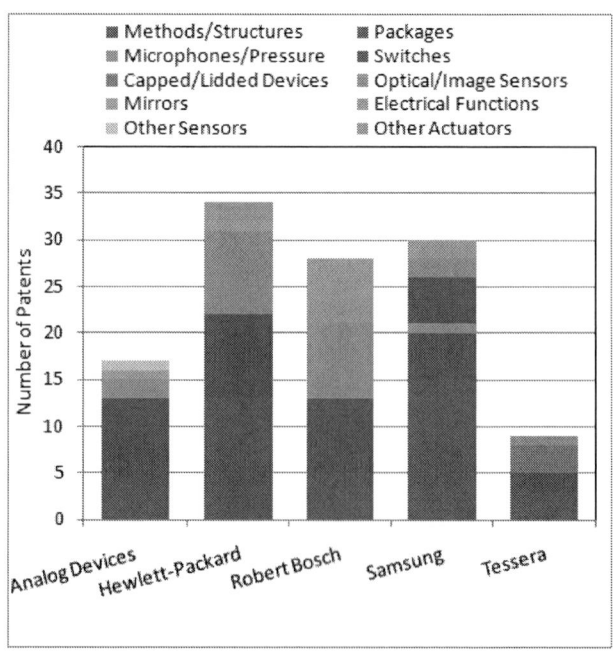

Figure 7. Distribution of market applications targeted in leading MEMs packaging portfolios (patents only)

Figure 7 refines the target market application analysis with a patents only view. The distributions of IP for HP, Bosch and Samsung change little from Figure 6 which reflects patents and pending applications because for these firms either issued patents dominate their portfolio or the focus remains equally balanced between patents and applications.

Figure 8 shows the distribution of market applications targeted in pending patent applications for the same set of portfolios, the complementary picture to Figure 7. HP and Bosch show little activity in pending MEMs packaging patent applications compared to Analog, Samsung and Tessera. Also observe the concentrations in packages by Analog and in lidded devices by Tessera. Samsung shows a broad distribution of target applications in both their pending patent applications and issued patents (Figure 7).

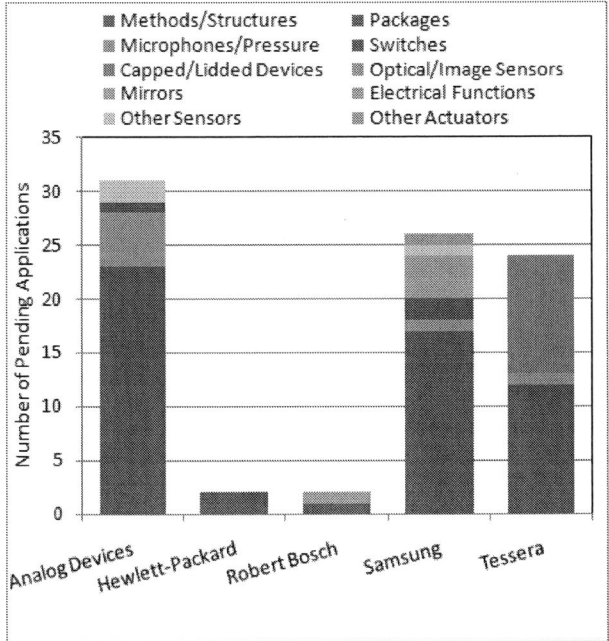

Figure 8. Distribution of market applications targeted in leading MEMs packaging portfolios (pending patent applications only)

Another view of the leading portfolios results from identification of key subject matter claimed in the patents and applications. Key subject matter groups found in these portfolios include (in order of frequency):

- MEMS device structures
- Lidding
- Sealing
- Cavity packages
- TSV or Through Silicon Vias
- Anti-Stiction technologies
- Electrical interconnect
- Bonding
- Encapsulation and molding
- Dicing or singulation
- Release methods

Figure 9 shows the distribution of these subject matter groups in patents and pending applications for leading portfolios while Figures 10 and 11 show the respective subject matter distributions for patents only and patent applications only.

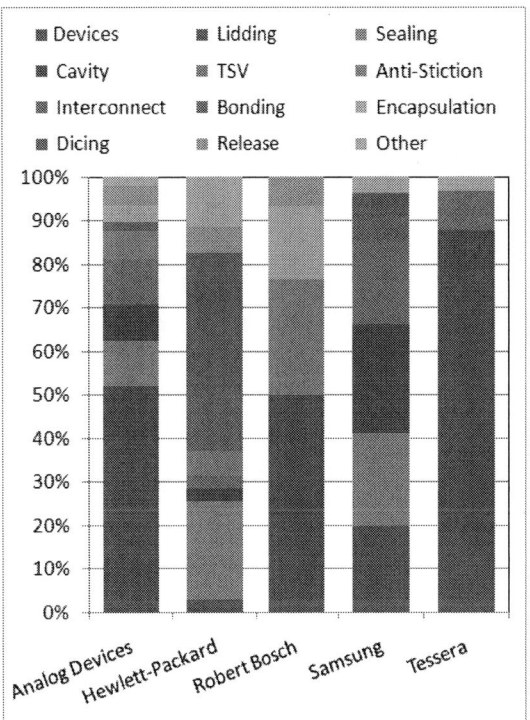

Figure 9. Distribution of key subject matter in leading MEMs packaging portfolios (patents plus pending applications)

This analysis echoes the patterns seen in Figures 6, 7, and 8. In particular we note a strong emphasis by Tessera in the lidding area while Samsung and Analog hold broad distributions of subject matter within their MEMS packaging portfolios. Half of Bosch's entire MEMS packaging portfolio relate to device structures.

While strongly represented in issued patents assigned to Analog, HP, and Samsung, patent applications for these firms deemphasize sealing technology possibly indicating a shift toward wafer-level processes as an alternative. The data also show a similar de-emphasis of cavity packages from issued packages to pending applications.

CONCLUSIONS

The authors demonstrated a methodology to analyze patent applications to provide insight into competitive strategies in emerging markets using MEMS packaging as an example. In emerging fields, patent applications typically out-number issued patents. Transient in nature, patent applications reveal the intentions and aspirations of the assignee, which often undergo substantial revision before finalization as an issued patent (or rejection altogether). Because early filing preserves IP rights, and because patent law requires publication of pending applications, patent applications provide a powerful window into competitive strategies in emerging markets.

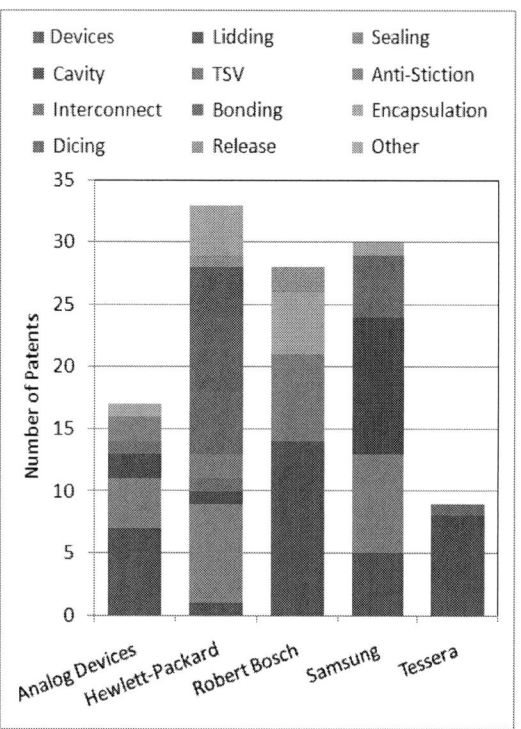

Figure 10. Distribution of key subject matter in leading MEMs packaging portfolios (patents only)

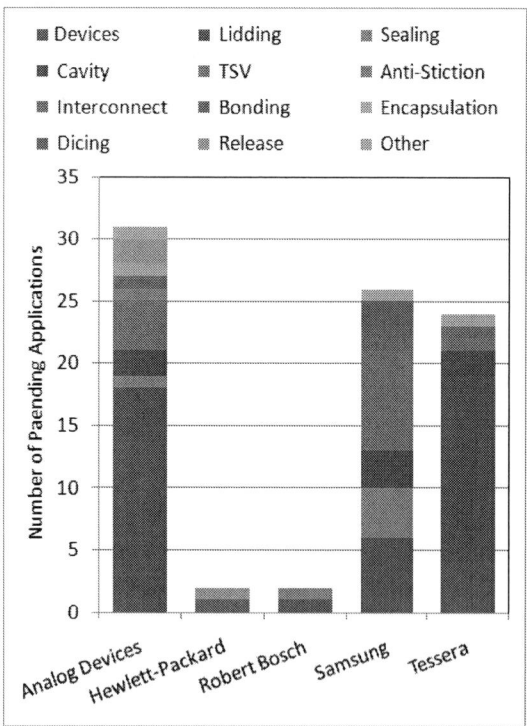

Figure11. Distribution of key subject matter in leading MEMs packaging portfolios (pending patent applications only)

REFERENCES

1. Harhoff, D., "Citation Frequency and the Value of Patented Innovation," August, 1997.
2. Trajtenberg, M., "Market Value and Patent Citations – A First Look," Dept. of Economics, UC Berkeley 2001.
3. Baron, R., "Better Accounting for Patent Portfolios," Intellectual Property Magazine, October 16, 2000.
4. Wirbel, L., "MEMS Market Demands New Foundry Strategies," Electronic Engineering Times, October 6, 2008.
5. Gibb, K., "Texas Instruments: The Evolution to a Lower Cost DMD," Advancing Microelectronics, March/April 2009.
6. Lin, C-W, "Implementation of three-dimensional SOI-MEMS wafer-level packaging using through-wafer interconnections," 2007 J. Micromech. Microeng. 17 1200-1205
7. IEEE MEMS 2009 Conference, Sorrento Italy, Jan 25-29, 2009.
8. METRIC 2009, Pittsburgh, Pennsylvania, March 25-26, 2009.
9. International Conference On MEMS 2009, Madras India, May 2009.
10. MEMS Industry Group (MIG), www.memsindustrygroup.org.
11. http://money.cnn.com/2009/12/11/news/economy/patent_filings/index.htm
12. Bauer, C., "The IP Landscape for Photovoltaics," ICEP 09, Kyoto, Japan, April 14-16. 2009.
13. Bauer, C., "The IP Landscape for Photovoltaics," ESTC 2008 – Electronics System-Integration Technology, Conference, Greenwich, UK, Sept 2008.
14. Bauer, C., "The IP Landscape for 3D Semiconductor Assembly; Chip Stacks, Origami, PoP," Proc. ICEP 2007, March, 2007

EMERGING SUBSTRATE TECHNOLOGIES FOR PACKAGING

Henry H. Utsunomiya
Interconnection Technologies, Inc.
Suwa City, Nagano Prefecture, Japan
henryutsunomiya@mac.com

ABSTRACT

This presentation will outline the market dynamics driving the development of advanced substrates in today's industry landscape. Technological advancements have shifted from Personal Computers to Mobile Applications like Cellular phones, MPEG players, etc. An overview of the technology revolution of substrates and their corresponding packages will be discussed in terms of technology supply chain matching: assembly technology, substrate technology and system technology. Advanced substrate technology roadmaps of leading Japanese suppliers will be reviewed. The evolution of buildup substrates, which have standardized under the direction of CPU & FPGA manufacturers will be explained. The two technological directions of substrates for advanced electronic packages will be mapped out: miniaturization and functional integration, and explained in terms of "More Moore" and "More than Moore", respectively. Fine pitch wiring on substrates can be combined with embedded active and passive devices and has been proven as a technology. Examples/ illustrations of advanced substrate applications in servers, hand -helds, and high end servers and communication equipments will be reviewed: PoP, Embedded Active Devices, Si interposers and Wafer Level Packages, etc.

Key words: Packaging substrate, Embedded Active Devices, Si Interposer, Wafer Level Package, PoP

INTRODUCTION

Worldwide printed wiring board production in 2008 was estimated 51,530 million US$ and Japan was produced 11,561 million US$ (22.4 %) by WECC (World Electronic Circuits Council) Global PCB production Report. And worldwide packaging substrate production in 2008 was estimated 8,925 million US$. In 2008, Japan produced 4,414 million US$ packaging substrate and total share was 49.5 %. Taiwan is second large production of packaging substrate with 2,316 million US$ and Korea followed Taiwan with 1,410 million US$ [1].

Japanese domestic packaging substrate production amount in 2008, buildup structure substrate shares 55.4 % and rigid structure substrate shares 20.1 %, while tape structure substrate including TAB (Tape Automated Bonding) and COF (Chip on Film/Flexible) shares 19.3 %. Package substrate production in 2009 is expected to be divided as follows; rigid substrates are estimated to be 35.6 Billion ¥, 11.7 % share, buildup substrates are estimated to be 190.0 Billion ¥, 62.5 %, and the tape substrate share is estimated to be 61.2 Billion ¥ or 20.1 % share [2]. In addition, Japanese packaging substrate manufacturers expect a CAGR of 11.9 % (2009 to 2013) for next 5 years. Figure 1 and figure 2 are illustrated Japanese packaging substrate production amount progress and Japanese printed wiring boards production amount progress respectively.

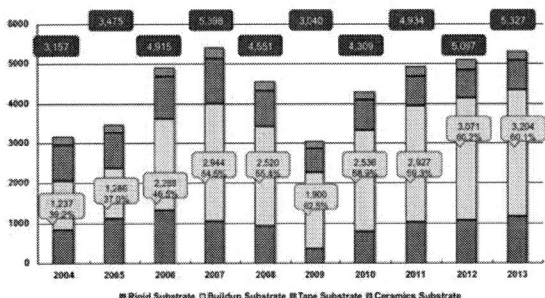

Figure 1. Japanese Packaging Substrate Production Amount Progress (Unit: 100 Million ¥) [2]

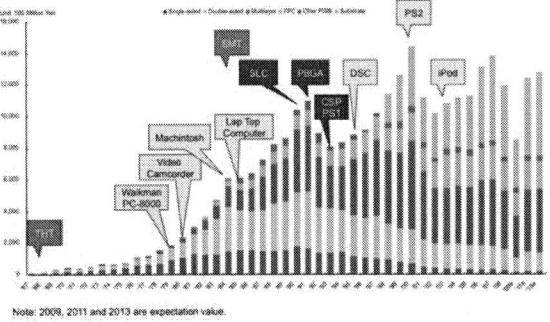

Figure 2. Japanese PWB Production Amount Progress (Unit: 100 Million ¥) [2]

151

Figure 3 and figure 4 are shown printed wiring board products production amount share in 2007 and average growth rate of 2005 – 2007, and PWB production amount share in 2008 and CAGR 2005 - 2008 respectively.

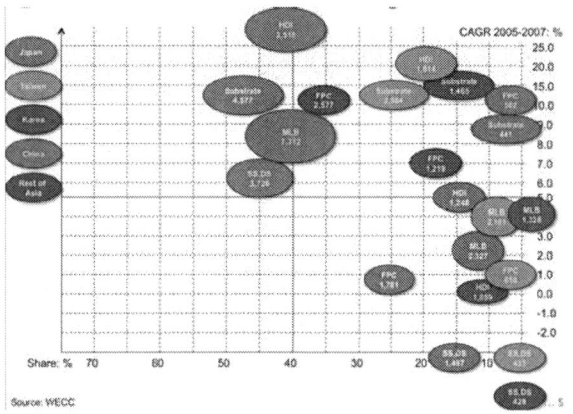

Figure 3. PWB Production Amount Share & CAGR in 2007 (Unit: Million US$) [3]

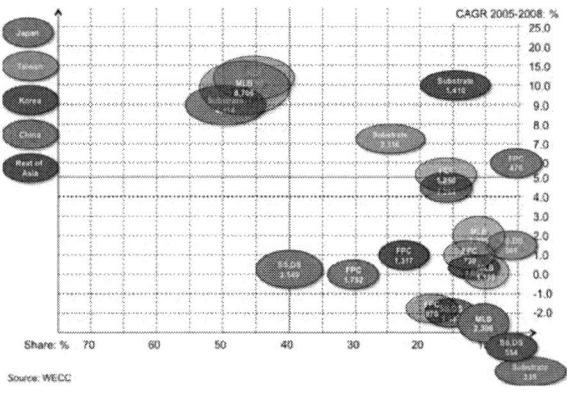

Figure 4. PWB Production Amount Share & CAGR in 2008 (Unit: Million US$) [3]

BACK GROUND OF PACKAGING SUBSTRATE ROADMAP

Actual scaling of semiconductor is more accelerated than ITRS roadmap and 32 nm technology nodes was announced by Intel and IBM such as SRAM using 19 billion transistors and 22 nm SRAM development in 2008 and 2009 respectively. To accomplish 32 nm technology nodes, several challenges on reducing power consumption, improving signal speed and external data transmission speed are required for reduced leakage current, reducing coupling for minimized signal propagation delay, and increased data bus width by increased number of I/O terminals. Scaling of semiconductor enable to reduce gate delay, while interconnect delay increasing exponentially due to

resistivity increased by narrower wiring and increased length of global wiring. In addition, I/O terminal pitch should be decreased to less than 125 µm to assure power integrity. Figure 5 is shown performance bottleneck by interconnect delay and figure 6 is shown technology innovation by data bus width.

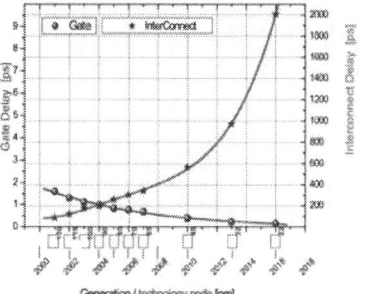

Figure 5. Performance Bottleneck by Interconnect Delay [4]

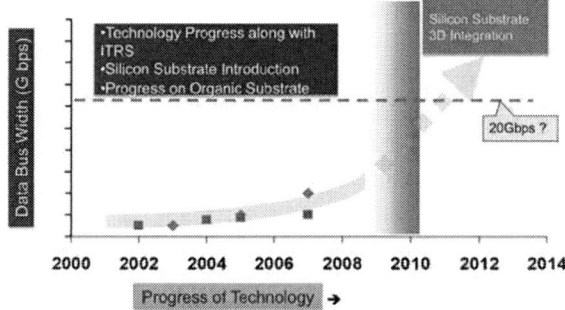

Figure 6. Technology Innovation by Data Bus Width [5]

For flip chip package, organic substrate is difficult to adopt high bump account and less than 100 µm bump pitch with economical cost due to difficulty on 15 - 10 µm via hole diameter generation with existed laser drilling process. More over difficulty on 5 µm line width and space generation with 50 µm via diameter by current CO_2 laser drilling. In addition, in small form factor portable products required 150 µm solder ball pitch for motherboard in 2011. For very fine pitch chip package bonding, alternative technology such as silicon interposer with TSV is one of candidate technology to solve finer wiring and via diameter for high-end applications. And fan-out CSP can ease to solder ball pitch for portable products. In addition, embedded devices into substrates, motherboards and module boards will enable to solve these issues. Table 1 is shown chip to package interconnect technology requirements and Table 2 is shown package to board interconnect technology requirement.

Table 1. Chip to Package Technology Requirements [6]

Item		Unit	2009	2010	2011	2012	2013	2014	2015	2018	2020
Au Wire Bond	Single in-line	µm	35	30	30	25	25	25	25	25	25
	Two-row Staggered Pitch		40	35	35	35	30	30	30	30	30
	Three-tier Pitch		50	50	45	45	40	40	35	35	35
	Wedge		20	20	20	20	20	20	20	20	20
Cu Wire Bond	Single in-line	µm	50	45	40	40	35	35			
	Dual-row		60	50	45	45	40	40			
Chip on Film		µm	25	20	20	15	15	10	10	10	10
Flip Chip	Area Array: Low End	µm	210	210	200	200	180	180	150	150	150
	Area Array: Mobile Products		150	135	120	110	110	100	100	95	95
	Peripheral: Mobile		60	50	50	40	40	40	40	35	35
	Area Array: Notebook	µm	150	135	120	110	110	100	100	95	95
	Area Array: High Performance		160	150	150	130	130	120	120	110	110
	Flip Chip: Harsh										

Table 2. Package to Board Technology Requirements [6]

Item		Unit	2009	2010	2011	2012	2013	2014	2015	2018	2020
Solder Ball Pitch	Low Cost, Handheld	mm	0.65	0.65	0.50	0.50	0.50	0.50	0.50	0.50	0.50
	Cost-Performance		0.65	0.65	0.50	0.50	0.50	0.50	0.50	0.50	0.50
	High-performance		0.80	0.80	0.65	0.65	0.50	0.50	0.50	0.50	0.50
	Harsh		0.65	0.65	0.50	0.50	0.50	0.50	0.50	0.50	0.50
Small Portable Products	Low Cost, Handheld	mm	0.65	0.50	0.50	0.50	0.50	0.50	0.50	0.50	0.50
	Harsh		0.65	0.65	0.50	0.50	0.50	0.50	0.50	0.50	0.50
	CSP Area Array Pitch		0.20	0.20	0.15	0.15	0.15	0.10	0.10	0.10	0.10
	QFP Lead Pitch		0.30	0.30	0.30	0.30	0.30	0.30	0.30	0.30	0.30
	SON Land Pitch		0.40	0.30	0.30	0.30	0.30	0.30	0.30	0.30	0.30
	QFN Land Pitch		0.40	0.40	0.40	0.40	0.40	0.40	0.40	0.40	0.40
	P-BGA Ball Pitch		0.80	0.65	0.65	0.65	0.65	0.65	0.65	0.65	0.65
	T-BGA Ball Pitch		0.65	0.50	0.50	0.50	0.50	0.50	0.50	0.50	0.50
	FBGA Ball Pitch		0.30	0.30	0.20	0.20	0.20	0.20	0.20	0.15	0.15
	FLGA Land Pitch		0.30	0.30	0.30	0.30	0.30	0.30	0.30	0.30	0.30

Generally, motherboard design rule is define one line between via holes and major mounted device package is fine pitch BGA, while packaging substrate design rule is defined two or three lines between via holes. For 0.15 mm pitch solder ball F-BGA requires 40 µm via diameter on 75 µm land diameter with 25 µm line width/space, generally. In case of flip chip bump pitch of 100 µm, 30 µm to 50 µm via diameter on 70 µm land diameter with 6 µm line width/space is needed for 2 lines between via holes and 3 µm line width/space is needed for 3 lines between via holes with same via land diameter. For 30 µm via formation, it is capability limitation of UV-YAG laser and it may require Excimer laser. However, through put of Excimer laser is not enough for huge number of via holes formation. Table 3 is shown general design rules on motherboard and substrate.

Table 3. General Design Rules for Motherboard & Substrate [7]

Item	Mounted Device	# of Lines/ Via	Pad Pitch (mm)	Space Width (µm)	Line Width (µm)	Space Width (µm)	Land Diameter (µm)	Via Diameter (µm)
Mother-board	FBGA (CSP)	1 Line/Via	0.50	75	75	75	275	150
			0.40	60	75	60	200	100
			0.30	50	50	50	150	75
			0.20	30	40	30	100	50
			0.15	25	25	25	75	40
			0.10	18	18	18	50	30
Substrate	FC	2 Lines/Via	0.18	18	18	18	90	60 - 50
			0.15	15	15	15	75	40 - 30
			0.13	15	15	15	50	30 - 20
			0.10	15	10	15	30	15 - 10
		3 Lines/Via	0.18	10	10	10	90	60 - 50
			0.15	8	10	8	75	40 - 30
			0.13	8	10	8	50	30 - 20
			0.10	5	5	5	30	15 - 10

▨ Semi-additive Process Required   ▨ UV-YAG Laser, Excimer Laser Required

SUBSTRATE TECHNOLOGY ROADMAP

For packaging substrate roadmap, we classified technology difficulties into 6 categories such as "Low Cost": P-BGA for conventional consumer products with conventional technology, "Hand-held": FBGA (CSP) for portable products, "Mobile Products": SiP and PoP for smart phone and mobile PC, "Cost Performance": Buildup substrate P-BGA for CPU, GPU and game processor, "High Performance High End" Buildup substrate P-BGA and Silicon interposer BGA for high end routers, and "High Performance": LTCC BGA for routers and servers.

Table 4 is shown typical package type of each application.

Table 4. Description of Typical Package Type by Application [6]

Category	Package Type	Max. Body Size mm	Typical Body Size mm	Max. # of Pins	Typical # of Pins	Typical Substrate Materials	Max. # of Layers
Low Cost	P-BGA	35×35	27×27 31×31	1000	676 900	High Tg FR-4	4
Hand-held	F-BGA	15×15	10×10 12×12	400	300 360	High Tg FR-4	6
Mobile Products	SiP (F-BGA)	21×21	10×10 12×12 15×15	800	300 360 400	High Tg FR-4	6
Cost Performance	P-BGA	42.5×42.5	30×30 35×35 37.5×37.5	1300	300 360 400	High Tg FR-4 + Epoxy	3+4+3
High Performance	P-BGA S-BGA	70×50	45×45 50×500	2500	2500	High Tg FR-4 + Epoxy Silicon	6+6+6 6 8 10
High Performance	C-BGA	51×51	45×45	2500	1800	Al₂O₃	20

In mobile products application, the combination of high glass transition temperature FR-4 and epoxy film is dominant today.

Minimum line width/space of 20 µm/20µm today will decrease to 15µm/15µm in 2012 and it will expect to less than 10 µm in 2018. However, micro via diameter scaling is moderate and 80 μm today will expect to down to 50 μm in 2018, due to economical reason such as maintain via formation process cost and save capital investment for new laser drilling system. Table 5 is shown mobile product substrate roadmap.

Table 5. Mobile Products Substrate Roadmap [6]

Mobile Products (SiP, PoP) Parameter	unit	2009	2010	2011	2012	2013	2014	2015	2018	2020
Chip to Substrate Interconnect Land Pitch	μm	50	50	50	50	50	50	50	50	50
Min. Finished Substrate Thickness	mm	0.2	0.2	0.2	0.2	0.2	0.2	0.2	0.2	0.2
Core Material Tg	°C	180	180	180	210	210	210	210	210	210
Core Material CTE (X-Y)	ppm/°C	14	14	14	13	13	11	11	11	11
Core Material CTE (Z)	ppm/°C	40	35	35	30	30	30	30	30	30
Core Material Dk@1GHz	-	4.2	4.2	4.2	4.2	4.2	4.0	4.0	4.0	4.0
Core Materials Df@1GHz	-	0.013	0.013	0.013	0.013	0.013	0.010	0.010	0.010	0.010
Core Materials Young's Modulus	GPa	24	24	24	24	24	24	24	24	24
Core Material Water Absorption	%	0.10	0.07	0.07	0.07	0.07	0.07	0.07	0.07	0.07
Buildup Material Tg	°C	156	156	156	166	166	166	166	177	200
Buildup Material CTE (X-Y)	ppm/°C	13	13	13	12	12	12	12	12	12
Buildup Material CTE (Z)	ppm/°C	46	40	40	40	40	40	40	30	30
Buildup Material Dk@1GHz	-	3.4	3.4	3.4	3.0	3.0	3.0	3.0	3.0	3.0
Buildup Material Df@1GHz	-	0.013	0.012	0.012	0.010	0.010	0.010	0.010	0.010	0.010
Buildup Materials Young's Modulus	GPa	4	5	5	5	5	5	5	5	5
Buildup Material Water Absorption	%	0.2	0.2	0.2	0.2	0.2	0.2	0.2	0.2	0.2
Min. Line width/Space	μm	20/20	18/18	18/18	15/15	15/15	12/12	12/12	8/8	5/5
Min. Conductor Thickness	μm	25	25	25	20	20	15	15	10	8
Min. Through Via Diameter	μm	100	100	100	80	80	80	80	70	70
Min. Through Via Land Diameter	μm	250	250	250	200	200	200	200	150	150
Min. Micro Via Diameter	μm	80	70	70	60	60	60	60	50	50
Min. Micron Via Land Diameter	μm	150	130	130	120	120	120	120	100	100
Min. Through Via Pitch	μm	300	300	300	275	275	275	275	250	250
Min. Solder Mask Opening	μm	80	80	80	60	60	60	60	50	50
Min. Solder Mask Opening Tolerance	μm	20	20	20	18	18	18	18	15	15

Most of CPU, graphic processing engine and game processor are using flip chip interconnect today and it expect to use flip chip until 2020 for cost performance application. Since flip chip bump pitch will decrease from 150 μm today to 80 μm in 2018 due to die size shrinkage are happen for each technology nodes, minimum line width/space have to reach 5μm/5μm with 30 μm via diameter with UV-YAG laser or excimer laser. One of biggest issue on buildup material is residues are remain on the bottom of via holes after desmear process and it affects interconnect reliability. To accomplish sub ten micron wiring together with finer via diameter, material improvement including lower CTE with less fillers, very smooth surface morphology such as less than 100 nm average flatness for finer circuit, lower dielectric loss and optimized Young's modulus are necessary within next 5 years. Table 6 is shown cost performance substrate roadmap.

Table 6. Cost Performance Substrate Roadmap [6]

Cost Performance (CPU, GPU, Game Processor) Parameter	unit	2009	2010	2011	2012	2013	2014	2015	2018	2020
Chip to Substrate Interconnect Land Pitch	μm	150	150	150	125	125	125	100	80	50
Min. Finished Substrate Thickness	mm	1.1	1.1	1.1	1.1	1.1	1.1	1.1	0.8	0.8
Core Material Tg	°C	180	180	180	210	210	210	210	210	210
Core Material CTE (X-Y)	ppm/°C	10	10	10	8	8	8	8	8	8
Core Material CTE (Z)	ppm/°C	20	20	20	10	10	10	10	10	10
Core Material Dk@1GHz	-	3.2	2.8	2.8	2.8	2.8	2.8	2.8	2.8	2.8
Core Materials Df@1GHz	-	0.007	0.007	0.007	0.007	0.007	0.007	0.007	0.007	0.007
Core Materials Young's Modulus	GPa	34	24	24	24	24	24	24	24	24
Core Material Water Absorption	%	0.10	0.07	0.07	0.07	0.07	0.07	0.07	0.07	0.07
Buildup Material Tg	°C	200	200	200	210	210	210	210	210	210
Buildup Material CTE (X-Y)	ppm/°C	13	13	13	12	12	12	12	12	12
Buildup Material CTE (Z)	ppm/°C	46	40	40	40	40	40	40	40	40
Buildup Material Dk@1GHz	-	3.4	3.0	3.0	3.0	3.0	3.0	3.0	3.0	3.0
Buildup Material Df@1GHz	-	0.013	0.013	0.013	0.013	0.013	0.013	0.013	0.013	0.013
Buildup Materials Young's Modulus	GPa	4	5	5	5	5	5	5	5	5
Buildup Material Water Absorption	%	0.2	0.2	0.2	0.2	0.2	0.2	0.2	0.2	0.2
Min. Line width/Space	μm	18/18	15/15	15/15	12/10	12/10	10/10	10/10	5/5	3/3
Min. Conductor Thickness	μm	25	25	25	20	20	15	15	10	8
Min. Through Via Diameter	μm	100	100	100	80	80	80	80	70	70
Min. Through Via Land Diameter	μm	250	250	250	200	200	200	200	150	150
Min. Micro Via Diameter	μm	60	60	60	60	50	50	50	30	30
Min. Micron Via Land Diameter	μm	150	130	130	120	100	100	100	70	70
Min. Through Via Pitch	μm	300	300	300	275	275	275	275	250	250
Min. Solder Mask Opening	μm	60	50	50	50	50	50	50	40	40
Min. Solder Mask Opening Tolerance	μm	10	10	10	8	8	8	8	5	5

With the semiconductor die size becoming larger, matching of thermal expansion coefficient between the die and packaging substrate becomes critical for high-end high performance application. At present, 8 - 10 ppm/°C is required for core material and 16 ppm/°C for buildup layer material with reinforcement material such as SiO_2 fillers and/or glass yearns. In 2010, most tightest flip chip bump pitch become 125 - 100 μm, silicon interposer with polyimide redistribution layers will introduce to this application. And dielectric material for redistribution layer and/or functional layer materials will change to meso-porous silicon dioxide materials to adopt smaller CTE, mechanical strength and finer circuit in 2015 for tera bit scale data transmission package. One of potential solution is existed for current ultra low κ material for less than 32 nm technology node semiconductors.

Table 7 is shown high-end high performance packaging substrate roadmap.

Table 7. High-End High Performance Substrate Roadmap [6]

High End High Performance Parameter	unit	2009	2010	2011	2012	2013	2014	2015	2018	2020
Typical Materials	-	FR-4	Silicon	Silicon	Silicon	Silicon	Silicon	Silicon	Silicon	Silicon
Typical Buildup Materials	-	Epoxy	PI	PI	PI	PI	PI	SiO2	SiO2	SiO2
Max. Layer Counts	#	6+6+6	6+2	6+2	6+2	4+2	4+2	4+2	4+2	4+2
Typical Layer Count	#	6+6+6	6+2	6+2	6+2	4+2	4+2	4+2	4+2	4+2
Min. Finished Substrate Thickness	mm	1.0	1.0	1.0	1.0	0.6	0.6	0.5	0.5	0.5
Typical Finished Substrate Thickness	mm	1.0	1.0	1.0	1.0	0.6	0.6	0.5	0.5	0.5
Core Material Tg	°C	180	1410	1410	1410	1410	1410	1410	1410	1410
Core Material CTE (X-Y)	ppm/°C	6	3.0	3.0	3.0	3.0	3.0	3.0	3.0	3.0
Core Material CTE (Z)	ppm/°C	10	3.0	3.0	3.0	3.0	3.0	3.0	3.0	3.0
Core Material Dk@1GHz	-	12	12	12	12	12	12	12	12	12
Core Materials Df@1GHz	-	0.002	0.0005	0.0005	0.0005	0.0005	0.0005	0.0005	0.0005	0.0005
Core Materials Young's Modulus	GPa	5.5	185	185	185	185	185	185	185	185
Core Material Water Absorption	%	1.40	0	0	0	0	0	0	0	0
Buildup/RDL Material Tg	°C	200	300	300	300	300	300	700	700	700
Buildup/RDL Material CTE (X-Y)	ppm/°C	16	16	16	16	16	16	3	3	3
Buildup/RDL Material CTE (Z)	ppm/°C	20	20	20	20	20	20	16	16	16
Buildup/RDL Material Dk@1GHz	-	3.3	3.3	3.3	3.3	3.3	3.3	2.3	2.3	1.8
Buildup/RDL Material Df@1GHz	-	0.038	0.038	0.038	0.038	0.038	0.038	0.003	0.003	0.001
Buildup/RDL Materials Young's Modulus	GPa	4	5	5	5	5	5	10	10	10
Buildup/RDL Material Water Absorption	%	2.0	2.0	2.0	2.0	2.0	2.0	0.01	0.01	0.01
Min. Line width/Space	μm	12/10	8/8	8/8	8/8	5/5	5/5	3/3	3/3	1/1
Min. Conductor Thickness	μm	10	10	10	10	8	8	3	3	3
Min. Through Via Diameter	μm	180	180	180	150	150	100	700	70	70
Min. Through Via Land Diameter	μm	350	300	300	250	250	200	150	150	150
Min. Micro Via Diameter	μm	60	60	60	50	50	50	30	30	30
Min. Micron Via Land Diameter	μm	100	100	100	90	90	90	60	60	60
Min. Through Via Pitch	μm	350	300	300	275	275	275	250	250	250

PACKAGING SUBSTRATE TECHNOLOGY ROADMAP

Packaging substrate manufacturers in Japan are expects minimum line width and space of 7.5 μm/7.5 μm volume production will start in 2010 and 5.0 μm/5.0 μm volume production will start in 2014 according to survey results of Jisso Technology Roadmap 2009 edition, as shown in figure 7.

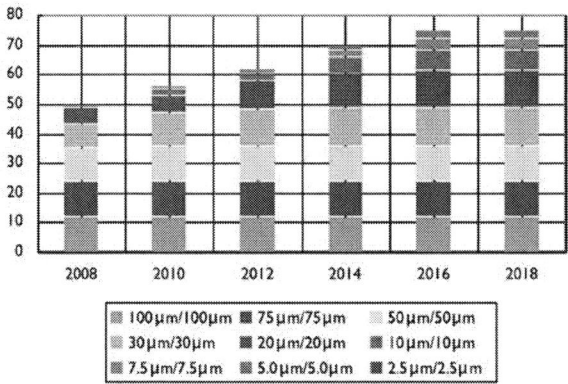

Figure 7. Minimum Line Width/Space for Packaging Substrate Volume Production Behavior (unit: μm) [7]

For printed wiring board roadmap, Jisso Technology Roadmap classified technology difficulties into 3 categories such as "Class A": conventional consumer products with conventional technology, "Class B": portable and cost performance products with leading edge technology and "Class C": highest performance independent cost with state-of-the-art technology.

In "Class C" buildup structure substrate, high-end Field Programmable Gate Array requires minimum line width/space of 5 μm/5 μm in 2010, and "Class B" for CPU and GPU will start volume production of minimum line width/space 5 μm/5 μm in 2014. Table 8 is shown minimum line width/space roadmap for substrate.

Table 8. Minimum Line Width/Space Roadmap for Packaging Substrate (unit: μm) [7]

Structure	Class	2008	2010	2012	2014	2016	2018
Rigid	A	35/35	35/35	35/35	30/35	30/30	30/25
	B	25/25	25/20	20/18	15/18	15/15	15/15
	C	20/15	20/15	15/10	15/10	15/10	12/7
Buildup	A	15/15	10/10	10/10	7/8	7/8	7/8
	B	10/10	7/8	7/8	5/5	5/5	5/5
	C	7/8	5/5	5/5	5/5	5/5	5/5
Tape	A	25/25	25/25	25/25	20/20	20/20	20/20
	B	20/20	20/20	20/20	15/15	15/15	15/15
Ceramics	A	50/50	30/30	25/25	25/25	25/25	20/20
	B	30/30	25/25	25/25	20/20	15/15	15/15
	C	25/25	25/25	20/20	15/15	15/15	15/15

Minimum via diameter below 30 microns is one of technical challenges for packaging substrate manufacturers. Conventional CO_2 gas Laser can fabricate via diameter around 50μm, and below 50μm via diameter UV-YAG Laser is adopts for fabrication. Below 20μm via diameter, Eximer Laser can form, however improvement of through put is required for commercial volume production. Figure 8 is shown volume production behavior survey results on Japanese substrate manufacturers for "Laser Drilling".

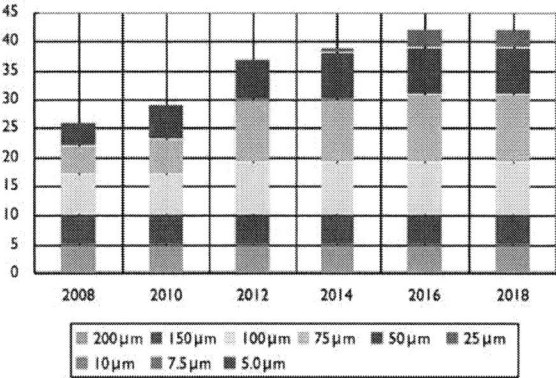

Figure 8. Laser Drill Volume Production Behavior [7]

As for rigid structure substrate, mechanical numerically controlled drilling is commonly uses for 100μm diameter. Incase of 50μm drill bit diameter, rotating speed must increasing to 500,000 rpm to 550,000 rpm while 100μm diameter hired 300,000 rpm. Thus capital investment for new mechanical drilling equipment is necessary to use less than 50μm drill bit diameters. For "Class B" substrate laser drilling, CO_2 gas laser will adopt major process technology toward 2018. Table 9 is shown minimum through via diameter/land diameter roadmap for "Class B" products.

Table 9. Minimum Through Via Diameter/Land Diameter Roadmap for "Class B" Substrate (unit: μm) [7]

Method	Structure	2008	2010	2012	2014	2016	2018
Mechanical Drill	Rigid	100/190	100/160	100/160	75/160	75/160	75/160
	Buildup	100/140	80/120	80/120	80/120	80/120	80/120
Laser Drill	Rigid	60/130	50/120	50/120	50/110	50/110	50/110
	Buildup	80/120	60/120	60/100	50/90	50/90	50/90
	Tape	80/160	60/110	60/110	50/100	50/100	50/100
Punching	Tape	150/200	150/200	150/200	150/200	150/200	150/200
Ceramics		30/30	25/25	25/25	20/20	15/15	15/15

In case of "Class C" buildup substrate, 20μm via diameter was in production in 2008 and 30 μm via diameter also in volume production in 2008 for "Class B" buildup layers. In 2014, 10μm via diameter will start production for high-end applications. Table 10 is shown minimum micro via diameter, land diameter and via pitch.

Table 10. Minimum Land Diameter, Land Diameter and Via Pitch Roadmap for Substrate (unit: μm) [7]

Structure	Description	Class	2008	2010	2012	2014	2016	2018
Buildup	Min. Micro Via/Land Diameter	A	60/100	50/90	40/80	40/80	30/70	30/70
		B	30/70	30/60	30/60	20/50	20/50	20/50
		C	20/60	20/50	20/40	10/40	10/30	10/30
	Min. Via Pitch	A	130	120	110	110	100	100
		B	100	90	80	70	70	70
		C	80	70	60	60	50	50

EMBEDDED ACTIVE DEVICES INTO PACKAGING SUBSTRATE

Embedded active devices and passive devices into packaging substrate as well as motherboard is one of potential solution to filled the gap between semiconductors and printed wiring board wiring density. And it is already start volume production in several manufacturers in Japan, Korea and Taiwan.

Figure 9 is shown categories of embedded passives and actives.

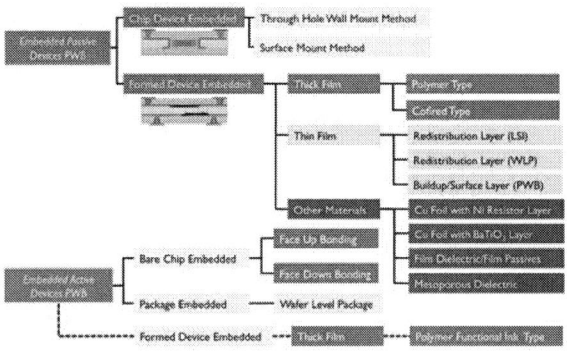

Figure 9. Categories of Embedded Passive Devices and Embedded Active Devices [7]

Regarding on embedded passive devices, there is two major technologies has been used in Japan. One is embedded discrete components such as 0603 (0.6 mm by 0.3 mm) resistors and capacitors with Cu plated terminals are mounted on to printed wiring board inner layer with surface mount equipment, then buried by prepreg or buildup layer materials. The other method is using printed wiring process technologies such as photolithography, etching and plating to form resistor and capacitor functions during PWB fabrication process. Embedded active device is hiring almost same technology of embedded discrete components, however placement of bare die has two directions such as face up and face down.

In the near future, formed active device by functional ink with carrier mobilities of 1 - 10 cm^2/Vs will be introduced. Thick film printed methods using gravure, flexography, screen, offset lithography, inkjet will be adopted for functional ink on a substrate. Table 11 is shown example of current status of active device embedded.

Table 11. Current Status of Embedded Active/Passive Devices [8]

Manufacturers	PWB Technology	Embedded Devices	Applications	Status
Dai Nippon Printing	B²it	LSI (Flip Chip), Passives (Chips)	Camera Module, One Segment Module, Finger Print ID Module	Volume Production
Toppan NEC Circuit Solutions	Buildup	WLP	One Segment Module	Volume Production
Casio Electronics	B²it	LSI (Flip Chip)	One Segment Module	Volume Production
Taiyo Yuden	EOMIN	Passives (Chip)	Power Supply Module	Volume Production
CMK	Buildup	WLP	TV Tuner Module	Volume Production
Oki Printed Circuit	B²it	WLP	RF Module	Volume Production
Denso	PALAP	LSI, Passives (Chips)	Car Navigation	Volume Production
Panasonic Electronic Device	SIMPACT	WLP, Film Passives	Car Navigation	Volume Production
Fujikura	Polyimide MLB	WLP, Film Passives	Camera Module, RF Module	Prototype
Meiko	Thin Core Buildup	WLP, Passives	RF Module, Camera Module	Production in 2010
Imbera DieDuck	Integrated Module Board (Imbera)	LSI (Flip Chip), Passives (Chip)	Cellular Phone	Volume Production
Samsung Electro Mechanics	Buildup	LSI (Flip Chip), Passives (Chips)	Cellular Phone	Volume Production

SUMMARY

In the next 5 years, the technology of packaging substrate will drive by flip chip interconnection technology with significant higher pin counts and embedded active and passive devices for System in Package. As a result, organic substrate will face technology limitation and silicon interposer or glass substrate will be introduced for high-end applications. In addition, organic substrate using buildup layers will approach minimum line width/space of 5 μm/5 μm with less than 30μm to 5μm via diameter.

Table 12 is addressed very advance PWB business structure comparison between last 5 years and next 5 years.

Table 12. Advanced PWB Business Structure [3]

Index of Change	2005-2010	2010-2015
R&D Theme (Min. Line Width)	30μm - 15μm, Flip Chip, EAD/EPD, WLP	10μm - 7μm - 5μm - 3μm, Flip Chip, EAD/EPD, Silicon Interposer
Via Diameter	75μm - 50 μm	30 μm - 5 μm - 3μm
Capital Investment	$0.15 Billion - $0.20 Billion	US$ 0.20 Billion - US$ 0.50 Billion
Design Parameter	Distributed Constant	Distributed Constant
Major Process	Semi Additive	Semi Additive, Silicon Technology, LCD Technology (Glass Substrate)
Bus Frequency	800 MHz - 1.2 GHz	> 10 GHz
Package Technology Driving Force	SiP/PoP, Embedded Actives/Passives, Wafer Level Package	Embedded Actives/Passives, Fanout WLP, TSV
PWB Type	Buildup PWB, FPC, FRPC	Silicon & Glass Interposer, Buildup PWB
Market Driver	Cellular Phone, Digital Home Appliances, Broad-band Internet	Digital TV, Super Smart Phone, Digital Home Appliance
Profit Structure	High Value Added & Differentiate	High Value Added & Differentiate
Business Model	Core Competence + Horizontal Division of Labor	Core Competence + Strategic Alliance + New Concept

The difficulty of advanced PWB business is capital investment that is tremendous huge for finer wiring board production such as 200 million US$ to 500 million US$ for 50,000 m^2 out put /monthly production factory from scratch. However, active devices embedded into printed wiring boards have potential solution to increase functional density without ultra finer wiring. Both finer wiring technology and integrated function into board level will be coexisted in next 5 years. To accomplish this scheme, infrastructural enhancement of materials, manufacturing technologies, quality assurance and modeling and simulation to advanced packaging substrate as well as embedded active devices are necessary.

REFERENCES

[1] WECC Global PCB Production Report

[2] Electronic Circuits Industrial Survey Report 2008 Edition, JPCA

[3] World Electronic Circuit Council Global PCB Production Report for 2008

[4] Fraunhofer IZM

[5] Toshihiko Nishio, Substrate Technology need by next generation high performance package, 28 Jan., 2009

[6] 2009 International Technology Roadmap for Semiconductors

[7] 2009 Jisso Technology Roadmap, JEITA

[8] H. Utsunomiya, KPCA International Symposium, 20 Nov., 2009

QUANTITATIVE ANALYSES OF INTERFACIAL PROPERTIES OF Cu-Cu BONDS FOR TSV INTEGRATION

Bioh Kim, Thorsten Matthias, Markus Wimplinger* and Paul Lindner*
EV Group
Tempe, USA and Florian/Inn, Austria*
b.kim@evgroup.com

Eun-Jung Jang, Jae-Won Kim, and Young-Bae Park
Andong National University, School of Materials Science and Engineering
Andong, Korea

ABSTRACT

We performed the quantitative analyses of the interfacial properties of Cu-Cu bonds with varying process parameters such as bonding temperature, chemical pre-treatment, pre-bond annealing and post-bond annealing. A 4-point bending method was used to measure the interfacial adhesion energy. Among all the tested parameters, bonding temperature and post-bond annealing have the most significant influence on the interfacial adhesion energy and interfacial voids. By adjusting those two experimental parameters, we could achieve even with a short bonding time (30min) the sufficient interfacial adhesion energies (≥ 5 J/m^2) with no interfacial seam voids. With increasing bonding temperature from 300ºC to 400ºC, the interfacial adhesion energy increases from 2.8 J/m^2 to 5.0 J/m^2 and the original bond interface tends to disappear. The post-bond annealing performed at 250-300ºC under N$_2$ atmosphere for 60min, when thermo-compression bonding was done at a low temperature (300ºC), drastically improves the interfacial adhesion energy from 2.8 J/m^2 to 8.9 – 12.2 J/m^2.

Key words: TSV, Cu-Cu bonds, interfacial adhesion energy.

INTRODUCTION

TSV integration enables the smallest form factor and highest performance due to the shortest and most plentiful interconnects between chips.[1] Due mainly to the thermal budget of CMOS devices, bonding processes compatible with TSV-interconnected CMOS wafers are limited only to direct oxide bonding, metal bonding (Cu-Cu or Cu-solder-Cu), adhesive bonding and various hybrids of those methods. Among all, metal thermo-compression bonding is the most intriguing because it can facilitate fine-pitch, high-density stacking of various devices leading to lower electrical resistance and higher mechanical strength.

The Cu-solder-Cu bond is formed at lower temperatures with lower manufacturing costs. However, the solder-based bonding has a weaker bonding interface and a lower electrical performance on account of the formation of brittle intermetallic compounds and Kirkendall voids.[2] Direct Cu-Cu bonding has several advantages over the solder-based bonding, including lower electrical resistivity, better electromigration (EM) resistance and more reduced interconnect RC delay.[3-4] However, the reliable Cu-Cu bonding for most industrial applications comes only from high temperature, high pressure and long process time mainly because of its tendency to form a native oxide which deadly impacts device reliability. As of today, high process temperature is one of the major bottlenecks of Cu-Cu direct bonding because it can negatively influence device reliability and manufacturing yield. The ultimate goal of this work is to develop the low temperature Cu-Cu bonding process while maintaining a strong bonding energy. As the first step, we investigated the impact of varying process parameters, such as bonding temperature, chemical pre-treatment, pre-bond annealing and post-bond annealing, on the interfacial adhesion energy and void formation.

4-POINT BENDING METHOD

The quantitative interfacial adhesion energy was measured by the 4-point bending test method with a modified tensile tester (LLOYD Instrument). Fig. 1 illustrates the schematic of the sample structure for the 4-point bending experiment. The sample was placed between four load pins. A notch with the depth of 450μm was made on the bottom Si sample with a diamond blade to uniformly initiate the crack propagation during the bending test. The 4-point bending test is based on the facture mechanics and used to calculate the interfacial adhesion energy between thin films, by measuring the energy release rate (G) which is required for a crack to propagate at the inner homogeneous material.[5] The sample has two elastic materials in a sandwich structure and the adhesion energy was measured during crack propagation at a constant moment, where a crack propagates along the weakest interface in multi-layers.[6]

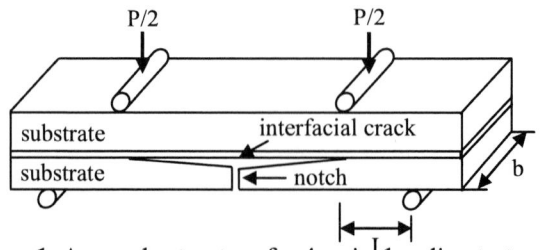

Figure 1. A sample structure for 4-point bending test

The interfacial adhesion energy, G (J/m^2), is calculated using the expression (1)

$$G = \frac{21(1-v^2)M^2}{4Eb^2h^3} = \frac{21(1-v^2)P^2L^2}{16Eb^2h^3}, \qquad (1)$$

where v is the Poisson's ratio for an elastic material (Si wafer : 0.28), E is the elastic modulus of the elastic material (Si wafer : 130 GPa), b and h are the specimen's width (3mm) and thickness (500μm), respectively, P is the measured load, L is the distance between the inner and the outer loading points (5mm) and M is the moment defined as $PL/2$.[7] In this experiment, the load cell was 20N, the loading speed was 0.08μm/s, and the distance between pins was 5mm.

INTERFACIAL PROPERTIES

All the bonding experiments with those wafers were done using an EVG520 bonder at 25kN, 10^{-3} Torr and N$_2$ atmosphere. The vacuum level was initially maintained at 10^{-6} Torr, but dropped to 10^{-3} Torr when N$_2$ was purged before the bonding process started.

Process parameters evaluated in this work include
- bonding temperatures; 300, 350 and 400°C,
- chemical pre-treatments with a diluted acetic acid (80 vol.%); 1, 5, 10 and 15min,
- pre-bond annealing; 100 and 200°C for 15min at forming gas (N$_2$+H$_2$) atmosphere, and
- post-bond annealing; 200, 250 and 300°C for 60min at N$_2$ atmosphere.

For the samples used for evaluating the effect of bonding temperature and annealing, wafer-level bonding was performed, whereas segment-level bonding was performed for the samples used for testing the chemical pre-treatment effect. Among all the tested conditions, bonding temperature and post-bond annealing have the most significant influence on the interfacial adhesion energy and seam void formation. By adjusting those two experimental parameters, we could achieve even with a short bonding time (30min) the sufficient interfacial adhesion energies ($\geq$ 5 J/m^2) with no interfacial voids.[8-9]

Bonding Temperature Effect

Wafers with 1.5μm thick, sputter-deposited Cu film on Si(100)/SiO$_2$/Ta(25nm) were used for this experiment. Neither pre-bonding surface treatment nor post-bond annealing was applied to the samples. The pressure applied to the stack was 25kN and the bonding time was 30min which is shorter than typical Cu-Cu bonding time (60-120min). The bonding temperatures tested for this evaluation were 300, 350 and 400°C. With increasing bonding temperature, the interfacial bonding energy increased as illustrated in Fig. 2. With the bonding temperature of 400°C, the interfacial adhesion energy was close to the minimal adhesion energy (5 J/m^2) required for the post-bonding processes such as grinding.[8-9]

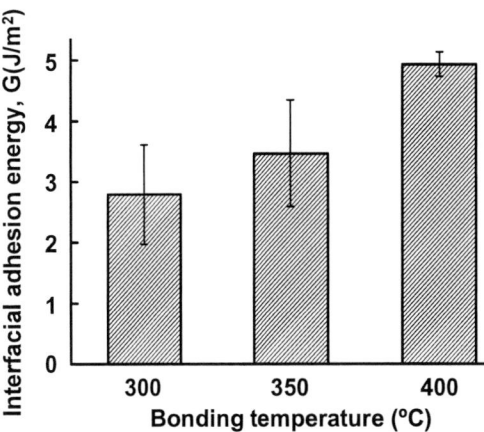

Figure 2. Interfacial adhesion energy as a function of bonding temperature

It was proven by FIB images that the original bond interface tends to disappear with increasing bonding temperature (Fig. 3), which leads to a higher interfacial adhesion energy. In all cases Cu layers were observed at both sides of the fractured samples, which means that the fracture occurred through the Cu-Cu interface.

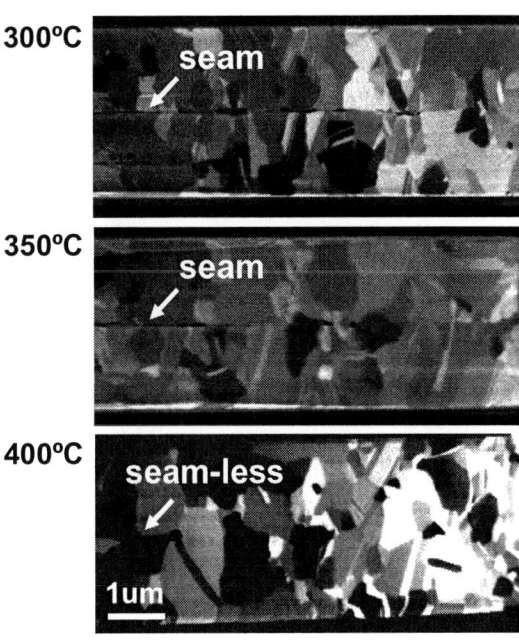

Figure 3. Microstructures of the samples bonded at different temperatures

Post-bond Annealing Effect

For the samples used for investigating post-bond annealing effect, bonding experiments were performed at 300°C and 25kN for 30min without any chemical pre-treatment or pre-bond annealing. Those samples were post-bond annealed at 200, 250 and 300°C at N$_2$ atmosphere for 60min. Fig. 4 compares the microstructures of the samples annealed at different temperatures.

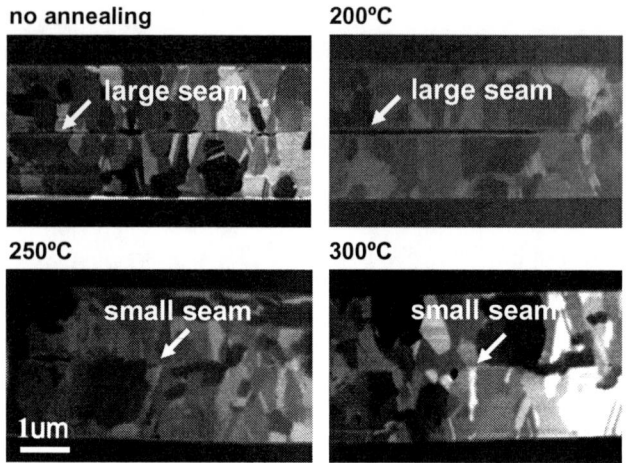

Figure 4. Microstructures of the samples post-bond annealed at different temperatures

In the case when no post-bond annealing was applied, the original interface or large seam was observed together with small voids, as this bonding was done at a low temperature with a short process time. With applying post-bond annealing at 200°C, no improvement in the interfacial properties was observed. After annealing either at 250°C or at 300°C, it was clearly observed that the interfacial seam between two layers decreased significantly. This means the post-bond annealing at 250-300°C can help Cu diffusion through the bond interface when the bonding was done at a low temperature (*i.e.*, 300°C in this experiment). Fig. 5 illustrates the change of interfacial adhesion energy with varying post-bond annealing conditions. A decrease in the interfacial seam size resulted in a higher bond strength with increasing annealing temperature. Regardless of annealing temperature, Cu layers were observed at both sides of the fractured samples.

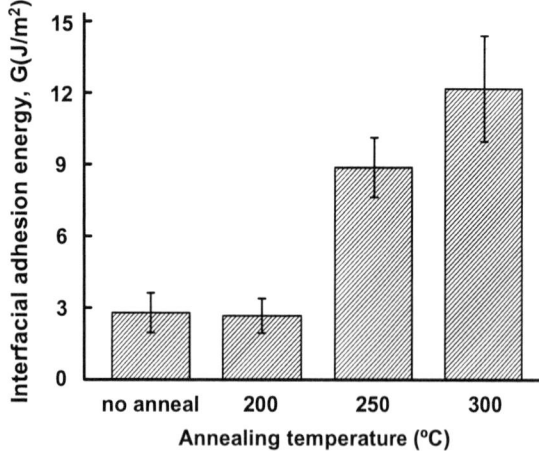

Figure 5. Interfacial adhesion energy as a function of post-bond annealing temperature

Post-bond annealing conditions evaluated was not strong enough to fully drive the recrystallization reaction at the interface between two Cu films, but drastically improved the interfacial properties by providing energy for Cu diffusion through the interface. Further optimization of subsequent post-bond annealing conditions can drastically improve the interfacial adhesion energy and thus lower the bonding temperature for Cu-Cu direct bonding.

CONCLUSIONS

The quantitative analyses of the interfacial properties of Cu-Cu bonds were performed with varying process parameters such as bonding temperature, chemical pre-treatment, pre-bond annealing and post-bond annealing. Among all the tested parameters, bonding temperature and post-bond annealing have the most significant influence on the interfacial adhesion energy and interfacial seam voids. By adjusting those two experimental parameters, we could achieve even with a short bonding time (30min) the sufficient interfacial adhesion energies (≥ 5 J/m^2) with no interfacial seam voids. With increasing bonding temperature from 300°C to 400°C, the interfacial adhesion energy increases from 2.8 J/m^2 to 5.0 J/m^2 and the original bond interface tends to disappear. The post-bond annealing performed at 250-300°C under N$_2$ atmosphere for 60min, when thermo-compression bonding was done at a low temperature (300°C), drastically improves the interfacial adhesion energy from 2.8 J/m^2 to 8.9 – 12.2 J/m^2. Further optimization of post-bond annealing conditions may improve post-bond interfacial properties.

REFERENCES

1. J.C. Eloy, *3D TSV Interconnects : Devices and Systems – 2008 Report*, Yole Development, Lyon, France, p. 5-13, July 2008.

2. J.M. Koo, *et al.*, *J. Electron. Mat.*, vol. 37, p. 118 (2007).

3. A. Fan, *et al.*, *Electrochemical and Solid-State Lett.*, vol. 2, p. 534 (1999).

4. S.I. Kim, *et al.*, *J. Korean Phys. Soc.*, vol. 50, p. 489 (2007).

5. R. Shaviv, *et al.*, *Microelectronic. Eng.*, vol. 82, p. 99 (2005).

6. H. Zhenyu, *et al.*, *Eng. Fracture Mech.*, vol. 72, p. 2584 (2005).

7. P.G. Charalambides, *et al.*, *J. Appl. Mech.*, vol. 111, p. 77 (1989).

8. R. Tadepalli, *Ph. D. Thesis*, "Characterization and requirements for Cu-Cu bonds for three-dimensional integrated circuits", Massachusetts Institute of Technology (2007).

9. B. Kim, *et al.*, "Effect of bonding process parameters on the interfacial properties of Cu-Cu direct bonds for TSV integration", in *Proc. SMTA-IWLPC* (2009).

To request a PDF version of this

Technical Conference paper,

Please contact info@smta.org

OPTIMIZATION OF COPPER PLATING PARAMETERS FOR FAST FILLING OF THROUGH SILICON VIAS (TSV)

Su Wang and S. W. Ricky Lee
Department of Mechanical Engineering, Hong Kong University of Science and Technology
Clear Water Bay, Kowloon, Hong Kong
rickylee@ust.hk

ABSTRACT

There is an increasing demand for electronic devices with smaller sizes, higher performance and increased functionality. The development of vertical interconnects or through silicon vias (TSV) may be one of the most promising approaches to provide the three-dimensional (3D) integration of integrated circuits (IC).

TSVs are typically filled with copper because of its high conductivity and wide applications in multilayer wiring. The filling is achieved with electroplating method. Even though the electroplating of copper for interconnections is well established for the copper damascene microfabrication process, it has been shown that the filling of TSVs with copper plating is a different situation due to the much larger dimensions of TSVs. Although quite a number of researchers have achieved void-free TSV filling, there still exist several technical issues to be resolved for the copper plating process. The typical issues include low throughput, high cost, heavy overburden and inconsistent void-free filling.

The plating of copper in TSVs is a very complicated electro-chemical process and has many critical factors. It is important to optimize each of these factors to achieve the desired plating results. In the present study, an electroplating process was developed to deposit copper in TSVs. The objective of this electro-plating process is to achieve a void-free copper deposited blind vias with aspect ratio 4 or higher. The ultimate goal is to achieve TSV filling with shorter plating time and better consistency. In order to achieve fully filled TSVs, chemical controlling parameters and physical parameters need to be optimized. Chemical controlling parameters typically involve copper salt, accelerator and suppressor. A series of experimental parametric studies were carried out to investigate the effects of these chemical parameters.

This paper describes the optimized plating process that results in void-free copper filling of blind via arrays. Successful plating results are demonstrated after optimizing each additive in the bath composition. The void-free copper deposition has been achieved for blind TSVs with diameters of 25 and 50 μm within four hours of plating time. All chemical process parameters and considerations will be discussed in detail in this paper.

Key words: Copper filling, electroplating, TSV, copper plating additives, void-free filling.

INTRODUCTION

Two-dimensional approaches have been traditionally applied for IC integration. However, consumer electronics market demand requires the semiconductor manufacturers to reduce the assembly size, cost and weight while functionality and performance need to be improved at same time. The continuous pressure for new IC packages has driven the semiconductor industry to develop more innovative packaging method using vertical interconnection for 3D integration. The general features of 3D packaging technologies include form factor miniaturization, integration of heterogeneous devices in a single package, replacement of long 2D interconnects with short vertical interconnects and reduction of power consumption. However, the use of spacers and wire bonds for electrical interconnection in system in packages (SiP) may not be suitable for all high performance applications. This is due to the high RC delay caused by long bonding wires. Therefore, 3D wafer or chip stacking with vertical interconnection by TSVs becomes potentially the best solution for semiconductor system integration [1-5].

Many cost models indicate that TSV filling process is one of the most expensive steps in the process flow [6]. TSV filling process also causes problems because of the lacking of a reliable measurement technique for online monitoring of TSV voids, which are one of the primary concerns of reliability issues.

Electro-copper plating deposition (ECD) for the TSV filling process is still an under-developing technology. Copper usually fills the deep blind via completely. The filling must be void-free to prevent reliability problems. Similar to the copper damascene process, TSV filling requires a bottom-up growth pattern to ensure the void-free filling. Both TSV and damascene ECD require inhibited deposition at corners and outer surfaces and accelerated deposition in the vias. However, due to the large differences in dimensions, the detailed parameters for these two microfabrication processes are not the same. Therefore, more efforts are required to optimize the ECD process for TSV filling.

ELECTRO-COPPER PLATING DEPOSITION FOR TSV FILLING

The basic copper plating process recovers copper ion (Cu^{2+}) in the solution to copper metal. The plating bath containing copper salt serves as the ion carrier. The work piece to be electroplated is placed at the cathode. The copper ions will be extracted from solution and deposited on the surface of the work piece at the cathode. The anode may be pure copper or other inert metals (e.g., platinum). Currently most electroplating processes in the semiconductor industry use pure copper as the anode.

The TSV filling process is a combination of chemistry, fluid boundary layer control and the timing of applied electrical current [7]. Due to the high aspect ratio of TSV, bottom-up filling is required for optimizing the plating results. Although pure bottom-up filling is not feasible, it would be ideal to combine conformal and bottom-up filling (see Figure 1) to achieve a better yield for void-free plating with low overburden.

The concentration and ratio of accelerator, suppressor and leveler will determine the features of plating, whether conformal or bottom-up filling. The functions of each additive are differentiated by their deposition behaviors on the cathode surface (see Figure 2). The accelerator and leveler will tend to go down to the bottom of the via and suppressor will more likely be attracted on the surface of the wafer. Therefore, the control of chemical additives is very important to achieve a void-free via filling. Without carefully optimized chemicals, the filling of copper may be sub-conformal and a big void may be formed in the via.

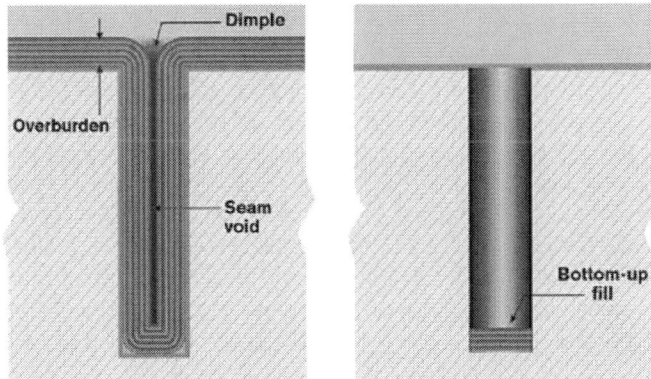

Figure 1. Pure conformal plating (left) and bottom-up filling (right) [7]

A = Accelerators   Cl = Chloride Ion
L = Leveler (weak suppressor)   Supp = Strong suppressor

Figure 2. Functions of additives on the cathode surface

EXPERIMENTAL PROCEDURE

A sample wafer with different sizes of TSVs was prepared. But this paper will focus only on the plating process and results of TSVs with 25 μm and 50 μm in diameter. The depth of TSVs is targeted at 200 μm and 300 μm for 50 μm diameter TSVs and 125 μm for 25 μm diameter TSVs. The TSVs were fabricated by the standard photolithography and deep reactive ion etching (DRIE) processes. A 5 μm thick insulation layer was deposited on the TSV walls by the PECVD method. Subsequently a barrier layer of 1000Å and a copper seed layer of 2000Å were formed by sputtering.

A cathode with plastic cover was prepared. The cover consists of two plastic plates. One of the plates has a center opening of 2x2 cm. A sample wafer with dimensions slightly larger than 2x2 cm was placed on center of the plastic cover. The sample was exposed due to the center opening of the plastic cover. A rim copper plate was prepared as well and placed on the peripheral of the sample. The copper plating was connected to the cathode by wiring around the whole plate. The two plastic cover were screwed together to hold all components together (see Figure 3). The function of plastic cover was to reduce the concentration of the electric field line on the peripheral of the sample by blocking the way of electric field lines. As the copper plate connects all peripheral of the sample, the contact point over-plating effect will be also minimized.

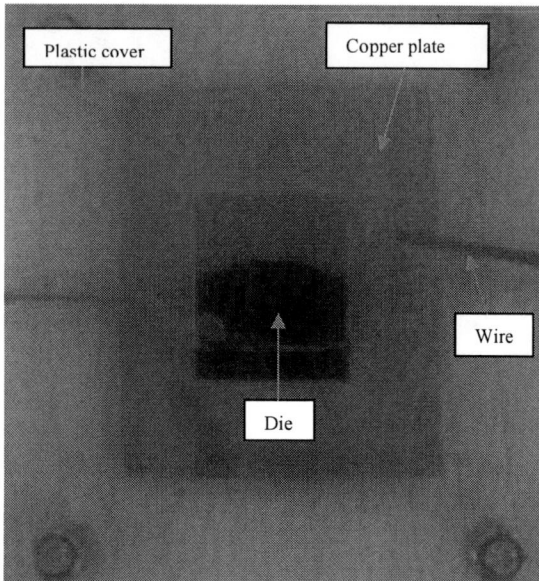

Figure 3. A sample of cathode setup

The anode was a pure copper plate which was placed in non-woven fabric bag. The non-woven fabric bag was to trap the anode copper dust in order to prevent the contamination of the plating bath. A direct current power supply with minimum current value 0.01 A was used. The minimum current density on the plating sample was 2.5 mA/cm².

A magnetic stirrer was used for solution agitation. The rotation speed of the magnetic stirrer was kept at about 300 rpm. Temperature of the solution was kept at room temperature of 23°C as most of the ECD process is carried out at room temperature. The cathode to anode distance was kept constant at 13 cm.

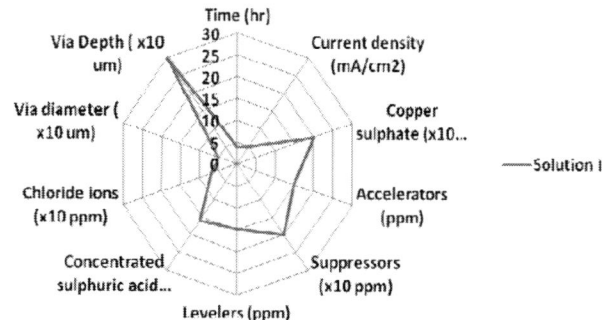

Figure 4. Radar graph for plating solution I

The initial experiments were done on vias with 50/300 μm (diameter/depth) dimensions. The result was made as the baseline for modifying the additives and optimizing the plating bath.

Figure 4 shows the basic plating bath composition and plating parameters (solution I). The basic bath consists of copper sulphate (200 g/L), sulphuric acid (16 ml/L). The additives include accelerator (SPS) (1-30 ppm), suppressor

(PEG) (100-300 ppm), leveler (10-30 ppm) and other chemicals.

Figure 5. Plating results with solution I

RESULTS AND DISCUSSION
The properties of electroplated copper in TSV depend largely on organic additives in the plating solution. Precise control of the electroplating process will require reliable monitoring of the organic additives in solution. The additives are generally presented as a mixture of complex organic compounds and the concentrations of the organic species can change rapidly as they react in the plating bath. Each component of the bath should be investigated carefully in order to achieve a desired void-free copper filling in TSV. From the preliminary results shown in Figure 5, we can see that very inconsistent TSV filling was obtained and many bottom voids were formed. Thus, modifications of the additives in this bath (solution I) were exercised in order to achieve optimum plating results.

Optimizing the Accelerator in the Bath
It is commonly believed that SPS accelerates copper reduction by forming an adsorption complex layer (RS-Cu+-Cl-) on the electrode surface, which will help the electron transport [8]. The functional group –S-S- is the active functional group of SPS during copper reduction process. Original SPS shows an effective accelerating effect on the copper reduction due to the existence of functional group –S-S- bond. Therefore, bottom-up filling can be obtained by adding SPS in ECD via filling solution during the via-filling process.

However, it is believed that dissolved oxygen can accelerate the oxidation process of SPS [9]. The disulphide (-S-S-) bond of original SPS can be oxidized to –S-SO- and can be further oxidized to –SO-SO- or S-SO₂- or even –S-SO₃-. The oxidized products do not have an accelerating effect on copper deposition. As a result, the accelerating functional group will be competing with non-accelerating oxidized group as plating goes on. Therefore, we need to adjust the level of accelerator so that the oxidized functional group will not overrule the accelerating effect of SPS too fast.

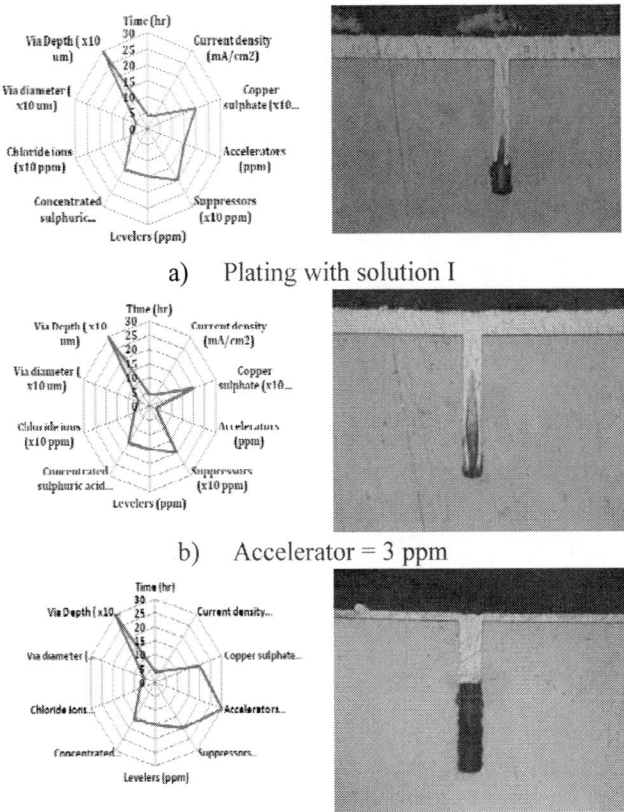

a) Plating with solution I

b) Accelerator = 3 ppm

c) Accelerator = 30 ppm

Figure 6. Plating results of different accelerator concentration

In Figure 6, we can see that when the amount of accelerator is too high, less bottom-up filling could be achieved. When accelerator was about 3 ppm, under similar plating conditions, bottom-up filling was most observable.

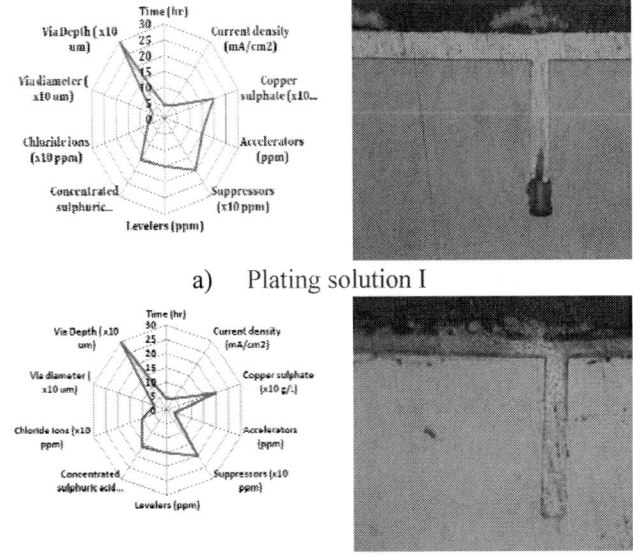

a) Plating solution I

b) Accelerator = 3 ppm, Chloride = 100 ppm

Figure 7. Plating results of different chloride ion concentration

Wei-Ping Dow *et al.* had also shown that the amount of chloride ions has an accelerating effect on via filling upon interactions with accelerator [10]. They pointed out that high accelerator concentration will favor more conformal deposition than bottom-up filling. However, the deposition behavior can be relieved with the help of extra chloride ions in the plating. They proposed that the formation of CuCl crystals in vias will favor more bottom-up filling. This behavior can be verified by the result shown in Figure 7.

Optimizing the Suppressor in the Bath
Suppressor such as PEG with high molecular weights is generally in a much higher concentration compare to accelerator. Immersion of the cathode in solutions containing PEG will result in the rapid formation of a thin passivating PEG-based film. Upon the addition of chloride ions in the bath, the combination of PEG and chloride ions will results in suppressing effect of the copper deposition reaction due to the formation of PEG-Cl film that blocks the access of copper ions to the surface [11-15].

We can see from Figure 8, the amount of suppressor does not make much difference when the potential is low. When potential goes up (always happen during plating), the suppressing effect is getting smaller. The amount of suppressor was adjusted and the amount of suppressor which works fine in current plating bath was studied.

In Figure 9, with the addition of suppressor, the closing up of vias opening was inhibited and more plating was done at the bottom of the via. However, suppressor did not have a significant effect on bottom-up filling. It supported more conformal filling. Note that the amount of accelerator has a significant effect on suppressor abilities [16]. Therefore, more experiment needs to be done to investigate the effect of suppressor as a function of accelerator.

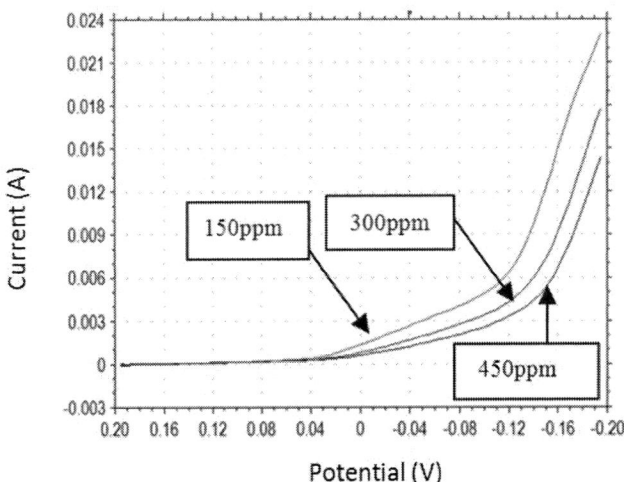

Figure 8. Effect of suppressor concentration on current density

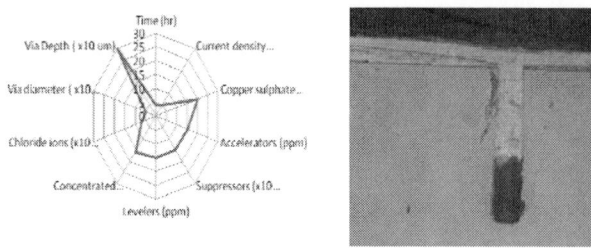

a)   Suppressor = 150 ppm

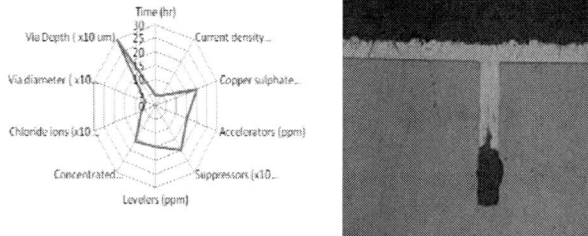

b)   Suppressor = 200 ppm

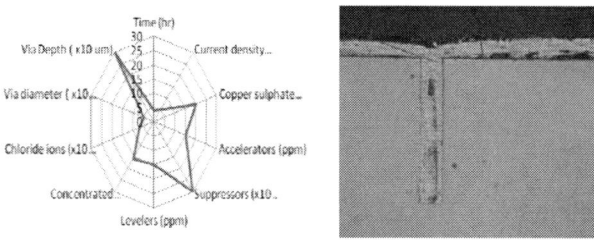

c)   Suppressor = 300 ppm

Figure 9. Plating with different suppressor concentrations

Optimizing the Leveler in the Bath

Leveler is a weak suppressor in the plating bath. It is used to inhibit the deposition of the copper at the corner of the trench to mitigate the over-plating. With the presence of leveler in the plating bath, the deposition rate will be suppressed, compared to the bath with just accelerator, suppressor and chloride [17]. However, the inhibition of the plating may largely depend on the concentration of JGB (a common leveler). Many mechanism predictions have been made on the function of leveler in the plating bath [18-20]. However, this mechanism is still not yet well understood.

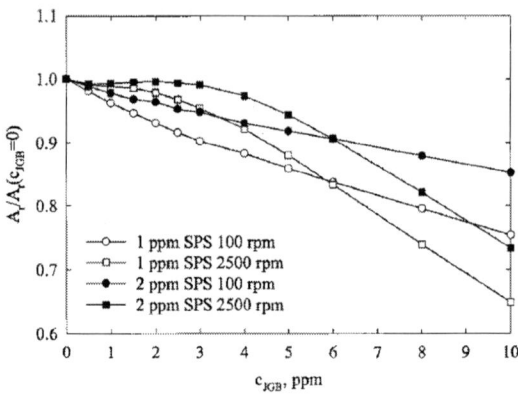

Figure 10. Effect of JGB concentration on different SPS concentration and rotation speed [17]

It has been also suggested by Yang *et al.* [17] that, the surface coverage will not be changed at high concentration of leveler (see Figure 10). Therefore, super-filling or bottom-up filling cannot be achieved at the high concentration of leveler.

Based on the results shown in Figure 11, increasing amount of leveler could improve the plating quality when the leveler was 20 ppm in concentration. Although a slim void was observed at the center of the via, the via filling quality was still relatively good. However, as the amount of leveler increased, the suppressing effect of the leveler dominated the bath and the plating rate dropped dramatically.

Optimizing the Wetting Agent in the Bath

Chemistries with poor wetting ability will result in bottom voids. The wetting problems can be addressed by adding surfactant with high wetting capabilities to pre-treat solution or electrolyte directly. The surfactant needs to be chosen such that it will provide extra wetting capabilities for the electrolyte without significantly affecting the electrochemical performance of the electrolyte. A PEG based wetting agent was chosen as it is from the same family of suppressor and will provide extra wetting ability to the bath.

In Figure 12, the adding of wetting agent eliminated the bottom voids formation. Although a slim void still existed in the via, we could confirm that the solution reached the via bottom more readily than before. The filling nature of the basic additives was not affected by the addition of wetting agent either.

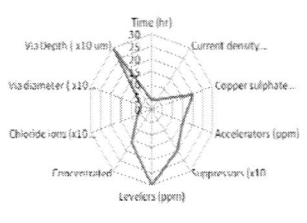 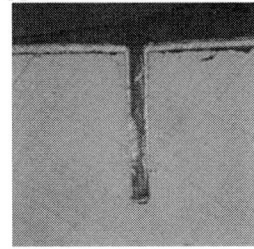

a)   Leveler = 20 ppm

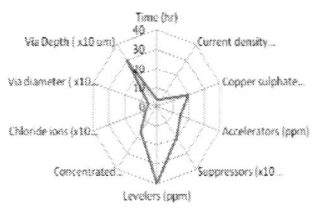 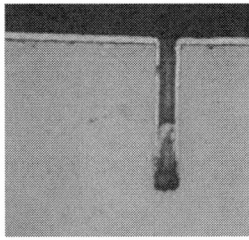

b)   Leveler = 30 ppm

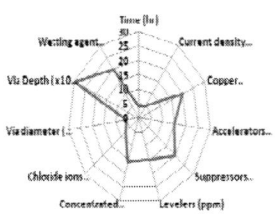 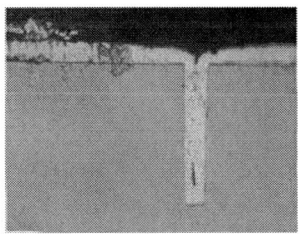

c)   Leveler = 40 ppm

Figure 11. Effect of different amount of leveler in bath

Wetting agent =20 ppm

Figure 12. Effect of wetting agent in the bath

Other Ways of Optimizing Wetting Conditions in the Bath

There are other ways to improve wetting conditions in the bath. One of them is by ultrasonic sound pretreatment. This can be done in either deionized water or copper electrolyte. In some cases, it can be done in pretreatment solution as well. The pretreatment solution normally includes copper salt, wetting agent and some additives like accelerators. Ultrasonic wetting will enable the plating solution to reach the bottom much more easily (see Figure 13).

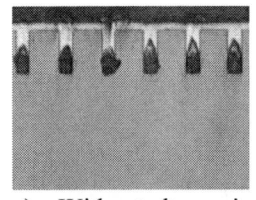 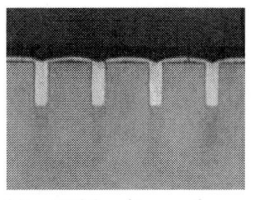

a)   Without ultrasonic   b)   With ultrasonic

Figure 13. Effect of ultrasonic wetting

Ultrasonic wetting can be a good way to wet the vias but there are some limitations. Firstly, the ability of wetting depends largely on the via aspect ratio and via diameter. The wetting will not be very effective for vias with diameter 10μm and below and the via aspect ratio is more than 10. Secondly, the ultrasonic wetting might be difficult to control on a larger scale, causes uniformity problems later. A combination of ultrasonic and chemical pretreatment wetting is generally the best way to deal the wetting problem of TSV.

Plating After Additives Adjustment

After investigating the effects of the main additives in the solution, the original plating solution I was adjusted according to the results. Better results were shown in the new solution II. Compared to solution I, solution II has less accelerator, more suppressor, more leveler, and the addition of wetting agent (see Figure 14).

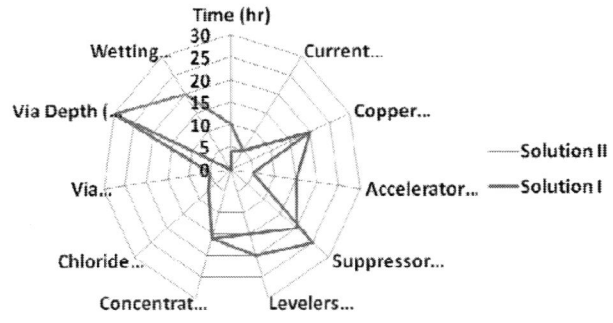

Figure 14. Radar graph for solution I and II comparison

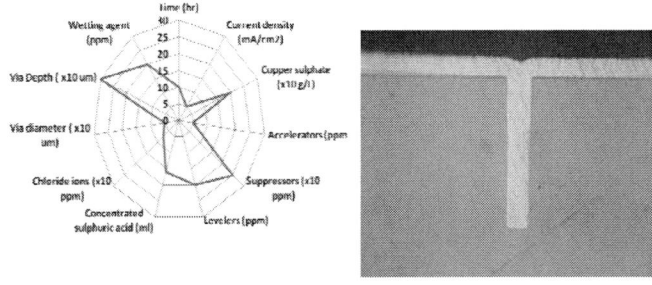

Figure 15. Plating results of 50/300 μm TSVs with solution II

Complete filling of 50/300 μm TSVs was achieved with plating solution II as shown in Figure 15. Plating done with solution II, complete fill was achieved with acceptable consistency. Bottom voids and center slim voids disappeared upon using plating solution II.

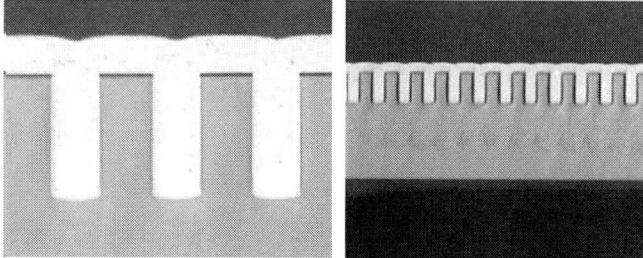

Figure 16. Plating results of 50/200 μm TSVs with solution II

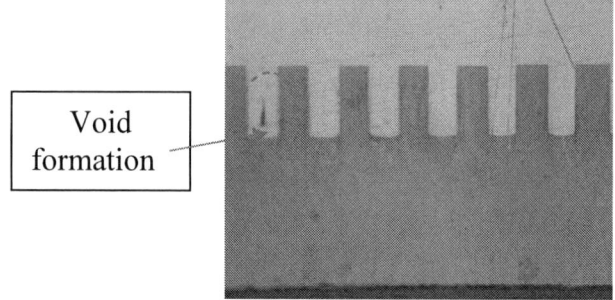

Figure 17. Voids formation in 50/200 μm TSVs with solution II

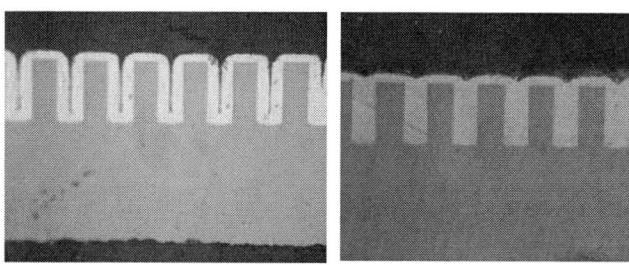

Figure 18. Plating for 3.5 hrs (left) and 4 hrs (right)

Plating of Different Via Profile
High aspect ratio vias of 50/300 μm should be much more difficult to plate compared to low aspect ratio via of 50/200 μm. Although successful plating results had been shown for 50/300 μm TSV filling, the consistency of the plating was still too low to be considered a stable and consistent TSV filling process. Plating solution II was studied on 50/200 μm in order to achieve higher consistency and shorter plating time.

From the results shown in Figure 16, as the aspect ratio of TSVs got smaller, much higher consistent plating was observed. The consistency of successful plating was over 90% of the total number of vias. However, there were still some voids observed (see Figure 17). As the over-burden of the plating was heavy, the plating time needs to be reduced in order to reduce the total over-burden on the sample surface. The minimum time required for plating 50/200 μm vias was studied and four hours plating was found to be the least amount of time required for filling this via profile with plating solution II (see Figure 18).

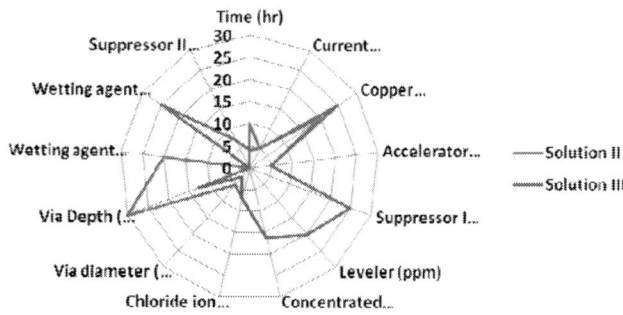

Figure 19. Radar graph presentation for solution III

Small Diameter Vias of 25 μm Via Diameter
25 μm diameter TSVs may be the most possible solution for 3D technology packaging for actual applications. It will require less time to fill due to the smaller volume of vias. However, as the diameter is smaller, it is getting more challenging to fill the vias. Because the solution is more difficult to reach the bottom of vias, plating results for 25/125 μm commonly shows a big bottom void in the via. In order to overcome this problem, higher concentration of copper ions was added into the solution. At same time, a new sulfonate based wetting agent (wetting agent 301 and purchased commercially) and a new PEG based suppressor were added into the bath as well. These modifications introduce the new plating solution III (see Figure 19).

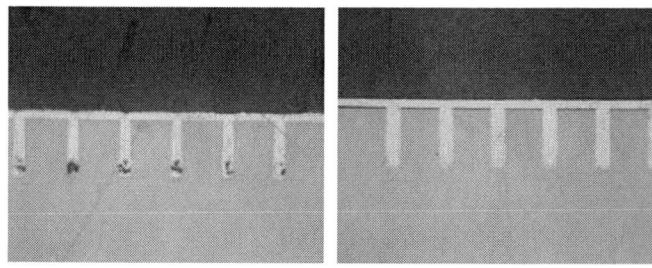

Figure 20. Plating results of 25/125 μm TSVs with solution II (left) and solution III (right)

In Figure 20, although some successful plating results were shown, the consistency of the successful plated via (80%) was still not in the desired range (>90%). There were still some vias showing bottom voids. The minimum plating time required for 25/125 μm via filling was about 1 hour and 15 minutes. Further investigation is on-going for better via filling consistency and shorter plating time.

25/65 μm TSVs are much easier to fill compare to higher aspect ratio 25/125 μm. It can be filled much faster and better consistency. In Figure 21, the successful plating via is nearly 100% and the filling time is about 35 minutes. It is

168

done with plating solution III and a higher current density of 10mA/cm².

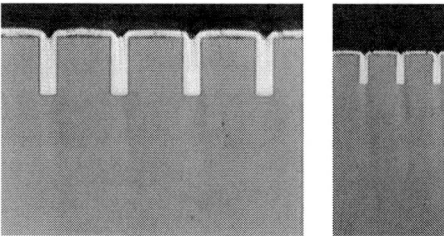

Figure 21. Plating results of 25/65 µm TSVs with solution III

Filling Mechanism of Solution III

A proper filling mechanism will ensure the complete filling of the vias. A bottom-up growth mechanism will be prefered to enable detect free filling. However, the filling rate for large TSV vias can be really slow which will affect the overall throughput for TSV device production. Therefore, a filling mechanism with both conformal and bottom-up effect is prefered. The filling of copper in the via is both upwards from bottom and inwards from side. In Figure 22, the filling mechanism of 25/65 µm via profile is shown upon increasing duration. With plating solution III, the filling of copper is faster at bottom of the via.

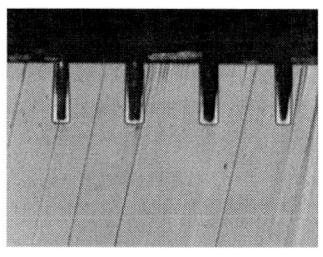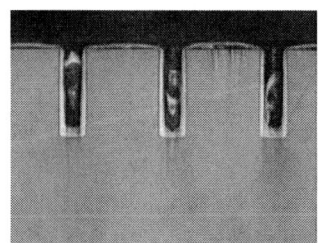
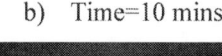

a) Time=5 mins   b) Time=10 mins

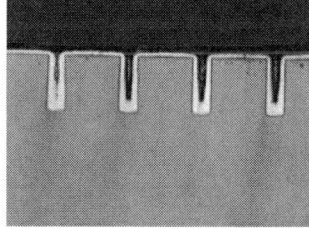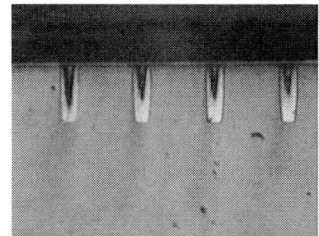

c) Time=15 mins   d) Time=20 mins

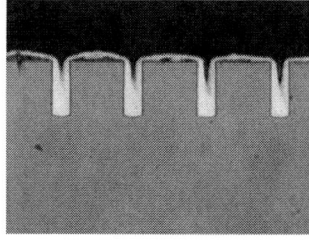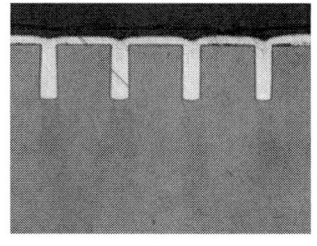

e) Time=30 mins   f) Time=40 mins

Figure 22. Filling mechanism of 25/65 µm TSVs with solution III upon increasing duration

CONCLUSIONS

In the current study, the effects of different chemical controlling parameters of the plating bath were investigated. Optimization of organic additive concentrations and ratios in the plating solution was performed in order to achieve good quality via filling. The accelerator, suppressor, leveler and wetting agent were adjusted to obtain better plating results. Complete filling of 50/300 µm vias was achieved with solution II within 4 hours. However, the consistency for this via profile was still low. 50/200 µm via profile was also filled completely with solution II. The consistency was much higher and could reach more than 90%.

Solution III with new wetting agent and suppressor were made to plate 25/125 µm diameter vias. Successful plating results have been shown. But the consistency of plating still needs to be improved. The plating was done within 1 hour and 15 minutes. Solution III is also tried on 25/65 µm vias. Plating time is only 35 minutes and plating consistency is 100%. A combination of conformal and bottom-up filling mechanism is shown with solution III.

In order to further optimize the copper plating process of TSVs, more studies need to be done to investigate the relations between different additives. Physical parameters such as periodic pulse reverse power supply, cathode rotation and agition of solution will have to be investigated as well. More experiments are under way to setup an optimized copper plating system for TSV filling with higher consistency and shorter time.

ACKNOWLEDGEMENTS

The authors would like to thank HKUST NFF and EPACK Lab staffs for their technical support. The material supply from Shanghai Sinyang Semiconductor Materials is also acknowledged.

REFERENCES

1. R. Beica, C. Sharbono and T. Ritzdorf, "Copper Electrodeposition for 3D Integration," DTIP 2008, (Nice : France 2008) (ISBN: 978-2-35500-006-5).

2. C. Laviron, B. Dunne, V. Lapras, P. Galbiati, A. Henry, B. Toia, S. Moreau, R. Anciant, C. Brunet-Manquat and N. Sillon, "Via First Approach Optimisation for Through Silicon Via Applications," in *Proc. 59th IEEE Electro. Compon. Technol. Conf.*, May 2009, pp. 14-19.

3. H. H. Chang, Y. C. Shih, Z. C. Hsiao, C. W. Chiang, Y. H. Chen and K. N. Chiang, "3D Stacked Chip Technology Using Bottom-up Electroplated TSVs," in *Proc. 59th IEEE Electro. Compon. Technol. Conf.*, May 2009, pp. 1117-1184.

4. F. Roozeboom *et al*, "3D Passive and Heterogeneous Integration Technology Options for System-in-

Package," *IEEE Workshop on 3D Integration*, 2007, Munich, Germany.

5. B. Kim, "Through-Silicon-Via Copper Deposition for Vertical Chip Integration," in. *Proc. Mater. Res. Soc. Symp.*, 2007, Vol. 970, Paper No: 0970-Y06-02.

6. A. Keigler, Z. Liu and J. Chiu, "Optimized TSV Filling Processes Reduce Costs," Semiconductor International, 5/1/2009.

7. A. Keigler, Z. Liu and J. Chiu, "Economical TSV Copper Filling," *Nexx System Presentation*, June 2008.

8. T. P. Moffat, D. Wheeler, D. M. Edelstein, and D. Josell, "Electrochemical Planarization of Interconnect Metallization," *IBM Journal of Research and Development*, Vol. 49, Issue 1 (2005), pp. 37 – 48.

9. A. Frank, and A. J. Bard, "The Decomposition of the Sulfonate Additive Sulfopropyl Sulfonate in Acid Copper Electroplating Chemistries," *J. Electrochem. Soc.*, Vol. 150, Issue 4 (2003), pp. C244-C250.

10. W. P. Dow, H. S. Huang and Z. Lin, "Interactions between Brightener and Chloride Ions on Copper Electroplating for Laser-Drilled Via-Hole Filling," *Electrochemical and Solid-State Letters,* Vol. 6, Issue 9 (2003), pp. C134-C136.

11. M. R. H. Hill and G. T. Rogers, "Polyethylene Glycol in Copper Electrodeposition onto a Rotating Disk Electrode," *Journal of Electroanalytical Chemistry,* Vol. 86, Issue 1 (1978), pp. 179-188.

12. R. Alkire and M. Verhoff, "The Bridge from Nanoscale Phenomena to Macroscopic Processes," *Electrochimica Acta,* Vol. 43, Issues 19-20 (1998), pp. 2733-2741.

13. D. Stoychev and C. Tsvetanov, "Behaviour of PEG during Electrodeposition of Bright Copper Coatings in Sulfuric Acid Electrolytes," *Journal of Applied Electrochemistry,* Vol. 26 (1996), pp. 741-749.

14. J. J. Kelly and A. C. West, "Copper Deposition in the Presence of Polyethylene Glycol," *J. Electrochem. Soc.*, Vol. 145, Issue 10 (1998), pp. 3472-3476.

15. T. P. Moffat, D. Wheeler and D. Josell, "Electrodeposition of Copper in the SPS-PEG-Cl Additive System," *J. Electrochem. Soc.*, Vol. 151 (2004), pp. C262-C271.

16. T. P. Moffat, J. E. Bonevich, W. H. Huber, A. Stanishevsky, D. R. Kelly, G. R. Stafford, and D. Josell, "Superconformal Electrodeposition of Copper in 500--90 nm Features," *J. Electrochem. Soc.*, Vol. 147, Issue 12 (2000), pp. 4524-4535.

17. P. Taephaisitphongse, Y. Cao, and A. C. West, "Electrochemical and Fill Studies of a Multicomponent Additive Package for Copper Deposition," *J. Electrochem. Soc.*, Vol. 148 (2001), pp. C492-497.

18. A. C. West, "Theory of Filling of High-Aspect Ratio Trenches and Vias in Presence of Additives," *J. Electrochem. Soc.*, Vol. 147, Issue 1 (2000), pp. 227-232.

19. Y. Cao, P. Taephaisitphongse, R. Chalupa, and A. C. West, "Three-Additive Model of Superfilling of

Copper," *J. Electrochem. Soc.,* Vol. 148, Issue 7 (2001), pp. C466-C472.

20. J. J. Kelly, C. Y. Tian and A. C. West, "Leveling and Microstructural Effects of Additives for Copper Electrodeposition," *J. Electrochem. Soc.,* Vol. 146, Issue 7 (1999), pp. 2540-2545.

SILICON INTERPOSER FOR HETEROGENEOUS INTEGRATION

M. Juergen Wolf, Kai Zoschke, Robert Wieland, Matthias Klein,
Klaus-Dieter Lang, and Herbert Reichl
Fraunhofer Institute for Reliability and Microintegration IZM
Berlin, Germany
wolf@izm.fraunhofer.de

ABSTRACT

Heterogeneous integration is one of the key topics for future system integration to address today's requirements of smart electronic systems in terms of performance, functionality, miniaturization, low production cost and time to market. The traditional microelectronic packaging will more and more convert into complex system integration. "More than Moore" will be required due to tighter integration of system level components at the package level. This trend leads to advanced 3D System in Package solutions (SiP).

One of the most promising technology approaches is 3D packaging which involves a set of different integration approaches. Silicon interposer with TSV's offers a new possibility to merge advanced devices e.g. high pin count ASICs, memories and MEMS using a silicon interposer with Through Silicon Vias (TSVs). Two applications using a Si Interposer are described in this paper.

INTRODUCTION

3D system integration is one of the most significant strategic key technologies in the field of microelectronic packaging. Solely a heterogeneous 3D system integration approach will be able to meet the requirements of future microelectronic systems in terms of performance, functionality and miniaturization. 3D silicon system integration offers the opportunity to use the advantages of silicon technology with a high flexibility to realize complex micro systems. Future applications will be found especially in the fields of transport and mobility, safety, energy and environment, information and communication as well as health monitoring. Examples for possible products are image sensors, fingerprint sensors, wireless sensor nodes, memory stacks, health monitoring devices and microprocessor modules [3-8].

Besides increasing integration level and functionality on chip level according to Moore's law, following "More Moore", different components, e.g. microprocessors, memory, passives, MEMS, etc., are integrated in one package as a system or subsystem. This technique is known as "System in Package" (SiP) – More than Moore" (Figure 1). In SiPs different technologies from wafer and board level and different materials are used, whereas organic material as a substrate carrier is of high importance because of the low cost.

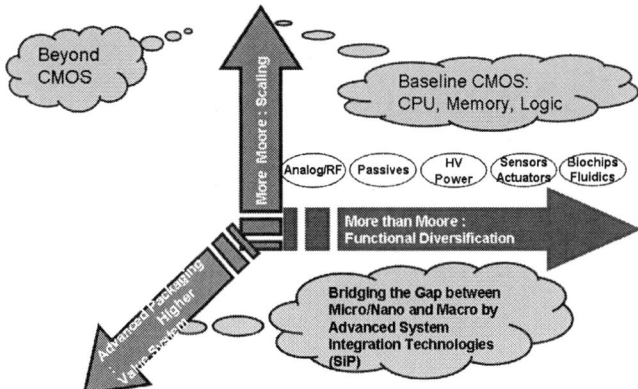

Figure 1. System Integration "More Moore" + "More than Moore" [1]

Present technologies enable the realization of organic substrates (PCB) with high density wiring and micro-vias. Passive and active components are assembled on both sides of the substrate [9-11]. The lateral required space can be reduced to a minimum by using CSPs (Chip Size Packages) or Flip Chips. Further miniaturization requires 3D integration of components with stacked dies.

The realization of advanced micro systems, e.g. 3D image processors, will be mainly driven by the enhancement of performance while low production cost have to be in focus.

3D integration technologies enable the combination of different optimized base technologies, e.g. MEMS, CMOS, with the potential of low cost fabrication through high yield and high miniaturization degree: Device stacks (e.g. controller and memory layers) fabricated with optimized 3D integration technologies will show reduced production costs in competition to monolithic integrated SoCs [3], [4], [5].

Silicon interposers with TSVs offer a new possibility to merge advanced devices e.g. high pin count ASICs, memories and MEMS using a silicon interposer with Through Silicon Vias (TSVs).

The components are electrically connected with the upper and/ or the lower wiring layer by flip chip interconnection or embedded devices. Furthermore, embedding of components allows the linkage of electrical and optical signal paths which in the future will be of main importance, especially for fast data transmission.

The general driving forces for 3D integration technologies into the production of microelectronic systems are:

- Form factor: reduction of system volume, weight and footprint,
- Performance: improvement of integration density and reduction of interconnect length leading to improved transmission speed and reduced power consumption,
- High volume low cost production,
- Applications: e.g. image sensors, memory stacks, processor/memory modules, sensor nodes.

3D INTEGRATION CONCEPTS

There are various kinds of advanced 3D SiP approaches. These include: Package-on-package (PoP), Package-in-Package (PiP), die stacking on PCB (wire bond and flip chip), stacking of flexible functional layers with embedded devices, advanced Printed Circuit Board (PCB) stacking w/ and w/o embedded devices, wafer level system integration based on Through-Silicon-Vias (TSVs) and device stacking. Figure 2 gives an overview about up-to-date categories of System in Package (SiP) defined by ITRS.

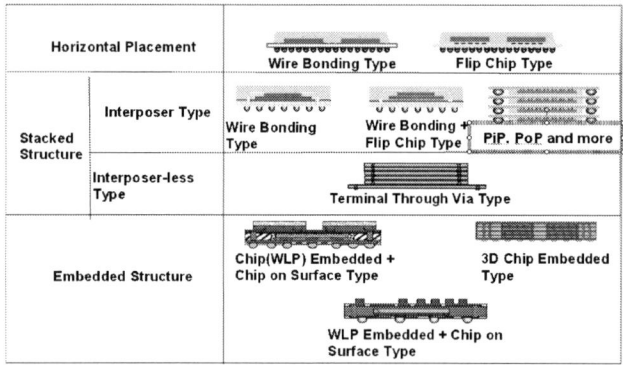

Figure 2. System in Package categories [2]

SILICON INTERPOSER WITH TSV

Wafer level packaging technologies are already used in a high volume production. Different WLP technologies [12], [13] are developed to satisfy the need to increase performance and functionality while reducing size, power and cost of the system.

Silicon interposers with TSV provide some additional features to address small sized 3D SiP. These include:

- Modularization,
- 3D stacking using Through Silicon Vias (TSVs),
- Integrated passive devices in RDL,
- Embedding of active devices into the RDL,
- Functional layer integration (actors, sensors, antennas),
- Integrated shielding (RF and power),
- Cooling,
- Integration of energy storage and converter,
- Optical chip to chip interconnects.

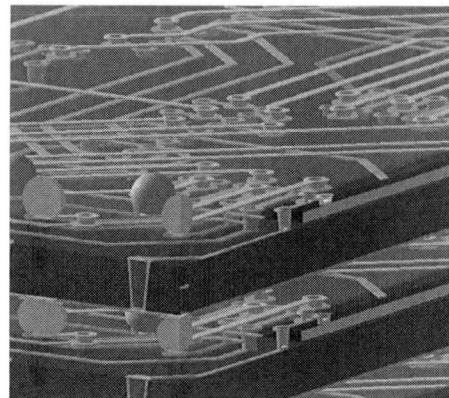

Figure 3. Schematic of a stackable silicon interposer with TSV and embedded active device

Stacking of such functional TSV interposers (Figure 3) allows the realization of modularized systems e.g. complex ultra-miniaturized sensor nodes, e.g. "e Grains", with testable sub-module layers for signal detection, data processing and transmission and power conversion at a high miniaturization degree.

Thin Chip Integration
Within Fraunhofer's "Wafer Level Thin Chip Integration (TCI) approach", thinned dies (thickness < 20 µm) are embedded and interconnected in a polymer dielectric layer with a multilayer thin film wiring or redistribution layer (RDL) on wafer level. A silicon interposer with or w/o trough silicon vias (TSV) can be used as a carrier. The thin chips are embedded face-up into the polymer layer. There is as well the potential for additional face-down mounted devices on top of the carrier (Figure 4).

One of the key advantages of this approach is the integration of passive components (resistors, capacitors, inductors, etc.) close to the active dice which results in minimal parasitic [14].

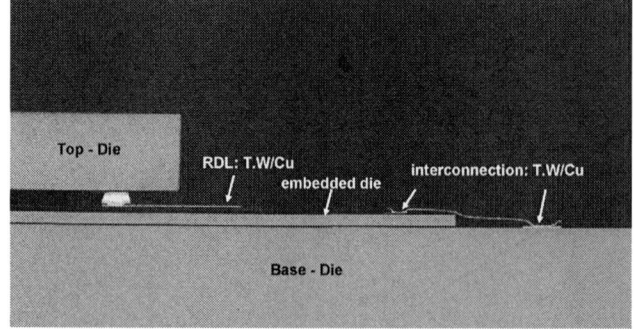

Figure 4. TCI-embedded die into RDL and interconnected by a thin film layer and a flip chip interconnect (source: European project "e Cube" [8])

Through Silicon Via Integration
Key process technologies are enabling 3D architectures with TSV interconnects included:

- High aspect ratio via formation,
- Isolation, barrier, and seed layer deposition,

172

- Via metal filling,
- Wafer thinning and thin wafer handling,
- Temporary bonding/de-bonding,
- Frontside/ backside multi level redistribution layers (RDL),
- Wafer/chip alignment, adjusted bonding (D2W, W2W).

Most of those 3D technologies are quite new to packaging providers and require a synergy between FE/ BE infrastructure as well as a well defined process integration.

Figure 5 shows a cross section of a 3D integrated test structure with W-TSV and CuSn-SLID interconnection after soldering. The tungsten filled ICVs are interconnected by Al wiring to the metallization of the top device and CuSn metal system to the metallization of the bottom device. The voidless metallization of the high-ratio aspect ICVs (diameter 2-5 μm) is realized by CVD of tungsten [15],[16].

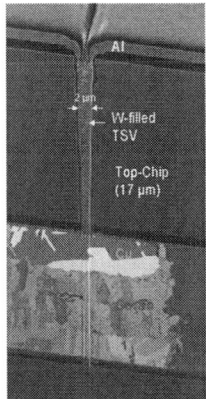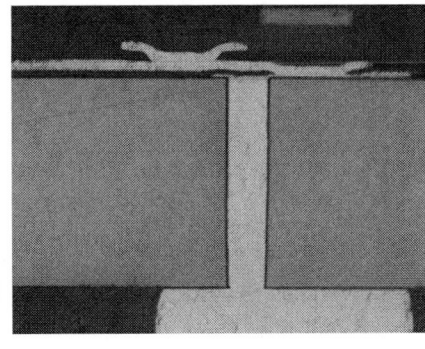

Figure 5. Cross-section of TSVs and interconnects
Left: W-TSV and SLID interconnection [15].
Right: Cu-TSV in silicon interposer with Cu-RDL and Cu-pillar interconnect

Copper TSVs are required in advanced systems because of the lower electrical resistivity. The Cu-TSVs (Figure 5) are filled by electroplating using a seed layer of sputtered TiW:Cu. Small sized vias are filled using Cu-CVD and TSV with larger diameters (>5 μm to 20 μm) are filled using electro deposition of copper [21] with an appropriate seed layer deposition process such as CVD or PVD [22]. A standard sputter PVD process can be used for aspect ratio in the range of 5 [22]. This allows the processing of metalized TSV in silicon substrate with a thickness of > 100 μm thickness today. Figure 6 shows a top view of a silicon interposer with Cu-TSVs (> 73150 TSV/cm²) and a multilayer Cu-RDL. The interposer acts as a high density wiring carrier for high IO count devices (Figure 7).

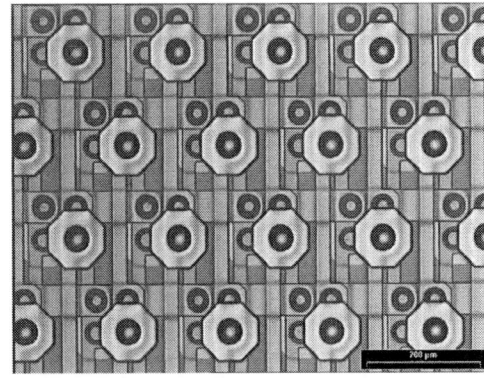

Figure 6. Detail of high density Si Interposer with Cu-TSVs and RDL

For the assembly and stacking of the devices, different interconnection technologies e.g. SLID [18] [19] or micro solder bumps are currently in evaluation. Most promising is wafer level assembly using micro bumps, e.g. SnAg, to deal especially with surface incoplanarity of the devices or warpage of the board/ substrate.

Figure 7 shows flip chip assembled devices with a high number of IOs an on a TSV silicon interposer.

Figure 7. FC- assembled devices on interposer

TSV Interposer for 3D Image Sensor Stack
A 3D image sensor with a TSV silicon interposer offers the possibility to implement a higher degree of vision-based system integration and to use standard components (not designed for 3D TSV) and gives more flexibility for the system design. The integration does not require especially designed circuits and components for the stacking. Thus different image sensor and processor which are already on the market today can be integrated by means of a TSV silicon interposer. The TSV interposer provides the electrical connection between image sensor and processor. The main challenge of the system concept was to enable a high density system integration of CMOS camera and processor. 321 IOs of the VIP-II SIMD core processor, 44 IOs of the camera chip as well as several voltage domains had to be routed on a TSV silicon interposer with minimized routing layers. For the next level interconnection the final 3D image sensor system had to provide 333 BGA balls with 540 μm pitch for PWB assembly and test.

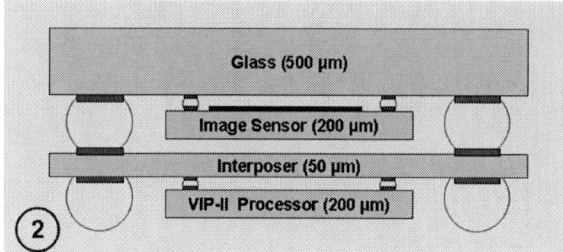

Figure 8. System concept for 3D Image Sensor Device [20]

The concept (Figure 8) is based on silicon interposer with TSV, which provides the electrical interconnection to the sensor die and the processor. The bumped image sensor is mounted face down onto a glass redistribution die which reroutes the IOs of the image sensor to a fan-out arrangement with larger pitch and size. These large contacts are then used to bridge the image sensor height to mount the glass die with the mounted image sensor flip chip wise onto the interposer top side. The processor is mounted on the interposer backside. For the assembly initial chip to wafer assemblies of image sensors dies to glass redistribution wafers as well as processor dies to interposer wafers followed by balling and dicing were used. Finally, glass carrier and interposer have to be assembled on single component level. The final interposer was designed to have a total size of 14 x 14 mm^2. It includes one topside and one backside routing layer as well as 3023 TSVs. Figure 9/10 show the final 3D stack according to the schematics in figure 8. [20] [23].

Figure 10. Backside view of 3D SiP Image sensor using Si-interposer (3023 TSV) [15].

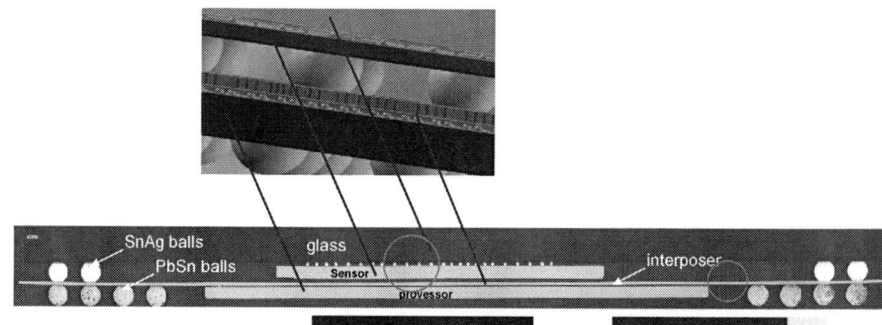

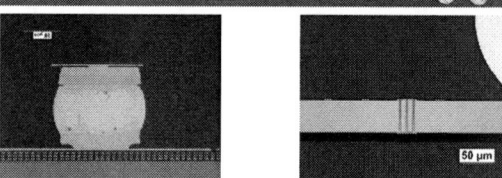

Figure 11. Cross-section of image sensor (Fig. 11) with Si-interposer (W- TSV) and SnAg bumps [20].

SUMMARY

Besides the progress in silicon technology following "Moore's law" there is an increasing demand for highly miniaturized complex system architectures which are based on 3D SiPs. Currently different approaches are in development, which are also combined in a new 3D SiP generation. The new 3D SiP solutions are driven by a wide range of applications.

Future advanced 3D solutions will result in complex 3D stacking approaches using TSV technology. Silicon interposer with TSV are one important element to combine different advanced devices into one miniaturized system with high functionality.

The main target today is to achieve cost reduction in TSV processing and stack assembly. But nevertheless 3D WL approaches technology is one of the main technology drivers in packaging and system integration.

Figure 9. 3D front side view of SiP image sensor using Si-interposer (3023 TSV) [15].

ACKNOWLEDGEMENTS

The authors would like to thank the cooperation with Infineon: Dr. Werner Weber and Ivan Limansyah. Furthermore thanks to the staffs involved in the 3D and Wafer Level System Integration program at Fraunhofer IZM especially Nils Jürgensen, Hermann Oppermann, Ingrid Kuna, Olaf Wünsch, Christina Lopper, Maria v. Suchodoletz, Karin Samulewicz and Julia Röder, T. Fischer, V Glaw. Also special thanks to the cooperation with the members of the EMC 3D consortium.

REFERENCES

1. ENIAC, ITRS 2008/9

2. A&P Roadmap ITRS, 2007/08 , white paper: System in Package - SiP, Semi, USA www.itrs.net/Links/2007ITRS/LinkedFiles/AP/AP_Paper.pdf

3. Reichl, H.; Wolf, M. J.: Hetero System Integration Challenges and Requirements for Packaging, MHSI 2006, Sendai, Japan, 6.-7. November 2006

4. Reichl, H.; Wolf, M. J.: "Potential Technologies for Wireless Sensor Nodes". Nano-Manufacturing Technology Pioneering Life Science for Health, Tokyo, Japan 2006

5. Reichl, H.: "Systems Integration – Requirements and Technical Solutions" 2007 IEEE European Systems Packaging Workshop, Como, Italy 29.1.2007

6. Wolf, M.J.; Reichl, H.: "The eGrain / e-Cube Concept" EWSN 2006, 13.-15.2. 2006 Zurich, Switzerland

7. Ramm P. and Sauer A., '3D integration technologies for ultrasmall wireless sensor systems – the e-CUBES project', Future Fab International, Issue 23 (2007) 80-82

8. www.ecubes.org

9. Utsunomiya, H. "Packaging Substrate technologies trend in Japan" Pan Pacific 2008, 22.-24.1.2008, Kauai, Hawaii, USA, Proc.

10. Boettcher L., Manessis D., Neumann A., Ostmann A., Reichl H.: "Chip embedding by Chip in Polymer technology", Device Packaging Conference 2007, Scottsdale, Arizona, Proceedings

11. Reichl, H.: „Potentiale der Leiterplatte für die Systemintegration - Multifunktionale PCB", Fachtagung Elektronische Baugruppen und Leiterplatten, 13.2.2008, Fellbach, Germany

12. Brunnbauer M., Fürgut E., Beer G., Meyer T., "Embedded Wafer Level Ball Grid Array (eWLB)," Electronics Packaging Technology Conference, EPTC 2007, Singapore, Dec. 2007.

13. Brunnbauer M. and Meyer T., "Embedded Wafer Level Ball Grid Array", 3rd Annual Device Packaging Conference IMAPS 2008, Scottsdale (Arizona, U.S.A.), March 2008.

14. Zoschke Kai, Reichl H.: „Herstellung integrierter passiver Komponenten auf Wafer Ebene", Mikrosystemtechnik Konferenz Dresden, 15.-17. Oktober 2007

15. Ramm P.: "3D System Integration: Enabling Technologies and Applications", International Conference on Solid State Devices and Materials SSDM 2006, Yokohama (2006) 318-319

16. Ramm P. and Buchner R.: "Method of making a vertically integrated circuit", US Patent 5,766,984, Sep. 22, 1994 [DE]

17. Wolf M. J., Reichl, H. "3D wafer level System integration" IMAPS Korea, 3.-4.9.2008, Seoul Korea

18. Ramm P., Wolf, M.J., Wunderle, B. 2008: „Wafer-Level 3D System Integration". In „Handbook of 3D Integration", Vol. 2, p. 289-318, Wiley Verlag, Weinheim

19. Klumpp et.al. "Chip to Wafer Stacking by using Through Silicon Vias and Solid Liquid Interdiffusion, 2nd Intern. IEEE Workshop on 3D System integration, Munich, Germany, Oct 1st, 2007

20. BMBF Projekt „KASS" FKZ: 01M3163A, 01M3163B

21. www.emc3d.org

22. M. Jürgen Wolf, Thomas Dretschkow, Bernhard Wunderle, Nils Jürgensen, Gunter Engelmann, Oswin Ehrmann, Albrecht Uhlig, Bernd Michel, Herbert Reichl „High Aspect Ratio TSV Copper Filling with Different Seed Layers". ECTC 2008, May 27-30, Orlando, FL, USA, proc.

23. M. Juergen Wolf, K. Zoschke, A. Klumpp, R. Wieland, M. Klein, L. Nebrich1, A. Heinig, I. Limansyah, W. Weber, O. Ehrmann, H. Reichl „3D Integration of Image Sensor SiP using TSV Silicon Interposer", EPTC 2009, Singapore

NANOTECHNOLOGY IS NOW STARTING TO FIND APPLICATIONS IN ELECTRONICS

Alan Rae
TPF Enterprises LLC
Wilson, NY, USA
alan@tpfenterprises.us

ABSTRACT

Four years ago we outlined some of the areas where nanotechnology might be applied to electronics. We are now seeing real examples of use, particularly in new areas such as photovoltaics, but there are applications in more conventional applications such as solder and surface finishes.

INTRODUCTION

The first presentation on nanotechnology at Pan-Pac predicted that nanomaterials would be entering the electronics supply chain and being assembled into products. Today, there are indeed products in use – not as many as anticipated and not necessarily the same products as predicted. We will explore some of these products and the areas in which they are being used.

We have seen controversy about nanomaterials in the workspace, we have seen completely new materials families such as graphenes emerge and we have seen new electronics markets such as solar photovoltaics where nanomaterials play a key role in their success. Other areas have taken longer to develop than expected.

THE iNEMI 2009 ROADMAP

The iNEMI roadmap (www.inemi.org) is a comprehensive document that reviews the key issues affecting the electronics supply chain. Gaps in the technology or infrastructure that can adversely affect NEMI members are identified, and the NEMI Research Committee was formed to prioritize and disposition the tasks and identify companies, universities and government laboratories that can address them for the mutual good. The results are published in the Research Priorities {1}, downloadable from the iNEMI web site.

Almost every roadmap chapter in the 2009 roadmap identifies aspects of nanotechnology that can enhance existing products or replace their structure or function.

Nanotechnology in semiconductors beyond the normal feature size shrinkage is raising interest in electronics circles. The novel 3-D structures known as FinFETs are predicted to be used in the next generation of semiconductor devices arriving in 2011-2012 and nano-wire based structures will be starting to appear in the same time frame. These are highly complex structures made by sophisticated

processes. Most nano applications that are reaching the market today are actually much more basic and many are concentrated in the area of improved materials.

Small size features – around the wavelength of light – can produce very interesting properties. Below the wavelength of light, nano structures can become invisible to the naked eye; band gaps in semiconductor materials can be modified to alter electrical and optical properties; metals can sinter and coalesce well below their melting temperatures and nanotubes and nano-wires can behave as individual transistors. Structured surfaces can be scratch-resistant, ultrahydrophobic or self-cleaning.

Nanotechnology has been described as a toolkit for the electronics industry in that it gives us tools that allow us to make nanomaterials with special properties modified by ultra-fine particle size, crystallinity, structure or surfaces. These will become commercially successful when they give a cost and performance advantage over existing products or allow us to create new products.

SEMICONDUCTORS AND PACKAGING

In 2009 we saw 45 nm node semiconductors deployed and real progress being made towards the next nodes. We haven't yet come to the "brick wall" predicting the end of Moore's Law... but it is still on the horizon.

Packaging continues to be a concern and we are starting to see a real emphasis on 3D packing to address integrated functionality, speed, form factor, and cost (yes, better-faster-smaller-cheaper still rules, especially for smart phones and netbooks). Nanomaterials use has generally been limited to evolution of filler systems used in packaging materials and underfills.

Replacement concepts for silicon semiconductor technology have focused for several years on carbon nanotubes. This technology has been slowed by the difficulties of separating metallic and semiconducting nanotubes produced together in the synthesis process which differ only in the stacking sequence of carbon atoms in the tubes. Better process control and new separation processes have allowed separation to take place but there are still cost and other issues involved.

Shepherding individual nanotubes into useful devices has also proved tricky. One approach is to create a random mesh (rather like a non-woven fabric) that is then cut to shape; the other approach, pioneered by the NSF Center for High-Rate NanoManufacturing, {2} is to assemble them into lithographically formed features. We are still some way from widely deployed applications of nanotubes in electronics although structural applications and those based on bulk conductivity are growing.

INTERCONNECTION

We predicted a widespread use in interconnection, which has proved rather more difficult to achieve in practice. In composites and coatings the law of mixtures is always difficult (conductivity is dominated by the less conductive material especially if conductivity is directional, as with carbon nanotubes) and so concentrations required to reach acceptable conductivity may be higher than is economic.

The iNEMI nano solder project showed that nano-sized SAC alloy could sinter at $180^{\circ}C$ and below but that flux formulation was problematic. Nano-sized metals can catalyze some flux ingredient decomposition as low as $120^{\circ}C$ and flux residues could really hinder solidification. Work is continuing on solder and solder replacement.

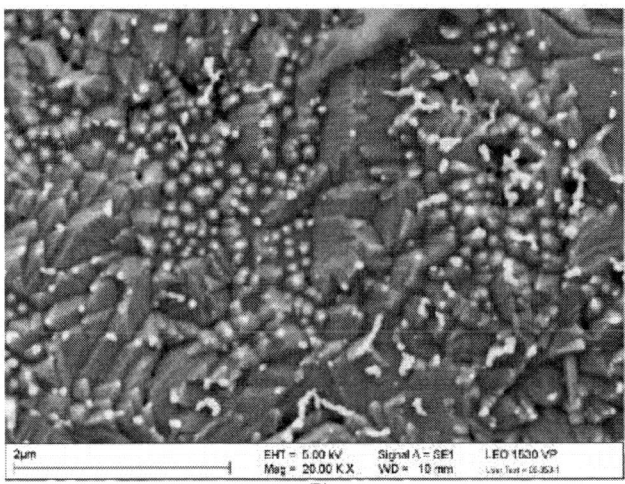

Figure 1 Enthone Nanofinish[R] (with permission)

Nano Surface finishes have been faster to market. Nano surface finishes have been commercialized by Enthone (Cookson Electronics Inc.) Their novel surface finish, originally developed by Ormecon in Germany, combines a conductive polymer with a nano silver to combine the best features of OSP and silver. This product is potentially a substitute for ENIG as well as other surface finishes due to its superior, aging, oxidation and solderability properties coupled to multiple reflow capability {3}

Printed electronics as a discrete industry has not developed as fast as the marketing reports suggested – except in the solar area (see below). Nano inks, particularly those based on silver, are compatible with polymer substrates but are in general waiting for the markets to develop.

CLEAN ENERGY

Clean tech represents a huge growth opportunity for electronics. For the first time ever, alternative energy sources outstripped nuclear electricity in the USA in the first half of 2009 - 0.7 quadrillion BTU January-May 2009, according to the Department of Energy's Monthly Energy Review {4}. Alternative energy depends heavily on nanomaterials used in electronics structures – in nano silver inks for current collectors for silicon cells, in printable thin film copper indium gallium selenide (CIGS) cell materials themselves as well as the conductors in the newer types of rigid and flexible cells that can lower cell costs below $1 per peak watt. They are an area of increasing focus for EMS companies, with Jabil, Flextronic and Celestica among others very active in the field.

Nanomaterials have many potential applications in fuel cells and alternative fuel catalysis but their first widespread application has been in solar photovoltaics.

In the "Front End" or cell formation process, reactive exothermic nano foils formerly produced by RNT Inc. are now supplied by Indium Corporation.{5}.This novel process uses a local exotherm to provide controlled heating and reliable mounting of sputter targets to backplates.

Printable CIGS (Copper Indium Gallium Sulpho-Selenide) cells are being produced using nano inks by NanoSolar Inc. {6}. The nano ink is coated onto a metal backplate to allow the formation of these flexible and potentially low-cost cells.

At the "Back End" of the process, metallization inks based on silver are used on the front surfaces of silicon solar cells. These metallizations together with back side aluminum and silver are co-fired at $800^{\circ}C$ causing cracking and distortion of the cells. Nano silvers processing at $200-300^{\circ}C$ are available but have yet to be widely accepted.

DISPLAYS

The use of carbon nanotubes as "self-sharpening" cold cathodes in plasma televisions has been limited by improvements in the competing LCD televisions and concerns over the energy consumption of large plasma TV sets, California has enacted ground-breaking legislation limiting the energy consumption of TV sets {7} and other states are likely to follow. Plasma TVs are typically less energy-efficient than LCD TVs which are becoming more efficient as LCDs replace fluorescent backlights.

Indium Tin Oxide (ITO) is used not only in TV sets but also monitors, phone displays and solar cells and is supply limited. Nano alternatives being given serious consideration include carbon nanotube and graphene containing films.

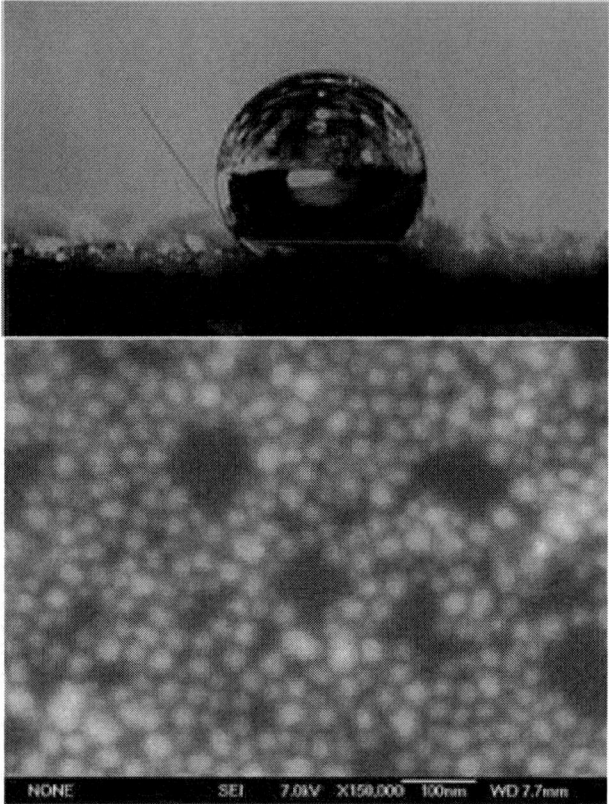

Figure 2 Hydrophobic and antibacterial coating of 80nm Ag in polyurethane (NanoDynamics)

Anti-smear nano coatings are being sold by suppliers including Aculon Inc {8}. Hydrophobic and oleophobic coatings of this kind will be increasingly used for screens and other surfaces.

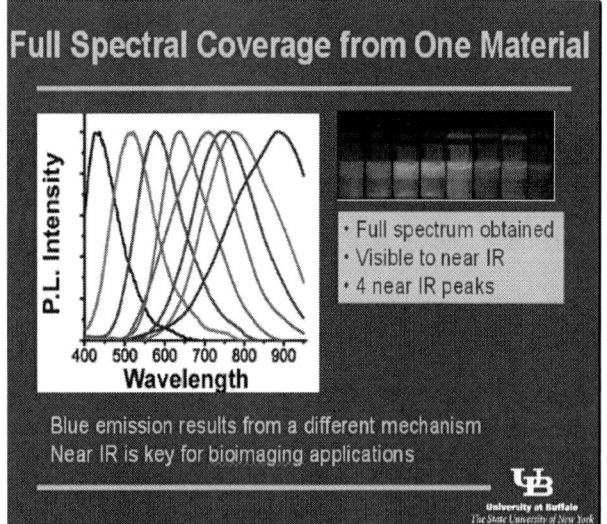

Figure 3. Spectral Coverage of Si quantum dots (University of Buffalo)

LEDs are replacing fluorescent tubes in TV and monitor screens to reduce power usage and mercury usage. Eventually they will start to replace fluorescent lighting more generally as costs start to fall. Efficiency and spectral

performance can be modified using quantum dots, nano-sized particles of silicon that absorb unwanted frequencies and re-emit at desired frequencies. Silicon quantum dots will replace the first generation Cd materials.

NEMS

NEMS – Nano-Electro-Mechanical Systems are the nano equivalent of the micro systems in MEMS. Two examples below are the NEMS nose and nanotube-based memory.

Figure 4. Schematic of a NEMS single molecule detector. Plan view is in the left, side view is on the right

The "NEMS Nose" is being developed by groups such as the Roukes Group at CalTech {9}. Here a nano-sized resonator can be used to detect molecular mass by alterations in the resonant frequency of a vibrating cantilever. Arrays with different surface treatments can be used for selective detection of chemical species and here we have created a NEMS analog of a mass spectrometer. This reduces the detector size from a device the size of a refrigerator to a NEMS device that can fit in a handheld detector. This has obvious applications for emergency services and medical use – and in addition, if the coating on the cantilever contained palladium, an extremely sensitive detector for hydrogen would be created – a sensor that will become vital if the hydrogen fueled vehicle becomes a reality. Hydrogen has an extremely wide explosive range – from 4% to 75% in air, much wider than gasoline which has a range from 1% to 7%. Hydrogen leaks in enclosed spaces are bad news!

Polymer based NEMS sensors such as those produced by NEMS AB {10} promise to be very inexpensive to produce and can be made selective to specific molecules.

Figure 5. Schematic of the two possible states of a carbon nanotube based memory cell (NRAM).

Nantero's NRAM {11}uses bundles of carbon nanotubes in arrays that can be switched to form a n array of 0's and 1s depending on whether they have moved position under an applied field. This non-volatile memory promises to be faster and more durable than flash memory and to use a lot less power then SRAM, fast and low power but can not hold a lot of data, or DRAM, slow and high power consumption but high capacity. Interestingly, as 2GB RAM memory is getting to be about the minimum memory requirement for a PC, cooling of memory is now an issue.

Totally new devices for new applications are quite rare and difficult to predict! These are totally new concepts giving us new tools to address opportunities that we don't know how to address just now. An example from the past was the AFM (Atomic Force Microscope) which allowed us to explore surfaces in previously unknown detail using a totally new principle. "Killer apps" like this may have shortcomings – there are a limited number of AFMs needed in the world and relatively few companies producing the detector heads.

NEMS can also enhance existing MEMS applications. These enhanced MEMS devices could detect smaller changes in acceleration, pressure, flow etc. Here the cost-benefit analysis is fairly straightforward if there is a clear market need for this type of measurement. This type of product will often be developed by established MEMS manufacturers

NEW MATERIALS REACHING COMMERCIALIZATION

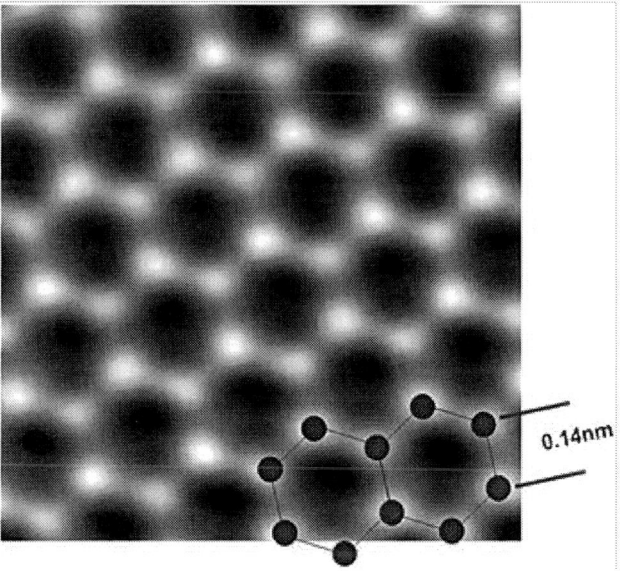

Figure 6. TEM of Graphene (Wikipedia)

One key new materials system has surfaced in the past 4 years. Graphene – composed of sheets of carbon atoms one atom thick – which seem to have many of the desirable properties of carbon nanotubes without the high cost and potential health and safety concerns. Potential applications include ITO replacement

Many nano materials have been developed because of their interesting properties and companies have been founded on products for which there is limited market demand (technology push). This tends to produce leading edge products with very limited immediate commercial potential. Work by iNEMI and others suggests the time for deployment in the electronics industry is typically 7 years for a new product that fits with the existing infrastructure and 15 years for a disruptive product. Doubters only need to look at the intensive phase of lead-free solder qualification and implementation (1999 to 2009…and still not complete for complex boards) or the implementation of MEMS

devices in accelerometer applications – 30 years! The electronics industry is fast moving in term of ultimate product development but very conservative when it comes to accepting new materials, devices and systems.

MANAGING HSE ISSUES

There continues to be a lively discussion internationally about whether nanomaterials should be regulated separately from traditional materials There are many opinions and the challenge is the diversity of types and diversity of applications – how do you equate nanosized silver, carbon nanotubes and liposomes, or nanomaterials in cosmetics, wind turbine blades, injected for medical image enhancement and automotive clear coat? Some nanomaterials behave differently at below 100nm – for example silver which can be sintered as low as 120^0C, well below its 961^0C melting point – and others such as sodium chloride (which you inhale every time you go to the beach) do not appear to behave differently.

At the moment products are registered under EPA's TSCA and, if appropriate, FIFRA programs. Only two materials have been singled out for special attention, carbon nanotubes because of their unique structure and properties, and silver because of the large number of antibacterial products being launched – not all of them properly registered – including computer keyboards and mice.

ISO TC 229 "Nanotechnologies" is at the forefront of the harmonization task. As with any new technology there is a "land grab" by national and international standards and other organizations; ISO is looking to harmonize nomenclature, metrology, health and safety guidelines and communication (data sheets, MSDS, labeling etc) to bring some order to the process.

It is highly likely that all nanomaterials presented to the electronics board fab and assembly industries will be in some way encapsulated as inks, as pastes, in resins or as coatings compatible with existing processes and probably will not be subject to specific regulations.

CONCLUSIONS

Nanomaterials are starting to be more widely used in electronics but it takes a significant time for these applications to take hold and for significant business to result. Most applications will be evolutionary rather than revolutionary but will still yield significant performance and economic benefits.

ACKNOWLEDGEMENTS

Past colleagues at NanoDynamics Inc; current colleagues and associates at Graphene Devices Ltd, University of Buffalo, Purdue University, Global Solar Technology, Cookson Electronics, iNEMI, ISO and the Inventures Group.

REFERENCES

[1] "Research Priorities", iNEMI Inc., http://thor.inemi.org /webdownload/RI/2009_Research_Priorities.pdf

[2] Center for High-Rate NanoManufacturing http://www.northeastern.edu/chn/

[3] Enthone Ormestar Ultra [tm] http://www.enthone.com/ pwb/index.aspx

[4] US Department of Energy Monthly Report http://www.eia.doe.gov/mer/

[5] "Indium Corporation Acquires Assets of Reactive Technologies" http://www.indium.com/news/ nanotechnologies/

[6] NanoSolar Inc. http://www.nanosolar.com/technology

[7} CA TV restrictions http://www.wired.com/gadgetlab/ 2009/03/california-tv/

[8}Aculon Inc. display coatings http://www.aculon.com/ displays.html

[9] Roukes Group CalTech http://nano.caltech.edu/

[10] NEMS AB http://www.nems.se/

[11] Nantero NRAM http://www.nantero.com/mission.html

SOLAR CELLS ON ULTRA-THIN CRYSTALLINE SILICON

Robert Mertens, Ph.D.
IMEC & K.U.Leuven
Leuven, Belgium
robert.mertens@imec.be

ABSTRACT

Possible avenues for cost reduction of crystalline silicon photovoltaics are reviewed. Driving forces in this process are the use of less grams of Si per Wp and the reduction of the manufacturing cost. A possible crystalline Si PV roadmap as viewed by IMEC is discussed. Two important developments are treated in more detail. The first development deals with the stress-induced lift-off method (SLIM) for kerf-loss free wafering of ultra-thin (40 to 50 micron) crystalline Si foils. A second lift-off method to create even thinner (a few micron thick) Si foils is the empty space in Si method. Both techniques require new techniques for handling these ultra-thin Si foils and new assembly methods for the photovoltaic modules.

Key words: crystalline silicon, solar cells, cost reduction, road map.

INTRODUCTION

Today crystalline Si (c-Si) based photovoltaics (PV) is by far the most important solar cell technology, taking more than 75% of the world wide market. During the last two years thin film PV, using materials such as amorphous Si, Cadmium-Telluride or Copper-Indium-diSelenide is drawing much attention and several thin film PV production plants have been commissioned or put in operation, claiming production costs considerably smaller than that for c-Si PV. Therefore in order to stay competitive with non c-Si thin film cells the production costs of c-Si PV have to be lowered. In that respect the following macro-trends in the production of c-Si based PV can be observed.

The Use of Less Grams of Si Per Watt Peak (Wp)

Today the Si wafer indeed represents 40% of the c-Si cell module cost. In order to decrease that cost the Si lost during wafering must be minimized (or in the limit completely avoided), the thickness of the active layer should be reduced and the conversion efficiency of the Si solar cells should increase up to or above 20%.

The Reduction of the Manufacturing Cost

Reduction of the c-Si PV manufacturing cost is possible by:

- equipment scaling and increase of the areal throughput
- the use of PV dedicated equipment
- upscaling of the fab size from MW/yr to GW/yr

- reduction or elimination of the use of expensive materials such as silver paste
- more standardization and vertical integration
- integration of the cell and module manufacturing.

The combination of these macro-trends results in a speeding up of the c-Si PV learning curve, strengthened by increased price competition from non c-Si thin film technologies and an accelerated reduction of the feed-in-tariffs e.g. in Germany. A critical aspect in this development is the industry-wide acceptance of a roadmap for c-Si based PV. In that respect IMEC shares the vision that the roadmap should be based on a sustained decrease of the Si consumption per Wp and an increase of the conversion efficiency.

A POSSIBLE C-SI PV ROADMAP VIEWED BY IMEC

Figure 1 shows a possible c-Si PV roadmap as viewed by IMEC. The essential characteristic of this roadmap is the sustained decrease in cell thickness. Starting from this technology we foresee two paths in our roadmap. The first one, represented by the upper part in figure 1, is evolutionary and includes a sustained decrease in cell thickness from the actual 200 μm in the standard cell to 40 μm (or even 3 μm) in the ultra-thin U-cell. We expect that by 2020 these ultra-thin cells will be the industrial standard. This evolution will however include two technology generations. In the near future a substrate thickness between 100 and 180 μm will be the standard and this will require modifications of the cell structure and the process sequence. At IMEC we have selected and further developed the industrial passivated emitter and rear cell concept, or i-PERC in short, as a quickly implementable industrial solution.

Although the i-PERC concept can be used with substrate thicknesses as small as 80 μm [1], difficulties with the assembly of such thin cells with contacts at both sides will probably necessitate a transition to cell structures having both contacts at the rear side. The i2-BC (industrial interdigitated back contacted) cell is such a cell structure that can be used for thicknesses of typically 80 to 120 μm. The technology generation, to be implemented in industry around 2020, is referred to as the U-cell (Ultra-thin) cell. In the first generation of the U-cell technology 40 to 50 μm thick Si substrates, prepared by the SLIM (Stress-Induced-Lift-off-Method) technique will be used. The second generation of the U-cells will use even thinner (a few microns thick) c-Si foils, made by the empty space in Si method. For both types of U-cells the Si foils will, prior to junction formation, be glued to a glass layer with the size of the module and processed in parallel using low temperature heterojunction formation processes.

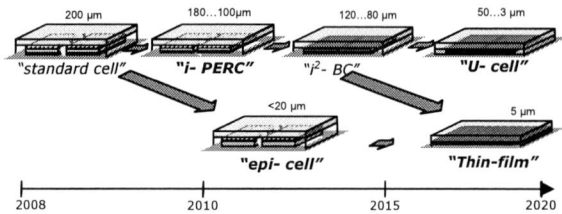

Figure 1. A possible c-Si PV roadmap viewed by IMEC

In parallel with this evolutionary path our road map includes a disruptive path, including epitaxial thin film Si solar cells, consisting of a thin (< 20 μm) high quality epitaxial layer on top of a low cost upgraded metallurgical grade Si substrate. This cell is referred to as the epi-cell. By 2020 the epi thin film cells will probably evolve in c-Si thin-film cells with thicknesses of only 5 μm, grown on glass or ceramic substrates.

It is important to note that several technologies will probably coexist and will be used in different applications. In this paper progress made at IMEC in these fields during recent years will be discussed.

INDUSTRIAL-TYPE PERC CELL

The fabrication of solar cells on thin (below 150 μm) Si wafers has received much attention in recent years due to the shortage and the high cost of Si feedstock. However, it is well known that two major challenges emerge when cells become thinner, i.e., the reduced long wavelength photon collection and the increased cell bowing, which respectively result in the loss of efficiency and production yield. The i-PERC is a quickly implementable industrial solution to meet these challenges. Figure 2 compares the structure of the i-PERC with that of a standard cell. Rather than the use of a full Al BSF (Back Surface Field) as in the standard cell, a passivation stack consisting of a layer of low quality oxide and a layer of SiNx is deposited on the rear side of the substrate. The dielectric stack is locally opened by laser ablation using a pulsed Niodymium-Yag laser. The resulting circular contact holes have a diameter of approximately 20 μm. An Al layer is then screen printed or evaporated on the backside and alloyed.

During the firing step the BSF is locally formed in the openings (Figure 3). It is found that the interface between the Si and AlSi alloy is primarily along {111} planes, as observed with SEM after the AlSi alloy is removed in a hot HCl solution (Figure 3 right, shown upside down).

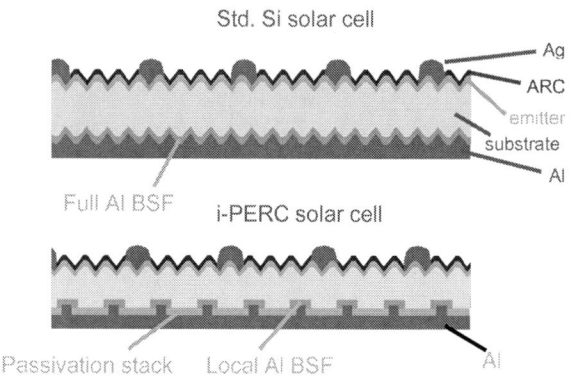

Figure 2. Design of the i-PERC cell

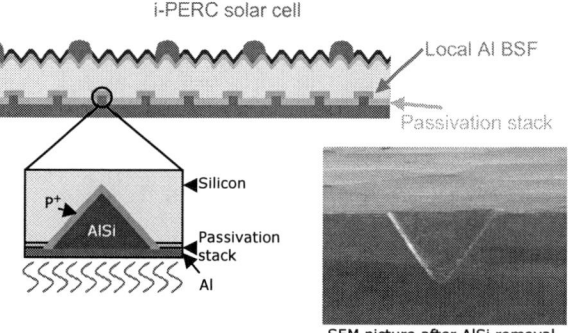

Figure 3. Formation of local back surface field

In the new i-PERC structure, most of the Si rear surface is separated from the electrode by the passivation stack, where the surface recombination velocity is lower than that at the local BSF passivated openings (typically ~700 versus ~1400 cm/s [2].

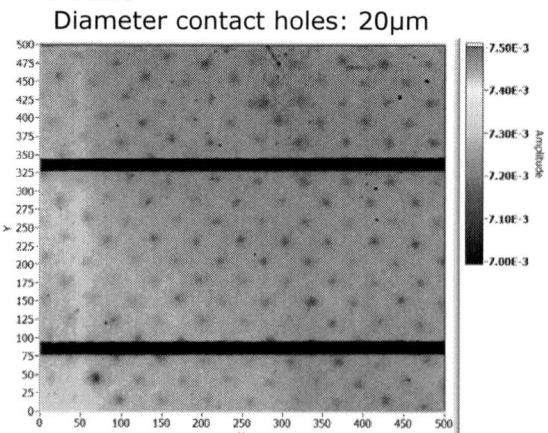

Figure 4. Laser beam induced current mapping of the i-PERC cell

Figure 4 shows a typical LBIC image (laser wavelength 850 nm) measured on the 80 µm thick i-PERC solar cell. The pattern of the rear local BSF can be clearly seen. The current generated in the regions above the passivation stack is higher than the one from the regions above the local openings, thus indicating a lower surface recombination velocity. The effectiveness of the passivation stack can be seen by comparing the open circuit voltage V_{oc} of standard cells with that of the i-PERC cells. As shown by figure 5 the V_{oc} of the i-PERC cells is almost independent of the thickness of the substrate; on the other hand the V_{oc} of the standard cells decreases for smaller substrate thicknesses. This clearly indicates the lower rear side surface recombination velocities in the case of the i-PERC cells.

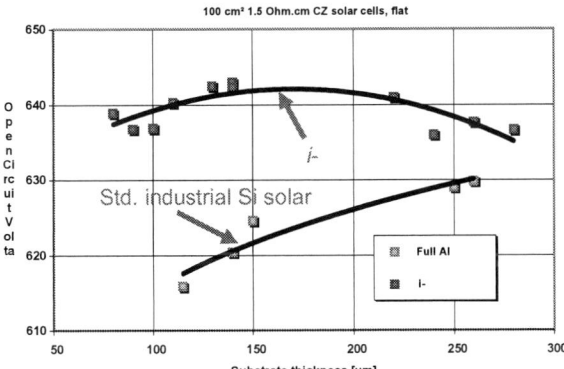

Figure 5. The open circuit voltage of i-PERC vs. full Al BSF

Some recent i-PERC results, obtained at IMEC, are shown in Table 1.

Material	Thickness (µm)	Area (cm^2)	Eff (%)	Ref
Cz	130	100	17.6	[3]
Cz	80	78	16.6	[1]
Multi	180	100	17.3	[4]
Multi	120	156	16.8	[5]
Multi	120	225	16.1	Priv.com.
EFG	170	100	16.0	[2]

Table 1. Best i-PERC results at IMEC

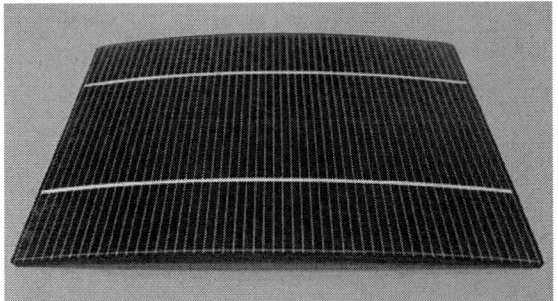

Figure 6. A 156 cm^2 130 µm multicrystalline Si full Al BSF cell showing bowing problems

Figures 6 and 7 compare a conventional large area (156 cm^2) 130 µm thick conventional full Al Back-Surface-Field cell with the new i-PERC cell on the same type of substrate with identical thickness and area. Comparison of these two figures clearly shows that the bowing

problem can be eliminated by using the new i-PERC structure.

Figure 7. A 156 cm^2 130 µm multicrystalline Si i-PERC cell where bowing is eliminated

INDUSTRIAL INTERDIGITATED BACK CONTACTED CELLS

The next technology generation on the roadmap are the industrial interdigitated back contacted cells (I^{2}BC) schematically represented on figure 8.

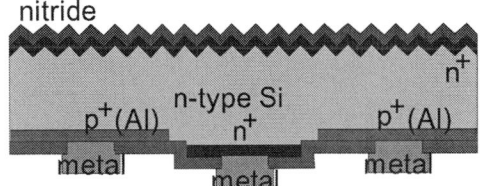

Figure 8. Schematic cross section of I^{2}BC solar cells on an n-type substrate

The wafer thickness for this type of cell is in the 80 to 120 µm range. Important characteristics of this type of cell are the texturing at the front side, the formation of the p-type emitter by the alloying of a screen printed aluminum layer and the passivation of both surfaces. As all contacts of this cell are at the rear side new module assembly techniques are required, as schematically shown by figure 9.

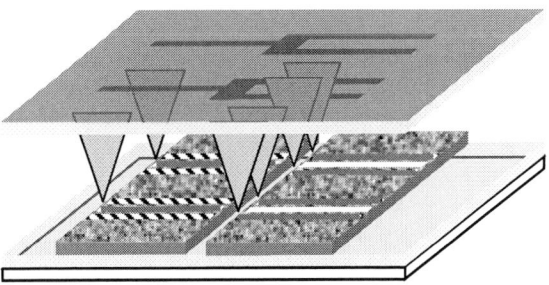

Figure 9. The assembly of a module with back side contacted Si solar cells

An advantage of back-contacted solar cells is that they allow a simplified interconnection and module manufacturing technology that can be easily automated. IMEC has proposed a new method that involves a fully automated pick-and-place of the cells and all the tabbing materials. In this concept the front cover glass (covered by a first sheet of encapsulant, e.g. EVA) stays at the same place while robot arms are laying out the cells, face down on top of it, at their final position. Conductive adhesive is then dispensed on the contact pads of all the cells. The triangles in figure 9 represent the dispensers, the solid lines the p-type contact channel and the dashed lines the n-type channel; both of which are covered by the conductive adhesive. The tabbing material is picked and placed at its correct position. The module assembling is finished when a second layer of encapsulant and the back sheet are positioned. This approach has led to prototype back-contact modules with high fill factors.

U (= ULTRA-THIN) CELL OF THE FIRST GENERATION

It is probable that by 2020 the thickness of crystalline Si solar cells will be reduced to 50 μm. This also corresponds to the optimal thickness of a crystalline Si cell, making the best trade-off between wafer cost and cell efficiency [6]. Moreover this will allow the Si consumption to be reduced to 2g/Wp.

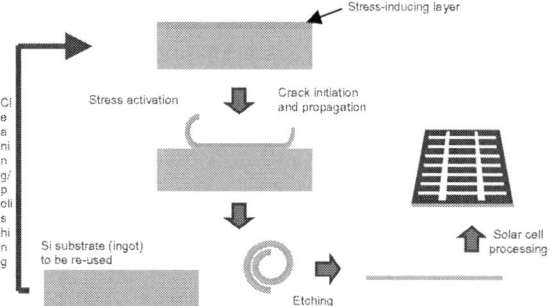

Figure 10. SLIM cut: experimental achievements

Recently [6] IMEC has presented a new wafering method for the production of ultra-thin crystalline Si. This new lift-off process named SLIM-cut (for Stress-induced LIft-off Method) requires only the use of a screen printer and a belt furnace; no ion-implantation, porous layer or additional thickening by epitaxy is needed to obtain high quality wafers in the thickness range of 50 μm without kerf-loss. As indicated in figure 10 the starting material is a Si substrate. A metallic layer is screen-printed on top of it and the wafer is annealed at high temperature in a belt furnace. Upon cooling down, the metal layer, as well as the Si substrate, undergo a thermal contraction, but the mismatch in coefficient of thermal expansion between the metal and the Si induces a high stress field in the substrate. To release the stress in the system the metal layer snaps off the parent substrate, peeling off at the same time a Si layer of approximately 40-50 μm. Figure 11 shows a photograph of the resulting substrate and of the Si layer that has been lifted off, still attached to the metal layer.

The substrate is cut along the whole surface but remains otherwise intact. It can be re-used for further additional layer lift-off. Starting the process from an ingot or from a very thick (several cm) substrate, the aim is to produce a big number of such thin daughter wafers from one mother substrate.

The metal layer is then removed in a metal-etching solution, resulting in a clean and stress-free Si layer. A SEM picture reveals a thickness of 40-50 μm relatively constant over the wafer (Figure 12).

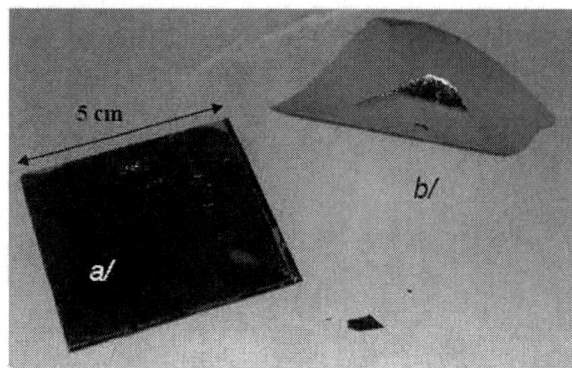

Figure 11. Photograph of the structure after the top layer (b) is peeled off the parent substrate (a) (25cm^2)

Since the process is purely mechanical, the characteristics of the resulting daughter wafer will depend on the mechanical properties of the parent substrate. In particular, since the mechanical properties of Si are highly anisotropic, the crystal orientation plays an important role. <111> is known to be the weakest plane in the Si crystal. As a consequence cleaving along this plane is energetically favorable and occurs with a higher probability. A fracture parallel to the surface will therefore be favored in a substrate oriented as such.

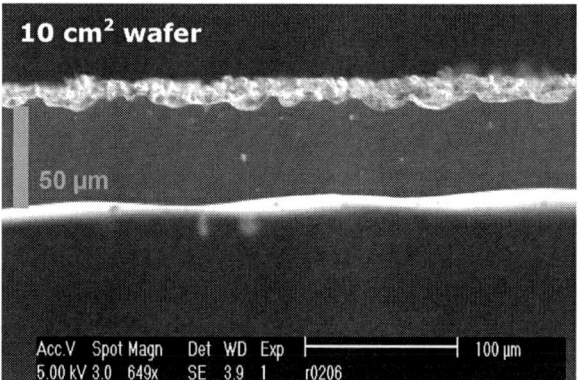

Figure 12. SEM picture of the film after flattening and metal-cleaning of the bi-metal

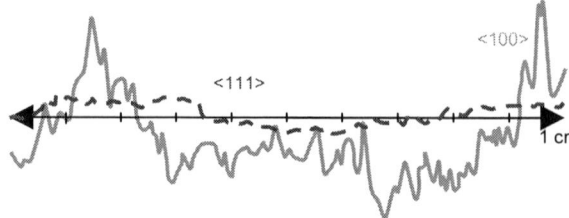

Figure 13. Comparison of the roughness (measured with a Dektak profilometer) of surfaces in the case of processing on a <100> oriented (solid line) and a <111> oriented (dashed line) parent substrate

Figure 13 shows the comparison of the profile of a surface in the case of a parent substrate oriented <100> and oriented <111> over a length of 1 cm. It appears very clearly that a <111> oriented substrate will produce daughter wafers with a much smaller roughness and is thus favorable for the SLIM-cut method.

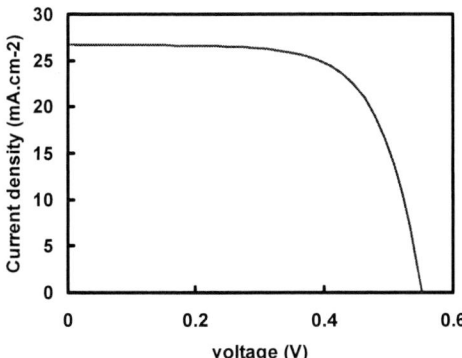

Figure 14. Current-voltage characteristics of a 1 cm² all made in a SLIM-substrate

In one of the resulting thin Czochralski daughter wafers a solar cell was made with a heterojunction emitter process described elsewhere [7]. The 1 cm² cell reached an efficiency of 10.0% (FF = 67.8%, J_{sc} = 26.7 mA/cm², Voc = 550 mV). There was no rear-surface passivation and no intentional surface texturing. The current-voltage characteristic of the cell is shown in figure 14. These results indicate that the electronic quality of the material is largely preserved during the lift-off process in spite of the high stress involved. Much higher efficiencies are expected when surface passivation and texturing are introduced.

In view of calculating the Si consumption of such device, we assume that 50 μm of Si are lost in a cleaning step to prepare the surface between two lift-off processes. Then we compare the Si consumption of one 40 μm cell of 10% with a 250 μm cell of 15.5% (obtained with 200 μm of kerf loss). We obtain a factor of 3.2. Assuming a Si consumption of 10 g/Wp for the standard technology, we reach already, for the unoptimized cell produced from the SLIM-cut method, a Si consumption of 3.1 g/Wp. This figure is expected to decrease even more as a more refined process is applied.

An important point is that for the SLIM-cut cells the 40 to 50 μm thick Si foils are bonded to a glass substrate before the junction formation process. Consequently, the junction formation must occur at low temperatures e.g. by the deposition of a heterojunction emitter using amorphous Si.

U (= ULTRA-THIN) CELLS OF THE SECOND GENERATION

Beyond 2020 the cell thickness will most probably further reduce to thicknesses of a few microns only. A concept that could provide such a thin monocrystalline silicon foil is the empty-space-in-silicon technique [8]. Following this method thin films of silicon can be formed by reorganization of regular arrays of cylindrical voids (macropores) at high temperature. A layer transfer process for thin-film Si cells based on the reorganization of such macropores has been developed and is schematically shown in figure 15. First macropores are formed by selective etching (step 1), the pores are annealed (step 2) thereby creating a thin (thickness of 1 to 3 μm) Si film,

detachment and transfer of the thin film to a low-cost substrate (step 3), and solar cell processing step 4.

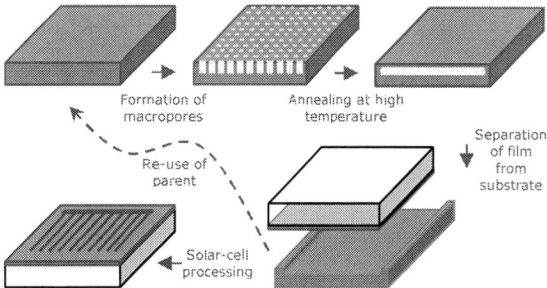

Figure 15. Layer-transfer process for thin-film silicon solar cells based on the reorganization of macroporous silicon at high temperature: formation of macropores (step 1), annealing (step 2), detachment and transfer of the thin film to a low-cost substrate (step 3), and solar-cell processing (step 4).

Figure 16 shows an example of a 1 μm thin film obtained after annealing in hydrogen at 1150°C. It can be clearly observed that the thin film is separated by a gap from the mother substrate. After detachment of the film the mother substrate can be reused again.

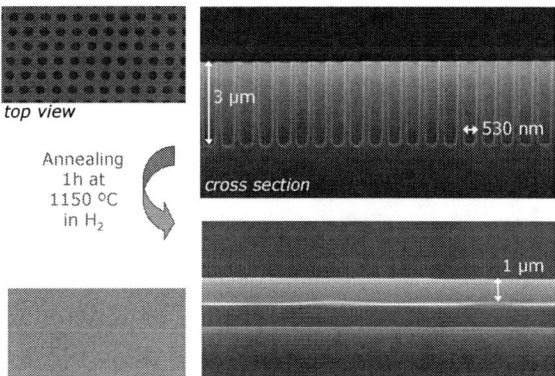

Figure 16. Example of a 1-μm- thin film obtained after annealing in hydrogen

Large-area thin films can be transferred to glass substrates and processed into proof-of-concept solar cells, demonstrating the feasibility of the concept for photovoltaic applications. The current voltage characteristic of such a cell is shown in figure 17, taken from [9].

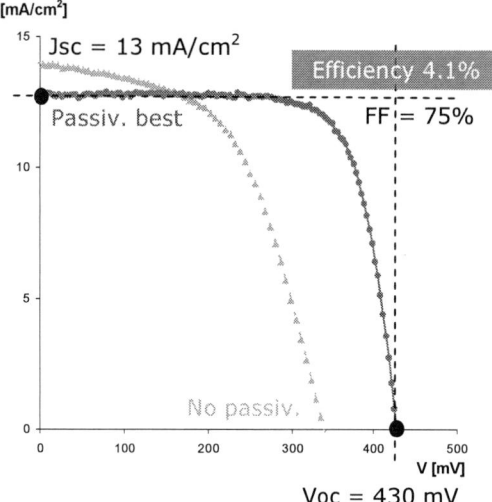

Figure 17. Current-voltage characteristic of first proof-of-concept cells on silicon films obtained by the empty-space-in-silicon technique. The cell thickness is 1 micron and no light trapping has been used. (after [9])

EPITAXIAL THIN FILM SOLAR CELLS

Epitaxial thin film Si solar cells, consisting of a thin (~ 20 μm) high quality epitaxial layer on top of a low-cost Si substrate (such as metallurgical grade Si) are an attractive alternative for bulk crystalline Si solar cells. Epitaxial solar cells usually have low short-circuit currents, due to the limited thickness of the epitaxial layer. This problem can be solved by introducing light trapping in the epitaxial layer. This can be achieved by a plasma texturing step to provide oblique light coupling and the use of a porous Si reflector between the substrate and the active layer as shown in the cross-section schematic in figure 18. Applying an electrochemical etching process prior to epitaxial deposition we can create a stack of alternatively low and high porosity layers, which act as a Bragg reflector and results in an internal reflectance of about 80% and in cell efficiencies of 13.5% on low-cost Si substrates [10].

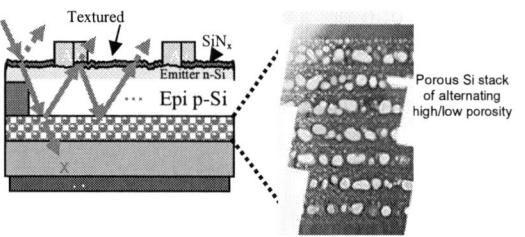

Figure 18. Cross-section schematics of a thin-film epitaxial solar cell with the porous Si Bragg reflector explicitly shown

Recently, we have improved the performance of the porous Si reflectors by chirping the porous Si structures [11]. A "chirped" structure refers to spatially periodic structures with the period changing across the structure. For this application, chirping the porous Si layer means that the periodicity in thickness of the different layers is varied across the stack as shown in figure 19. This variation is easy to achieve by programming the current source to the desired current density versus time during

electrochemical etching. Reflectors including such chirped structures and epitaxial layers were experimentally characterized by analysis of the reflectance spectrum. The wavelength band with high reflectance was broadened by at least 5% (Figure 20).

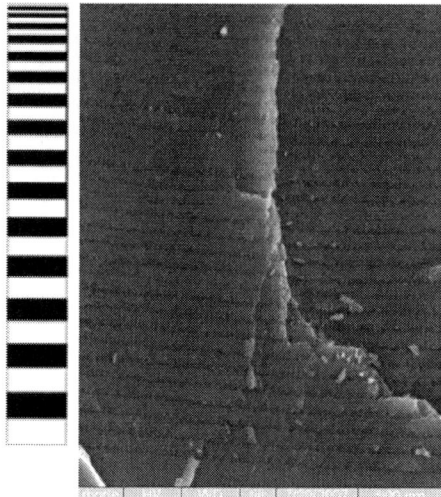

Figure 19. SEM cross-section of a chirped porous silicon reflector. Note that the period of the structure is lower close to the top surface in comparison with the bottom surface

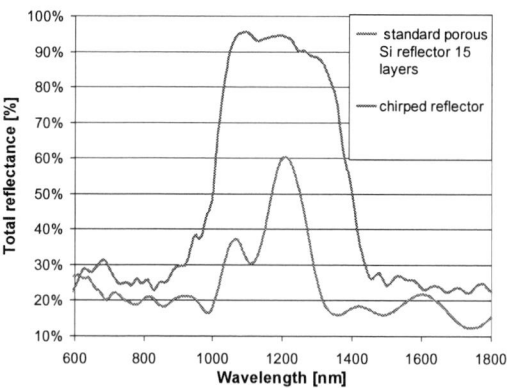

Figure 20. Experimentally measured reflectance curves of samples with Bragg reflectors, with and without chirping the structure

The total thickness of the chirped reflector used in figure 14 is approximately 7 micrometer, which took less than 300s to be etched. The refractive index ranges from 2.5 for the high porosity layers to 3.2 for the low porosity layers. In solar cell structures, the enhanced internal reflectance from chirped reflectors results, as expected, in an increased short-circuit current density and efficiency as shown in Table 2.

Reflector	Jsc (mA/cm²)	Voc (mV)	FF (%)	Eff. (%)
16 sub-layers, Fixed thickness	27.8	603	78	13.1
Chirped reflector, 40 sub-layers	28.7	598	77.6	13.3
Chirped reflector, 60 sub-layers	29.1	606	77.9	13.7
Chirped reflector, 60 sub-layers +DARC	29.5	605	77.5	13.9

Table 2. Solar cell results for thin-film epitaxial cells on low cost highly doped Si with "conventional" multilayer porous Si reflector and improved "chirped" reflector (screen printed, cell size 70 cm²)

THIN FILM CRYSTALLINE SI SOLAR CELLS ON GLASS OR CERAMIC SUBSTRATES

The cost of photovoltaic electricity could be lowered substantially if efficient solar cells could be made from polycrystalline Si thin films on inexpensive substrates. IMEC recently presented promising solar cell results that were obtained on polycrystalline Si films made by aluminum-induced crystallization (AIC) of amorphous Si followed by high-temperature epitaxial thickening [12]. The AIC process leads to very thin polycrystalline Si seed layers with a typical grain size in the range of 5 to 20 µm [13].

Thin film polycrystalline Si films can be prepared on alumina substrates (CoorsTek ADS 996R) by epitaxial thickening of AIC seed layers. The substrates were covered by spin-on flowable oxide (Fox – 25 from Dow Corning) to reduce their roughness. Next, double layers of Al and amorphous Si were deposited on these substrates in an electron-beam high-vacuum evaporator. In between both depositions, the aluminum was oxidized by exposure to air for two minutes. The nominal thickness of the Aluminum and amorphous Si layers was fixed at 200 nm and 230-250 nm respectively. After deposition, the samples were annealed in a tube furnace under nitrogen ambient at 500°C for 4 hours. During this annealing, the amorphous Si crystallized into polycrystalline Si and both layers exchange places [13]. Finally, the top aluminum layer was removed by selective wet chemical etching. Absorber layers were deposited on the AIC layers by thermal CVD. The depositions were performed in a single-wafer epitaxial reactor (ASM Epsilon 2000) under atmospheric pressure, at a temperature of 1130°C. double layers of p+ and p Si with variable thickness ratios were made. The p+ layer acts as a back surface field (BSF) while the p layer is the actual absorber layer. The total epitaxial layer thickness was always between 2 and 6 µm.

After epitaxial deposition, the samples were plasma textured in a prototype reactor from Secon using microwave antennas positioned above the substrates, with SiF_6 and N_2O as precursor gases.

Emitters (p-type) were formed in two different ways. The first type of emitter was obtained by phosphorous diffusion at 860°C from a doped pyrolithic oxide. This homojunction emitter was typically around 600 nm thick. The second type of emitter was formed by deposition of thin double layers of undoped and P-doped a-Si using plasma enhanced vapor deposition (PECVD) at 180°C. The total thickness of this heterojunction emitter was around 15 nm. Defect passivation of the layers was performed by plasma hydrogenation in a PECVD system at 400°C. Homojunction cells were hydrogenated after emitter formation, while heterojunction cells were hydrogenated before emitter formation.

To complete the cells, an antireflection coating (ARC) was deposited and metal contacts were formed. A SiN_x layer deposited by PECVD was used as ARC for the homojunction cells, while a conductive indium tin oxide layer deposited by rf-sputtering was used for the heterojunction cells. The contacts were formed by photolithography and wet chemical etching in combination with metal evaporation. All cells have an aperture area of 1 cm².

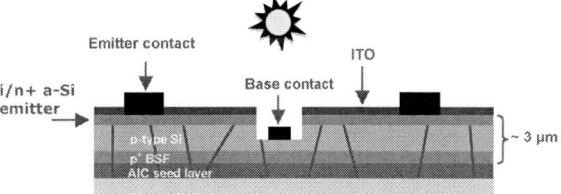

Figure 21. Schematic cross-section of a polycrystalline Si solar cell with heterojunction emitter on an alumnia substrate (not in scale)

Figure 21 shows a schematic cross section of such a polycrystalline Si thin film cell with a heterojunction emitter. The illuminated characteristic of such a 1 cm² cell is shown in figure 16. An efficiency of 8% is reached. This results from the use of a heterojunction emitter, yielding an increase in open circuit voltage, and from the efficient plasma texturing, causing a lower reflectance and better light trapping. The average grain size in this cell was around 5 µm.

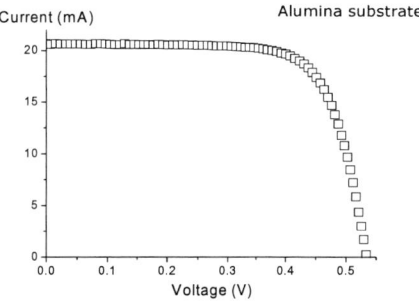

Figure 22. AM1.5 illuminated IV curve of a 1 cm2 size heterojunction thin film crystalline Si cell on an alumina substrate

The efficiency of the cells is actually mainly limited by

the large intragrain defect density of $10^9 \, cm^{-2}$. TEM studies reveal that most of the intragrain defects are already present in the AIC seed layers and got reproduced into the absorber layers during epitaxial growth. To reduce the intragrain defect density in our absorber layers, the AIC process needs to be optimized.

CONCLUSIONS

Several avenues for cost reduction of crystalline Si based solar cells exist. A possible roadmap up to the year 2020, as seen by IMEC, includes several technology generations. Important developments are the i-PERC cell, the ultra-thin cell, the epitaxial cell and the thin crystalline Si cell deposited on ceramic or glass substrates.

REFERENCES

[1]   Y. Ma et al., 23rd EU PVSC (2008).
[2]   P. Choulat et al., 23rd EU PVSC (2008).
[3]   G. Agostinelli et al., 21st EU PVSC (2006).
[4]   P. Choulat et al., 22nd EU PVSC (2007).
[5]   Y. Ma et al., 17th PVSEC (2007).
[6]   F. Dross et al., 33rd IEEE Photovoltaic Spec. Conf. (2008).
[7]   L. Carnel et al., J. Appl. Phys. 100 (2006).
[8]   T. Sato et al., Jpn. J. Appl. Phys., part I 39, 5033 (2000)
[9]   V. Depauw et al., Mat. Science and Eng. B 159-160, 286 (2009)
[10] F. Duerinckx et al., IEEE Electron Device Letters, Vol. 27 (2006).
[11] J. Van Hoeymissen et al., 23rd EU PVSC (2008).
[12] I. Gordon et al., 21st EU PVSC (2006).
[13] O. Nast et al., Applied Physics Letters 73 (1998).

AUTHOR INDEX

Abbott, Donald 14
Aguilar, Mario 121
Anderson, Richard 14
Aschenbrenner, Rolf 46
Barth, S. 58
Bauer, C. E. 145
Bennin, Jeffry S. 38
Bernard, David 31
Berndt, Hartmut 1
Bickford, Charlie 14
Bielick, Jim 99
Boger, Wendi 14
Bok, Zeger 78
Boutry, Hervé 64
Brun, Jean 64
Caleffii, Alessandro 133
Capra, Eugenio....................127, 133
Chapman, Brian 99
Cheung, Dason 121
Coronado, Juan 121
De Torres, H. Bartsch 58
Degan, Ken 14
Ebner, Karen 14
Evans Jr., Daniel D. 78
Farrell, Bob 14
Feinstein, Louis 14
Feldmann, Klaus 71
Feng, Zhen (Jane) 121
Ferrill, Mitchell 99
Fillion, R. A. 145
Fischer, M. 58
Fisher, Michael 99
Fragoza, Deb 14
Francalanci, Chiara 127
Franke, Jörg 71
Garcia, Omar 121
Goth, Christian 71
Gray, Brian......................................8
Haberland, Julian 46
Hsu, Allen 94
Isaacs, Phil 99
Jang, Eun-Jung 158
Jiang, Thompson 94
Kallmayer, Christine 46
Kim, Bioh 158
Kim, Jae-Won..................................158
Kim, Seung-Ho 105
Klein, Mathias 171
Kobeda, Eddie 99
Koshi, Masuo 21
Krastev, Evstatin 31
Kumbhat, Nitesh 52
Kurwa, Murad 121
Lang, Klaus-Dieter 171
Lee, Kiwon 105

Lee, Ning-Cheng 109
Lee, S. W. Ricky 162
Lewis, Theron 99
Liao, David 94
Lim, Andy 94
Lin, Charles 94
Lindner, Paul 158
Longworth, Don 14
Masuda, Junya 21
Matthias, Thorsten 158
McMahon, John 8
Mertens, Robert 181
Miller, Michael 14
Molnar, Ron 84
Morose, Greg 14
Müller, J. 58
Neuhaus, H. J. 145
Nishimura, Tetsuro 21
Nozu, Takashi 21
Paik, Kyung-Wook 105
Park, Young-Bae 158
Pasquito, Helena 14
Pawlowski, B. 58
Pinsky, David 14
Poupon, Gilles 64
Quinones, Horatio 90
Rae, Alan 176
Ratledge, Tom 90
Reichl, Herbert 46, 171
Ren, Eric 14
Robles, Carlos Alberto 121
Rodas, Fernando 121
Ru, Heinz 94
Sankaran, Nithya 52
Santos, Daryl L. 116, 139
Santos, Frederico 121
Sarkhel, Amit 14
Semmens, Janet E. 25
Shina, Sammy 14
Siblerud, Paul 161
Sillon, Nicolas 64
Souriau, Jean Charles 64
Sundaram, Venkatesh 52
Sweatman, Keith 21
Tan, Jerry 94
Tiller, Michael R. 38
Tummala, Rao R. 52
Utsunomiya, Henry H. 151
Valle, Jorge 121
Wang, Nick 94
Wang, Su 162
Weiland, Robert 171
Wimplinger, Markus 158
Wise, Jeff 84
Wolf, M. Juergen 171

AUTHOR INDEX

Zagordi, Daniele..127
Zamora, Luis Manuel 121
Zazzera, Alex...127
Zoschke, Kai...171